China's Space Programme

S. Chandrashekar

China's Space Programme

From the Era of Mao Zedong to Xi Jinping

 Springer

S. Chandrashekar
J R D Tata Chair Professor
National Institute of Advanced Studies (NIAS)
Bengaluru, India

ISBN 978-981-19-1506-2   ISBN 978-981-19-1504-8 (eBook)
https://doi.org/10.1007/978-981-19-1504-8

This Springer imprint is published by the registered company Springer Nature Singapore Pte Ltd.
The registered company address is: 152 Beach Road, #21-01/04 Gateway East, Singapore 189721, Singapore

Acknowledgements

This book is dedicated to Amma and Appa, whose belief and conviction that I could do difficult things well, helped me navigate the turbulent waters of the times we live in.

The 4 months in San Francisco spent with my younger sister Yoga and my dear friend Subbu, provided me the space needed for stringing together all the micro details of the Chinese Space Effort to create a coherent story. Without their help and support, this book would not have been written.

To my elder sister Shobana and my dear friend and guide Balu, a big thank you for their unwavering and steadfast faith in me and my abilities. I do hope that this book will live up to their expectations.

This book has gone through several iterations over the last 5 years before being published. My colleagues and dear friends, Lalitha and Viji at NIAS, have had the onerous task of looking through and editing these versions. They have also been of immense help in the collection and organization of all the data that forms the core part of this tome. A big thank you to them.

The source of inspiration for this book has been the time that I spent at ISRO working under the great architect of the Indian Space Programme, Satish Dhawan. Almost everything that I have learnt about Space, technology, society, and life came through my interactions with him. I wish he was here to read and critique this work.

Y. S. Rajan and V. Siddhartha have not only been colleagues and friends at ISRO but also mentors throughout my working life. Thank you for your help not only with this book but also for the many things that you have done for me over the years.

Two anonymous reviewers took time to patiently look through and review my manuscript. Their comments have been most helpful in improving the contents. I hope that they will be satisfied with the results. I would also like to thank Brian Harvey the author of two path breaking books on China's space programme for his generous gesture in providing me photographs of Chinese missiles for use in the book.

I would also take this opportunity to thank NIAS and my NIAS colleague, Ms. Hamsa Kalyani, for help in getting this book published by Springer.

A special thanks to Ms. Priya Vyas and Ms. Sushmita at Springer for their patience and support during COVID times.

Finally, I cannot forget my feline and canine friends, Zilly, Felix, Oscar, and Caesar, for enriching life and for being what they are—a source of joy and hope for all around them.

Bengaluru, India   S. Chandrashekar
October 2021

Preamble

The rise of China and its transformation into a major global power is one of the sagas of the twentieth century. After centuries of turmoil, humiliation, and a major civil war, the People's Republic of China (PRC) came into being in 1949. Since then, notwithstanding major domestic upheavals and a hostile international environment, it has achieved remarkable progress on all fronts. It is now one of the world's largest economies and is poised to overtake the US as the world's largest economy soon.

The Chinese model of development is a unique one that combines Marxist Leninist thought with the pragmatism of state supported capitalism and market forces. The organizational and institutional mechanism through which this blending has taken place for delivering results that no country has ever achieved, appears to be unique to China. This dynamic adaptation to the internal and external forces that aim to shape the national agenda by the political elite has created a country that is poised to become a major force in the geopolitical landscape of the twenty-first century.

The Chinese Space Programme easily stands out as one of the jewels in the Chinese crown. Its origins go back in time to the return of the well-known rocket and missile pioneer, Quian Xuesen from the US to China in 1955. This was also the time when after a period of consolidation, China's leaders were looking at an agenda for development with key roles to be played by science and technology. Since those early days China has gone from strength to strength and is now a major player in the global space arena.

A study of China's Space programme and how it has evolved and changed from its turbulent origins provide a lens through which one can understand the larger enigma that is China. This narrative about the origins and the evolution of the Chinese Space Programme attempts to provide an Indian perspective on the Chinese Space Programme. The study draws upon the vast literature on China and the Chinese Space Programme. However, its interpretation of the evidence on China's space capabilities and aspirations tries to integrate the various diverse and often contradictory strands into a coherent understanding of its strategy.

The book is divided into two parts. The first part provides an overview of how the space programme has evolved from its beginnings in the 1950s to the present. It uses the major political turmoils within China and the external forces that influenced

Chinese decision making, as the basis for dividing the period into four phases. The major achievements in each period, the internal and external forces acting on the programme as well as the major organizational and institutional changes that were affected during each phase are threaded together for providing an integrated picture. The last section of this first part provides an overall assessment on China's space strategy and how it has evolved in response to the internal and external forces shaping it. Recent developments in the military uses of space and how these are driving an increasingly hostile China-US dynamic are also covered.

The second part of the book addresses each major functional area of space in greater depth. This section covers Recoverable Satellites, Communications Satellites, Weather Satellites, Remote Sensing Satellites, Navigation Satellites, the Human Space Flight Space Programme, and Space Sciences. A separate section covers the launch vehicles that China has developed to place the various satellites it needs in their proper orbit. In each of these assessments Chinese capabilities are benchmarked against global standards to provide a flavor of China's competitive position in the different functional areas.

A special Section on Military Uses highlights how military uses have influenced the direction and thrust of the programme throughout this period. These space capabilities are also connected to China's War and War deterrence strategies.

The creation of the infrastructure that is required for the programme is covered in a separate chapter. This also includes the major organizations involved in the space effort and their relative standing in China's space hierarchy.

The final section of the second part of the book looks at China's International Cooperation efforts. The commercial, political, regional, and global dimensions of this cooperative effort are used to get an understanding of China's space strategy and its role in China's grand strategy.

Annexure 1 provided at the end of the first section of the book provides details of the satellites launched by China since the inception of the programme.

Other Annexures Tables and Charts in the second part of the book elaborate in greater detail the capabilities and achievements in each major functional area of space.

Contents

About the Author

S. Chandrashekar is the J R D Tata Visiting Professor at the National Institute of Advanced Studies. Before joining NIAS he was a Professor of Corporate Strategy at the Indian Institute of Management Bangalore (IIMB). His work at IIMB involved linking technology especially high technology with business, corporate and national strategy. Prior to his stint at IIMB he spent more than 20 years working at the Indian Space Research Organisation (ISRO) on satellites, rockets as well as the applications of space technology especially remote sensing. As Deputy Director of Earth Observation Systems he was also involved with all international matters related to space. He has been a member and leader of Indian delegations to the UN Committee on the Peaceful Uses of Outer Space (UNCOPUOS). He was the focal point at ISRO for all matters related to the militarisation of space. At NIAS his research is focused on missiles, nuclear weapons and space. Recent work includes the use of satellite imagery for identifying Uranium mills, the role of space in safeguarding India's National Security interests, a critical assessment of China's Space Programme and North Korea's recent missile and nuclear weapon tests. His current research critically reviews China's Anti-Access Area Denial military strategy directed against the US and its allies in the Asia Pacific Region.

Abbreviations

A2AD	Anti-Access Area Denial
AALPT	Academy of Space Propellant Technology
AASPT	Academy of Aerospace Solid Propulsion Technology
ABM	Apogee Boost Motor
APSCO	Asia Pacific Space Cooperation Organization
ASAT	Anti-Satellite
ASBM	Anti-Ship Ballistic Missile
ASEAN	Association of South-East Asian Nations
AVIC	Aviation Industry Corporation
BACCC	Beijing Aerospace Command and Control Centre
BLMIT	Beijing Landview Mapping Information Technology Company
BMD	Ballistic Missile Defence
C4	Command Control Communications Computers
CAAET	China Academy of Aerospace Electronics Technology
CAC	China Aerospace Corporation
CALT	China Academy for Launch Vehicle Technology
CAS	Chinese Academy of Sciences
CASC	China Aerospace Science and Technology Corporation
CASIC	China Aerospace Science and Industry Corporation
CAST	China Academy for Space Technology
CBERS	China Brazil Earth Resources Satellite
CD	Conference on Disarmament
CEODE	Centre for Earth Observation and Digital Earth
CETC	China Electronics Group Corporation
CFOSAT	China France Oceanography Satellite
CGWIC	China Great Wall Industry Corporation
CLEP	China Lunar Exploration Programme
CMA	China Meteorological Administration
CMC	Central Military Commission
CMI	Civil Military Integration
CMIPD	Civil Military Integration Promotion Department

CMSA	China Manned Space Agency
CNPGS	China North Polar Ground Station
CNSA	China National Space Agency
COPUOS	Committee on the Peaceful Uses of Outer Space
COSTIND	Commission on Science Technology and Industry for National Defence
COSTND	Committee on Science and Technology for National Defence
CPCC	Communist Party of China Congress
CRESDA	China Centre for Remote Sensing Data and Applications
CZ	Chang Zheng
DAMPE	Dark Matter Particle Explorer Satellite
DF	Dong Feng
DFH	Dong Fang Hong
DMC	Disaster Management Constellation
DSLWP	Discovering the Sky in Long Wavelengths Pathfinder
DSP	Defence Support Programme
ECT	Equatorial Crossing Time
ELINT	Electronic Intelligence
EO	Electro Optical
ESA	European Space Agency
FB	Feng Bao
FSW	Fanhui Shi Weixing
GAD	General Armaments Division
GLD	General Logistics Department
GLONASS	Global Navigation Satellite System
GOES	Geostationary Operational Environment Satellite
GPD	General Political Department
GPS	Global Positioning System
GSD	General Staff Department
GSO	Geostationary Orbit
GSSAP	Geosynchronous Space Situation Awareness Programme
GTO	Geostationary Transfer Orbit
HXMT	Hard X-ray Modulation Telescope
IAF	International Astronautical Federation
ICBM	Inter-Continental Ballistic Missile
IGSO	Inclined Geosynchronous Orbit
IRNSS	Indian Regional Navigation System
IRSA	Institute of Remote Sensing Applications
ISR	Intelligence Surveillance and Reconnaissance
ITAR	International Traffic Arms Regulation
ITU	International Telecommunications Union
LEO	Low Earth Orbit
MAI	Ministry of Astronautics Industry
MASI	Ministry of Aerospace Industry
MBB	Messerschmitt Bolkow Blohm

MCCC	Dongfeng Mission Control and Command Centre
MEO	Medium Earth Orbit
MIIT	Ministry of Industry and Information Technology
MIRV	Multiple Independent Reentry Vehicles
MMH	Mono Methyl Hydrazine
MOST	Ministry of Science and Technology
N2O4	Nitrogen Tetroxide
NASA	National Aeronautics and Space Administration
NDC	National Data Centre
NFIRE	Near Field Infrared Experiment
NOAA	National Oceanic and Atmospheric Administration
NOAC	National Astronomical Observatories of China
NORAD	North American Aerospace Defence Command
NOSS	Naval Ocean Surveillance System
NSSC	National Space Science Centre
OROS	Operationally Responsive Space Satellite
P&T	Posts and Telegraph
PAM	Payload Assist Module
PAROS	Prevention of an Arms Race in Space
PCC	Party Central Committee
PKM	Perigee Kick Motor
PLARF	Peoples Liberation Army Rocket Force
PRC	Peoples Republic of China
PSLV	Polar Satellite Launch Vehicle
QKD	Quantum Key Distribution
QSS	Quantum Science Satellite
QZSS	Quasi Zenith Satellite System
RA	Right Ascension of the Ascending Node
RADI	Remote Sensing Satellite Ground Stations
RFNA	Red Fuming Nitric Acid
RSGS	Remote Sensing Satellite Ground Station
SAAT	Sichuan Academy of Aerospace Technology
SAR	Synthetic Aperture Radar
SAST	Shanghai Academy of Space Flight Technology
SASTIND	State Administration for Science Technology and Industry for National Defence
SBIRS	Space Based Infra-Red System
SBSS	Space Based Surveillance Satellite
SLBM	Submarine Launch Ballistic Missile
SOA	State Ocean Administration
SSA	Space Situational Awareness
SSF	Strategic Support Force
SSN	Space Situation Network
SSO	Sun Synchronous Orbit
SSTL	Surrey Space Technologies Ltd.

STARE	Space Based Telescope for Actionable Refinement of Ephemeris
STSSATRR	Space Tracking and Surveillance System Advanced Tracking Risk Reduction Satellite
TDRS	Tracking and Data Relay Satellite
TLE	Two Line Element
TT&C	Tracking Telemetry and Command
UCS	Union of Concerned Scientists
UDMH	Unsymmetrical Di Methyl Hydrazine
UNCOPUOS	United Nations Committee on the Peaceful Uses of Outer Space
UNOOSA	United Nations Office for Outer Space Affairs
USSR	Union of Soviet Socialist Republics
XPNAV	X-ray Pulsar Navigation Satellite

Part I
Origins and Evolution

Chapter 1
The Origins of China's Space Programme (1956–1976)

1.1 The Early Years and the Focus on Missiles

The return of Quian Xuesen from the US to China in 1955 also coincided with Chinese efforts to modernize their programme of economic development in which science and technology inputs were seen as key ingredients. Quian made a major contribution to the formulation of a 12-year plan in the aviation, missile, and space domains[1] (Chang 1995).

In January 1956, the Chinese government established the Institute of Applied Mechanics in Beijing headed by Quian. The main aim of this Institute, which functioned under the Chinese Academy of Sciences (CAS), was to carry out research on Applied Mechanics and High-Speed Aerodynamics for defense purposes. The political leadership also recognized that missiles had a crucial role to play for ensuring national security and protecting its international interests. To take care of these needs, Chinese leaders created the Fifth Academy for taking the lead role in the development of missiles. Quian was the founding director of this institute though it was headed by a Peoples Liberation Army (PLA) veteran at the Ministry level. By the end of 1958, the Fifth Academy had established its first set of ten laboratories dealing with different aspects of missile development. Figure 1.1 provides an overview of the organization structure of the early Fifth Academy.

The missile programme enjoyed top level support from major political figures that included Mao Zedong, Zhou Enlai, and others. The main spearhead of the effort was Marshal Nie Rongshen, a veteran of the Long March. He headed an early Aviation Committee that formally became the Committee for Science & Technology for National Defence (COSTND) in 1958. Initial efforts involved a major cooperative programme with the Soviet Union who helped China by supplying them with

---

[1] Quian Xuesen was a Chinese student of Theodore Von Karman the US aerospace pioneer. He became a US citizen and made a major contribution to the early aerospace efforts in the US. During the McCarthy era, he was accused of being a spy and deported to China. He went on to become the "Father of China's Space Programme".

© National Institute of Advanced Studies 2022
S. Chandrashekar, *China's Space Programme*,
https://doi.org/10.1007/978-981-19-1504-8_1

Fig. 1.1 Organization structure of early fifth academy

missiles, consultancy as well as with the training of Chinese engineers in the Soviet Union. They also provided them with the required blueprints for manufacturing the missiles and helped them with the establishment of production facilities.

In April 1958, construction activity for the Jiuquan Missile Test facility also began. This would become the launch center for the first Chinese satellite in 1970.

The early focus on missiles did not deter Quian from proposing a parallel project for launching a Chinese satellite in January 1958. This thrust towards satellites may have also been triggered by the launch of Sputnik in October 1957. The proposal put forward through the CAS, received support at the highest level. Mao himself wanted this programme to go ahead as a part of China's Great Leap Forward effort. A programme for the development and launching of sounding rockets as a part of the satellite development project was also cleared by a Committee in June 1958.

However, given the resource constraints of that time, China's capabilities to carry out both missile and satellite development simultaneously appeared to be unrealistic. Deng Xiaoping, as Party General Secretary, decided that China's focus should be directed towards the development of missiles and informed the CAS that the satellite programme had to be deferred[2] (Chandrashekar et al. 2007).

---

[2] The reference provides a technological and organizational perspective on missile developments in China.

Case Study 1

The Early Satellite Efforts and the Shanghai Connection[3] (Kulacki and Lewis 2009)

Immediately after the launch of Sputnik in 1957, five scientists, Qian Xuesen, Zhao Jiuzhang, Qian Sanqiang, Chen Fangyun, and Cai Xiang, tried to sell the idea of China building and launching its own satellite. As mentioned, the Soviet Union was actively supporting the PRC in its efforts to build a nuclear and missile arsenal. These ongoing collaborations could have also played a part in Chinese decision-making[4] (Lewis and Xue 1989, 1994). If needed, development problems in modifying the missiles to become space launch vehicles could be resolved with help from Soviet engineers and technologists.

Qian Xuesen was a well-known rocket scientist of international repute, Zhao Jiuzhang was a meteorologist and geophysicist trained in Germany, and Qian Sanqiang was a nuclear physicist trained in France. Chen Fangyun was educated in England and had carried out research on ground survey systems for Space applications. He later became an Academician of the International Academy of Astronautics.

The proposal went to the Chinese Academy of Sciences (CAS). Zhang Jingfu was its Vice Director and the Party Secretary. It was then sent to Mao who approved the programme in May 1958.

CAS then formed Group 581. The objectives of founding this group were:

- Development of Sounding rockets,
- Launch of a 200 kg satellite, and
- Launch a satellite with a mass of several thousand Kg.

This also led to the setting up of three design academies under the CAS.

The First Design Academy headed by Guo Yanghuai and Yang Nansheg would be responsible for the overall design of the rocket and satellite.

The Second Design Academy headed by Lu Quiang and Chen Yuanjiu would be responsible for the control systems. The Third Design Academy headed by Zhao Jiuzhang and Quian Ji would be responsible for environment testing of the various packages and subsystems developed for the satellite.

To consolidate all its satellite related work, the CAS created the Shanghai Mechanical and Electronics Institute with Wang Xiji as its Director in 1958. The scientists and engineers for this Institute came from the CAS in Beijing, CAS

---

[3] pp. 4–14.

[4] Together the two references provide a fascinating account of the development of the missile and nuclear development programmes from their start to the late 1980s.

branch institutes in Hebei, Hunan, Sichuan and Shanghai, Jiao Tong University, Harbin Polytechnic University, Jiangnan Shipyards, the Radio Electronics factory, and the Textile machinery factory.

Though Mao had supported the early launching of a satellite, the economic problems created by the Great Leap Forward resulted in the deferring of the satellite programme. The missile programme was accorded the immediate priority.

However, the CAS Institutes did not completely give up their efforts. By 1959, their goal was to focus on starting a programme of building and launching sounding rockets. This was independent of the missile program that China was already pursuing.

The first two stage rocket system had no control system and combined liquid and solid fuels; it weighed 190 Kgs, was 5.3 m tall and could reach an altitude of 8–10 Kms. The launch took place on February 19, 1960. The second launch of the T-7 M sounding rocket took place on September 13, 1960.

Following this the cooperation between the three CAS institutes, (Geophysical Institute, Bio physical institute and the Shanghai Mechanical and Electronics Institute) became stronger. The focus was on R & D programmes in Space Medicine, Aerodynamics, Orbital Mechanics, Rocket Engines, Propellants, Attitude Control Technology, and environmental sensing missions in the upper atmosphere. Zhao Jiuzhang who was then the Director of the CAS Geophysical Sciences Institute led this effort.

In anticipation of future successes in the missile area, Quian once again started the push for the launching of a Chinese satellite. In 1962, he recruited four engineers from the Shanghai Mechanical and Electronics Institute to serve as the torchbearers for the satellite effort. These scientists went back to their Institutes and in turn trained others creating a core cadre of trained personnel. R & D efforts to build a satellite continued in the CAS institutes but there was no significant push from the CPC.

All this changed when China successfully tested the Dong Feng-2 (DF-2) missile on June 29, 1964. This was witnessed by Zhao Jiuzhang and a small delegation of CAS scientists who then decided that it would be appropriate for China to take the next step of launching a satellite. In December of that year, Zhao had the opportunity to put across the idea of satellite launch to Zhou-Enlai. Since the ballistic missile technology was reasonably well developed in China, Zhao felt that a missile could be modified to launch a satellite. The question then was to build the satellite.

1.2   The New Push for a Satellite

The missile development programme got reformulated after the withdrawal of support from the Soviet Union. After several setbacks, the DF-2, the DF-3, the DF-4 missiles as well as the sounding rocket programme achieved successes. The DF-5 ICBM programme also showed significant progress. Towards the end of 1964, the Fifth Academy became the Seventh Ministry of Machine Building responsible for all missile and space projects. Its various sub-academies all became full-fledged Academies. Figure 1.2 provides a snapshot of the structure of the Seventh Ministry of Machine Building in 1965–66.

In January 1965, after the missile efforts had shown significant successes, the plan for the satellite was once again revived. Two separate parallel requests for a satellite made their way into the higher levels of decision-making within the Chinese political establishment.

Two of the original scientists who spearheaded the satellite effort within the CAS, Zhao Jiuzhang and Lu Quiang, made one proposal. This was routed through Zhang Jingfu, the Party General Secretary, to the CAS Interplanetary Science Committee. The proposal was approved and passed on to the Party Central Committee for a final decision.

Simultaneously another proposal prepared by Quian Xuesen was discussed in a CSTND meeting. The participants at this meeting consisted of Gen. Zhang Aiping (Deputy Director of CSTND and member of General Staff Department), Zhao Jiuzhang and his team as well as 30 technical experts from CAS. The meeting ended with no decision being taken.

To break the impasse, Zhao Jiuzhang then approached yet another Deputy Director of COSTND, Gen. Luo Wuchu, CAS party Secretary Zhang Jingfu and Qian Xuesen to resolve the various issues.

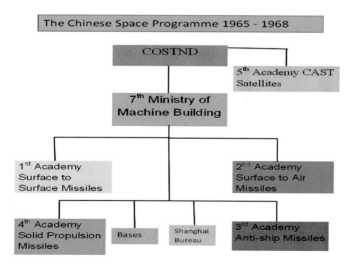

Fig. 1.2   The Chinese space programme 1965–1968

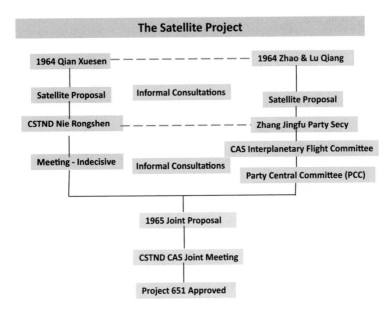

Fig. 1.3 The satellite project

Following these consultations on April 29, 1965, the Party Central Committee (PCC) adopted a proposal to put a 100 kg satellite in orbit by 1971. The Party Central Committee then sent the proposal to the Special Committee which guided national S & T Policy. This Committee included this proposal in the National Plan during its 12th Meeting in Beijing on May 4–5, 1965. It also directed the CAS to submit a plan for implementation of the proposal by July 1965. On receiving this plan, the Special Committee approved the proposal in its 13th meeting held on August 1965. The comprehensive National Satellite Program was called Project 651.

Figure 1.3 illustrates how the two separate proposals got merged to create the first satellite project—Project 651.

The first meeting of Project 651 was held in NW suburbs of Beijing and chaired by the CAS Vice-Director, Pei Leishen. Almost all institutes and organizations with an interest in space took part in this meeting. These included representatives from CSTND, the Committee on Industry for National Defense, National Committee for S & T, General Staff of the PLA, the Air Force, The Navy, the Second Artillery, the PLA Signal Corps, the First, Fourth, and Seventh Machinery Bureaus, Ministry of Post & Communications, Academy of Military Sciences as well as 13 research institutes under CAS.

In early 1966, a Satellite Academy was also established within the CAS. Work on a recoverable satellite also began in the Seventh Ministry. In May 1966, just as the Cultural Revolution was beginning, a committee that included Quian, finalized the requirements and the names for the first Chinese satellite. The satellite would be named Dong Fang Hong (DFH) translating simply into "the East is Red". The launcher would be called Chang Zheng (CZ) which means "Long March".

The original aim of Project 651 was to build a satellite with experimental instrument packages that would help design satellites for earth observation, weather monitoring, and communications. The CAS was given the responsibility for the satellite and the associated packages. The CAS initiated several mission-oriented research projects across China for the realization of the satellite. The Satellite Academy with Zhao Jiuzhang as the head, was responsible for the satellite design. A separate project, the 701 Project Bureau, was set up for managing the Tracking stations for the satellite mission. Within the Seventh Ministry, a design Institute (the Eighth Design Institute) was created for developing the CZ 1 launcher. The CAS Academy for Military Science and the CAS Academy for Medical Science were also set up for research on the space environment and space medicine.

1.3 The Cultural Revolution and the Creation of CAST

The beginning of the satellite project also coincided with the launching of the Cultural Revolution by Mao. This led to serious divisions within the various work centres responsible for the space and missile effort. Rebel factions led by the Red Guards created an atmosphere that seriously threatened activities within workplaces. The authority of COSTND and the Politburo above it was seriously undermined. At one stage all the Deputy Directors of COSTND and many scientists including Quian became victims of Red Guard actions[5] (Chen 1999).

Urgent action was needed to negate the impact of the Cultural Revolution on various projects and programmes of strategic importance. Within 2 months of the warring factions taking control of various units within the Seventh Ministry, Zhou Enlai along with Nie Rongzhen, Ye Jianying, and Li Fuchun met with all the political factions. They firmly indicated that they would not be allowed to take control of the key ministries of the Central Government.[6] They also imposed martial law within all defence-related ministries. A PLA General became the Director of the Committee of Military Administration that was responsible for the running of these key Ministries. In addition to proclaiming martial law, Zhou also managed to put all the defence-related industries within the ambit of Commission of Science & Technology for National Defense. This effectively placed COSTND directly under the military with the powers to approve and oversee all defence projects within the various defence ministries. By December 1967, the immediate problems seemed to have been resolved.

These political actions effectively transferred all the responsibilities of the satellite project from the CAS to the Seventh Ministry. The Shanghai-based Mechanical and Electronics Institute as well as some other institutes and work centers where expertise had been created in earlier years also came directly under COSTND.

---

[5] See p. 93 for details.

[6] Zhou's actions covered all the critical ministries including the second, third, fourth, fifth, and seventh ministries.

In early 1968, an Academy for satellites was also created under COSTIND that assumed direct responsibility for the satellite project. This is also sometimes referred to as the Fifth Academy or as the China Academy for Space Technology (CAST). Quian Xuesen became the first Director of CAST with Sun Jiadong as his second in command. Some institutes and personnel working on satellite-related projects within the CAS were also taken over by CAST[7] (Chen 1999). The original objectives of Project 651 were toned down to accommodate the political interests of the Cultural Revolution. Instead of the scientific packages, the satellite would broadcast the song "The East is Red" to the whole world. At 173 kg, China's first satellite was the heaviest first satellite put into orbit in accordance with Mao's wishes.

Immediately after its creation CAST put in place plans and proposals for the development of various kinds of satellites that included weather, telecommunication, and navigation satellites. The Plan also proposed the initiation of a human flight programme[8] (Chen 1999). A Space Flight Engineering Institute that dealt with the medical aspects of human space flight was also created as a part of CAST.

CAST continued to function directly under COSTND till July 1973 when it was transferred to the Seventh Ministry and became its Fifth Academy. However, because of the delays in the human space flight programme, the Space Flight Engineering Institute remained with COSTND.

1.4   The First Satellite and Its Aftermath

On April 24, 1970, the first "East is Red" Dong Fang Hong (DFH) 1 satellite was successfully placed in orbit through a Chang Zheng (CZ) 1 launcher. The success was repeated the following year when a CZ 1 launcher placed the Shijian 1 satellite in a low earth orbit.

The turmoil of the Cultural Revolution continued till the end of 1976 and officially ended only in 1977. During this time, China launched a total of 11 satellites on 11 rockets using three different kinds of launchers[9] (Chen 1999). Apart from the CZ 1 launcher used for the first two satellites, the CZ 2 launcher developed by the Beijing-based First Academy and the Feng Bao (FB) Launcher also derived from the DF-5 missile developed by the Shanghai-based space Group (the later Eighth Academy) were used for placing these satellites in low earth orbit. These satellites included recoverable satellites with film capsules, Electronic Intelligence (ELINT) satellites

---

[7] See p. 177.

[8] See p. 180.

[9] According to Chen Yanping, the Feng Bao (Storm) Launcher was launched by China in August 1972. This carried an ELINT package. Other open sources suggest that this launch was a partial failure. Since most publicly available data bases do not include this launch, it has not been included in this assessment. See p. 86.

Table 1.1 Creation of industrial infrastructure by the 7th Ministry

Industrial base	Academy affiliation	Location (province)	Start year	End year
061 base	2nd academy	Guizhou province	1965	1975
062 base	1st academy	Sichuan province	1965	1975
063 base	4th academy	Shaan'xi province	1965	1977
064 base	3rd academy	Shaan'xi province	1967	1978
065 base	2nd academy	Guizhou province	1968	1970 merged 061
066 base	3rd academy	Hubei province	1966	1978
067 base	1st academy	Shaan'xi province	1969	1976
068 base	2nd academy	Hunan province	1970	1977

as well as some other experimental satellites. The launch data indicates that both the Beijing-based and the Shanghai-based groups received approximately equal shares of the space activity.

In addition to the satellite effort highlighted above, the transition from the Fifth Academy to the Seventh Ministry also resulted in an organized effort to create an industrial base for producing missiles and satellites[10] (Chen 1999).

The two main industrial centres of China at Beijing and Shanghai were of course the main locations around which research and production activities were initially organized. However, even in those early days, based on security considerations, decisions were taken at the highest levels to create parallel facilities in China's interior.

To accelerate space- and missile-related efforts, many research Institutes and several factories producing defence products were transferred to the Seventh Ministry. These included factories producing electronics, semiconductors, optics, chemicals, and precision engineered products. Vocational training schools were also identified for training workers needed for the space and missile programmes. After training, these workers were transferred to the ministry. The labor force needed for the missile effort was also augmented by the addition of trained workers as well as retired soldiers with special skills. By the end of 1970, the Seventh Ministry had about 250,000 people working for it. That this happened during the height of the Cultural Revolution is indeed a remarkable achievement.

The various bases created in China's interior during this period and their affiliations to the academies are provided in the Table 1.1 [11] (Chen 1999).

During this turbulent period, the space programme withstood the buffeting it received from different political groups within China and came out relatively unscathed. That it was able to do this successfully was in no small measure due to the support that it received from the top political leadership that included Zhou Enlai and Nie Rongzhen. Lin Bao, the Defence Minister and at one time Mao's designated successor, was another powerful player whose alliance with Mao's wife

---

[10] See pp. 181–184.

[11] p. 184.

and the Shanghai-based Gang of Four could have also created problems. The Gang of Four and the Red Guards also exerted a powerful influence on the happenings inside China.

The emerging space and missile programme was able to respond and adapt to these challenges in a variety of ways.

Though the original satellite project was under the CAS, the Cultural Revolution forced the top leaders to intervene and place all the defence industries including the Seventh Ministry under military protection. This in a way strengthened the influence of the military over the space programme. Scientists and engineers, though influential, were now directly under the military. In organizational terms, this can be viewed as a diminution in the power of the CAS vis-a-vis the PLA.[12]

One major loss that the Seventh Ministry sustained during this period was the transfer of its Third Academy (tactical naval missiles) and all its related activities to the newly created Eighth Ministry. Though it would eventually come back to the Seventh Ministry after Deng's assumption of power, the missile and space programmes did suffer a setback from this decision taken during the Cultural Revolution.

The creation of two parallel launcher and satellite groups, one based in Beijing and the other in Shanghai, to accommodate the interests of the Shanghai-based power center appears to be a very creative response to the political challenges faced by the nascent space and missile programmes.

The eventual integration of the Shanghai space interests with the space programme by making it the Eighth Academy of the Seventh Ministry, reflect the political and strategic acumen of the top political leaders Zhou Enlai and Nie Rongzhen.

Though China could ill afford to create parallel capabilities in space due to resource constraints, this accommodation of different political interests was possibly the only way to insulate the emerging space programme from the trials and tribulations of the Cultural Revolution.

In retrospect, it created parallel competing groups and capabilities that China could leverage to foster change and innovation as its space and launcher programmes matured.

Case Study 2

Power Struggles, the Loss of the Third Academy, and the Emergence of Shanghai

At various stages during the Cultural Revolution, the nascent missile and space programme had to find various ways to cope with the problems posed by a wide spectrum of political interests.

---

[12] In more recent times as China allocates more resources for science, including space-based science, the CAS appears to have become more influential in the space ecosystem.

Though each power center saw the programme differently, they all seemed to think that the space programme could be used to further their interests.

The declaration of martial law and the placing of the various defence ministries directly under the PLA was the first step that Zhou, Nie, and other top-level decision-makers took to insulate the space programme from the political power struggles going on inside China. During this process, the CAS that was originally in charge of the satellite project, had to cede control to the Seventh Ministry that directly came under the PLA. This control over the space programme continues even today as a legacy of the Cultural Revolution. Though scientists and engineers are influential, control is still with the military.

In the early days of the Cultural Revolution, Lin Bao, the Defence Minister, became very powerful. He placed a lot of loyalists in key posts within the Seventh Ministry. He also wanted an accelerated programme to establish Chinese missile and space capabilities to be on par with the US and the Soviet Union. However, he was soon ousted and died while trying to flee to the Soviet Union.

Shanghai, as one of China's leading industrial centres, had always been a development and production center for the space and missile programme. A Shanghai base had always been a key element in the activities of the Fifth Academy as well as the Seventh Ministry.

One of the major power blocs that emerged during the heydays of the Cultural Revolution was the Gang of Four headed by Mao's wife Jiang Qing. One of their objectives was to shift the center of space activities from Beijing to Shanghai. By establishing the Shanghai Institute of Satellite Engineering, the Shanghai Institute of Launch Vehicles, and the Heavy Rocket Test Centre, they created a second research and production base for both satellite and launch vehicle activities[13] (Chen 1999).

Both Zhou Enlai as well Nie Rongshen recognized the importance of accommodating some of these interests to protect the nascent missile and space programme from the vicissitudes arising from the Cultural Revolution.

The transfer of designs from the first academy and the development of the parallel DF-5-based Feng Bao launcher were the compromises made to accommodate the special interests of the Shanghai Group.

Even at the third level of the Chinese hierarchy, scientist like Quian Xuesen seemed to have identified the need to accommodate the political influence of the Shanghai group. As early as 1962, he took recruits from the Shanghai-based Institute for Mechanical and Electrical Design and provided them

---

[13] pp. 100–101.

with direct training under him. These four scientists then went on to train others in space engineering and technology[14] (Chang 1995).

The record also indicates the critical role played by Zhou Enlai in protecting the interests of the space programme. His periodic visits to different work centers, his personal interest in protecting key scientists such as Quian, his dealings with the various political factions to get across the message of his personal commitment to the space programme, were all aimed at ensuring that the space programme remained largely insulated from the power struggles going on within China[15] (Chen 1999).

Nie Rongzhen, as the head of COSTND, also played a part. This informal alliance of political, military, and scientific interests at different levels in the Chinese hierarchy can be seen in the special relationships between Zhou, Nie, and Quian. These informal ties shaped the space programme during this critical period in China's history.

Though the Seventh Ministry did survive and grow, it did have to yield its Third Academy to accommodate the political interests of the Cultural Revolution. As indicated in the main text, its transfer to a newly created Eighth Ministry show that its leaders did make needed compromises in dealing with the political turmoil arising from the Cultural Revolution. It had to wait for the return of Deng for the restoration of the Third Academy.

1.5 The Initiation of a Human Space Flight Programme

As in the case of Sputnik, the Yuri Gagarin human space flight of April 1961, triggered a response from China. Within 2 months of Gagarin's flight, a Quian effort via the CAS led to the holding of 12 seminars between 1961 and 1963. The CAS also set up a Human Space Flight Committee that carried out preliminary research and formulated plans for a human space flight effort. A China Space Flight Engineering Institute was also set up in 1968. It continued to function even during times when there were no concrete plans for human space flight.

COSTND formally proposed a human space programme (Shu Guang) in 1970. This involved the launch of two astronauts into space on an 8-day mission. The first unmanned module was to be launched in 1973 followed by a human flight in 1974. Resource constraints prevented this programme from taking off. However biological experiments related to human space flight were carried out using sounding rockets

---

[14] p. 226.
[15] pp. 96–98.

developed as a part of this programme. The recoverable satellite programme (FSW) series was also linked to the manned programme as precursors for eventual human space flight.

Two dog flights on sounding rockets took place in July 1966 and October 1966 as a part of this activity. Though the programme was not fully taken up, low-key activities related to human space flight continued. There seemed to be unanimity within the scientific and political communities in China that this was a goal that China would eventually have to reach.

1.6 A Critical Appraisal and the Role of Key People

The nascent phase of China's space programme that stretched over two decades from its inception in 1956 was the most turbulent period in China's recent history. The growing rivalry between the Soviet Union and China, a hostile US-led Western alliance as well as two major internal power struggles[16] initiated by Mao, created almost impossible conditions for any hi-tech programme to take off. Despite these hurdles, the Chinese Missile and Space Effort was able to go ahead and make significant progress.

China launched 11 satellites, developed 3 different launchers, and created the necessary launch bases and industrial infrastructure needed for the space programme during this period. All these achievements were made possible because of the continuous high-level political support that the programme was able to access as the different political factions fought among themselves for power and influence.

The political leadership including Mao himself, saw satellites and space achievements as a tool to project China's power on the global stage. This desire to project China as a major power often resulted in political considerations dominating decisions irrespective of whether the results could be realized with the resources available.

Lin Bao, one of the aspirants to become Mao's successor, wanted targets and timeframes that could not have been realized even under the best of conditions. The technical leadership of the space programme had to successfully navigate these waters to build the required capacities and capabilities. In such negotiations, scientific and technical considerations often took a back seat to political directions.

The premature initiation of a human space effort even before the launch of a first satellite and the way in which the payload choices were made for the first Chinese satellite provide clear evidence of the dominant role played by Mao and other leading political figures.

---

[16] The Great Leap Forward and the Cultural Revolution both of which were initiated by Mao created a lot of political and economic turmoil that made it difficult for any entity to carry out any organized activity.

Major compromises on the technology development agenda often had to be made to cater to the interests of different political groups. The parallel initiation of two launcher development programmes (the CZ 2 and the Feng Bao) during a period when resources were scarce, the transfer of the Third Academy from the Seventh Ministry to a new Ministry provide illustrations of such compromises.

Through all these ups and down and the internal and external turmoil that China experienced, the space programme was fortunate to have an exceptionally capable and powerful political elite who provided the continuity and stability to build the capabilities for an ambitious and aspiration driven programme. Figure 1.4 provides an overview of this network of power and influence that steered the programme through its early years.

The political support of Prime Minister Zhou Enlai, the political and military connections of the Long March veteran General Nie Rongshen, and the technical leadership provided by the US-trained Quian Xuesen were the pillars around which the programme was built.

There is little doubt that Zhou Enlai provided the top support for the programme from its inception. His personal interventions, his periodic visits to various work centers, and the exercise of state authority through his position as prime minister, ensured that the missile and space programmes came through the various political crises relatively unscathed.

At the operational level, Nie Rongzhen, a veteran of the Long March and a member of the Politburo steered the programme through its most turbulent times. Though often

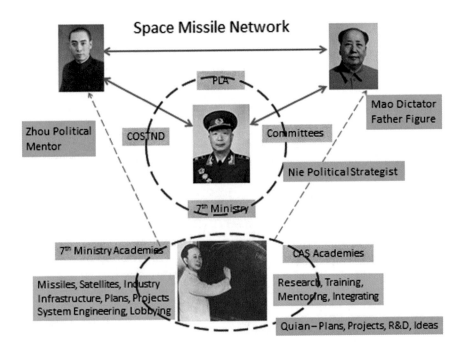

Fig. 1.4 The space missile network

under attack from several quarters, by using his connections with Zhou and other Party higher ups, he made sure that the interests of the space and missile programmes were protected.[17]

Quian Xuesen also played a key part in creating the technology base for running a hi-tech Missile and Space Programme. Through his dual role both in the CAS as well as in the Fifth Academy, he brought about the necessary integration between research and development teams. His proposals on missile development, satellite development as well as human space flight received ready hearings at the highest political levels.

His Directorships of key CAS and Seventh Ministry Academies helped him link the scientific and technical people with decision-makers at the top. Without his support and guidance that came out of his deep understanding of all matters related to missiles and satellites, the technology base that China needed would not have been created so quickly.

Though he was personally involved only in one ICBM project that had to be closed because of technology and infrastructure problems[18] (Chang 1995), he provided the vision and the strategy for China's space and missile programmes.

Quian played a crucial liaison role between the Political, Military, and Scientific communities to get the programme moving. **From this record of his involvement, he can certainly be called the "The Father of China's Space Programme"**.

References

Chandrashekar S, Gupta S, Nagappa R, Kumar A (2007) An assessment of China's ballistic and cruise missiles. National Institute of Advanced Studies (NIAS) Report R4–07. http://isssp.in/wp-content/uploads/2013/04/An-Assessment-of-Chinas-Ballistic-and-Cruise-Missiles.pdf

Chang I (1995) Thread of the Silkworm. Basic Books

Chen Y (1999) China's space activities policy and organization 1956–1986. Dissertation, Faculty of the Columbian School of Arts and Sciences, George Washington University, ProQuest Dissertation and Theses

Kulacki G, Lewis JG (2009) A place for one's mat: China's space program 1956–2003. American Academy of Arts and Sciences

Lewis and Xue, 1989 Lewis JW, Xue L (1989) China Builds the Bomb. Stanford University Press

Lewis JW, Xue L (1994) China's strategic seapower: the politics of force modernisation in the nuclear age. Stanford University Press

---

[17] The connections between Zhou and Nie go back to when they both studied in Europe especially Paris. Nie was apparently a protégé of Zhou.

[18] A 10,000 km Missile Project that closed in two years had Quian as the Chief Designer. See p. 219.

Chapter 2
Economic Reform and the Deng Era-1977–1990

2.1 The Early Years of Reform

The end of the Cultural Revolution and the emergence of Deng Xiaoping as China's leader created a new set of challenges for China's space Programme. It will be clear from the evidence presented in this section, that during the early years of this transition, the space programme lost some of its special status. It was viewed as just another programme that had to be assessed in terms of its contribution to China's development agenda. The first 2–3 years of transition were particularly painful. However, the space programme soon adapted to these new conditions and within a few years once again became a showpiece[1] (Chen 1999).

While the changes introduced were traumatic and involved a significant refocus of efforts, the Political Leadership did not lose sight of the strategic and critical role of missile and space capabilities. The recoverable satellite effort for, e.g., was not touched and launches continued through the transition period and well into the 1980s and early 1990s.

Based on a critical review of capabilities and China's needs the major tasks entrusted to the Seventh Ministry under the Deng Regime were the following:

Development of an ICBM, launching it, tracking it until impact in the southern Pacific Ocean to demonstrate its capabilities;

Development and launch of a Geostationary Communications Satellite to provide China with a modern satellite-based communication and TV broadcast system.

Launch of an indigenous Submarine Launched Ballistic Missile (SLBM).

Deng was not against science and technology. He however wanted them to be oriented towards the realization of economic and societal benefits. In August 1978, he asked the Seventh Ministry to focus on using space for adding economic value, especially satellites for telecommunications and remote sensing. Deng also wanted

---

[1] Chen Yanping's thesis on the early days of the space programme are particularly insightful. See pp. 106–123, 189–213.

© National Institute of Advanced Studies 2022
S. Chandrashekar, *China's Space Programme*,
https://doi.org/10.1007/978-981-19-1504-8_2

all the defence programmes (including the space programme), to develop and produce civilian goods for meeting domestic needs, a practice that continues within the space programme. Though Deng had been involved with the space programme and even helped it on occasions, he did not think of it as a privileged programme that required special treatment. He did not want China to become a part of a space race. He also ruled out any grandiose ambitions to go to the moon. He wanted the space programme to build and launch satellites that would benefit the country.

Deng did not rule out importing satellites for China to realize immediate economic benefits. In 1979, during a visit to the US, he signed a space cooperation agreement with President Carter that covered China's possible imports of US-made Communications satellites.[2]

In response to these initiatives, the space programme diversified by using some of the capacities built up for various space-related systems to produce goods needed in the domestic market. Through this, they were able to raise the resources needed for the continuation of the space programme as well. They also sought to improve their resource position through the export of space products and services. The Chinese satellite building capabilities were still far behind those required by international customers. However, they had made considerable progress in developing launch capabilities.

The reforms initiated by Deng were also accompanied by major restructuring exercises that sought to bring in greater organizational cohesion within the Defence Industry complex. While on one hand, resources were constrained because of programme focus, restructuring brought in additional resources and capabilities that could be harnessed to further both strategic and domestic commercial interests. The Third Academy that had been moved out of the Seventh Ministry to a new Eighth Ministry was brought back into it in a merger of the Eighth Ministry with the Seventh Ministry in 1981. The name of the Seventh Ministry was also changed to the Ministry of Astronautics Industry (MAI).[3]

The structural changes also affected the oversight commissions that were responsible for the programmes. The Commission on Science & Technology for National Defence (COSTND) was merged with the Commission on National Defence Industry into a Commission on Science Technology and Industry for National Defence (COSTIND). These changes reveal Deng's commitment to directly link Science and Technology research with industry to meet China's societal needs.

---

[2] However, these initiatives did not fructify. Chinese insistence on shared manufacture of satellites rather than outright purchase could have been one reason for this. China did tie up with MBB, a German company for the supply of critical components for its first communications satellite.

[3] The restructuring exercise also directly affected the specific projects undertaken by the various academies. The FB-1 parallel launcher being developed by the Shanghai Group was terminated. The rationalization of the work between the First Academy (CALT) and the Eighth Academy (SAST) would have been difficult without Deng's insistence on a new orientation. China seems to have in place an arrangement that seems to work and as in the case of the human space flight programme, used these competitive capabilities to foster original ideas for new programmes.

Table 2.1 Market orientation—changing names

Old name	New name
7th ministry of machine building	Ministry of astronautics industry
The first academy	China Academy of Launch Vehicle Technology (CALT)
The second academy	China Chang Feng mechanical & electronics technology academy
The third academy	The sea eagle electronic-mechanical technology academy of China
The fourth academy	The Hexi industry corporation
The fifth academy	Chinese Academy of Space Technology (CAST)
The seventh academy	China space civil engineering design & research institute
The eighth academy	Shanghai Academy of Space Flight Technology (SAST)
The ninth academy	China academy of basic technology for space electronics
Guizhou base	China Jiangnan space industry company group
Sichuan base	Sichuan aerospace industry company group
Hubei base	China Sanjiang space group
Shaanxi base	Shaanxi Lingnan machinery company
New creation	China Great Wall Industry Corporation (CGWIC) - Export Arm

To provide visible evidence that the space and missile programme was responding to market forces a renaming exercise was also implemented. The new names were chosen to reflect the main function that was being carried out by an Academy or an organization. Thirteen Academies and entities under the Ministry of Astronautics Industry (MAI) were renamed. Table 2.1 provides the new and old names of these entities.

Along with these internal changes, the restructuring also created new organizational mechanisms for promoting exports. In 1979, as a part of this export promotion exercise, the Great Wall Industries Corporation (GWIC) was created within the Seventh Ministry for the promotion of exports.

On the international front following Deng's visit to the US in 1979, China became a member of the International Astronautical Federation (IAF), the International Telecommunications Union (ITU), and the Committee on the Peaceful Uses of Outer Space (COPUOS) in 1980. It also participated in the second United Nations Conference on the Peaceful Uses of Outer Space in 1982.

These domestic and international actions paved the way for China's foray into the international market for space launch services.

Case Study 1
What did The Deng Reforms Achieve?

The Deng reform introduced into China's space and missile programme has also been termed "the one third system"[4] (Chen 1999). According to this mode of operation, one third of the income of the Seventh Ministry would come from directly supported R&D projects that were relevant to the defence and space industry. This was the share of the total budget that COSTIND would provide. Another one third of the resources would come through central government contracted projects that would be cleared by the State Council. These could be satellite or other projects of relevance to various other ministries that would have their support for using the service. This measure would ensure that the application end of the space programme would get priority. The last one third of the resources had to come from the efforts of the Seventh Ministry itself by making and selling products required by domestic consumers and industry.

After these reforms, the power of COSTIND over the Seventh Ministry came down with some of it shifting to the State Council that supported projects relevant to other government departments. It also provided a large degree of autonomy to the various entities within the Seventh Ministry to raise money through relevant domestic production of goods and services. They were also free to use these resources in a way that was appropriate for achieving their goals.

This model of a publicly owned entity operating in this complex mode seems to have a particularly Chinese character. In the west, corporations operate mainly on market dynamics. The aerospace sector is focused largely only on the production of hi-tech products. Though there have been periods when some of them become conglomerates operating in several diversified product markets these have largely failed the test of sustainability over longer periods of time.

The hybrid orientation that Deng provided, seems to have withstood the test of time at least as far as China is concerned. Though the financial performance of Chinese entities (Return on Investment, etc.) may be lower than their western counterparts, when the overall societal benefits are considered, these models of industry performance may be quite robust and relevant for a country like China.

---

[4] pp. 210–212.

2.2   Early Outcomes

By 1982, the space programme had delivered on two of the three major missions entrusted to it by the Chinese political leadership. On May 18, 1980, a DF-5 ICBM after traveling 9079 km, landed within its impact zone in the South Pacific. The missile was tracked over land as well as through its trajectory over the oceans through several ship-borne tracking terminals. In October 1982, a JL-1 missile was launched from a submarine.

The task that still had to be completed was to place a Chinese communications satellite in Geostationary Orbit.

2.3   The Communications Satellite Project

The origins of the Communications Satellite project predate the Deng reform era. There are several accounts of its origins and how it evolved till the first launch took place in 1984[5,6] (Harvey (2013), Kulacki and Lewis (2009). The Box item below presents an overview of the major events that took place extracted mainly from the above references along with some other data compiled from other sources.

Case Study 2
Decision-Making on The Communications Satellite Project

In March 1974, three Chinese telegraph workers of the Post & Telegraph (P&T) Department write directly to Zhou Enlai that China should make a claim with the ITU for slots in the GSO for its communications satellites. They also state that if China did not act quickly, they might lose their entitled slots.

Two months after receiving the letter on May 19, 1974, Zhou sends the letter on to the Politburo members—Li Xiannian, Zhang Chunqiao, Ye Jianying, and Wang Hangwen—for consideration. One of the Politburo members Li Xiannian, who is also the Vice-Chairman of the State Council, passes the letter on to the National Planning Commission.

On May 21, a meeting is convened to discuss this matter. The P & T Ministry, the Radio & Television Department and CSTND all take part in this meeting.

Four months later in September 1974, the report on the communications satellite project based on the discussions in the meeting is sent to the Party Central Committee.

---

[5] pp. 135–141.
[6] pp. 14–19.

On March 31, 1975, Ye Jianying places the proposal before the CMC. Deng acting on behalf of the terminally ill Zhou Enlai pushes the proposal and persuades the CMC to clear the project.

A dying Mao formally clears the project immediately afterwards and the 331 Project to build and launch a communications satellite becomes a reality.

Though China notified the ITU about its decision to place a satellite in GSO, the internal politics particularly immediately after the end of the cultural revolution delays progress.

The project involved three major sub-projects. The first was to build the communications satellite. To reach the satellite to a Geostationary Transfer Orbit (GTO) a new rocket based on the CZ 2 had to be built and qualified. This involved adding a third cryogenic stage to the CZ 2. The cryogenic engine needed for this stage had to be developed. Finally, a new launch complex optimized for launch to the Geostationary Orbit (GSO) had to be established at Xi Chang.[7]

After Deng became the leader in August 1977, he makes his ally Zhang Aiping, the head of CSTND. The communications satellite project becomes one of three priority projects of the Seventh Ministry.

In 1978, after Deng became the Chairman of the CPCC National Committee, he reinforces the role of satellites, especially for distance education[8] (Kulacki and Lewis 2009). In order to accelerate progress on this front, Deng is not averse to importing a satellite. Ren Xinmin is appointed as the Project head for the communications satellite project.

After Deng's visit to the US and the signing of a space cooperation agreement with President Carter the import of a US satellite to accelerate the use of satellite applications becomes an option. However, negotiations breakdown and China decides to go ahead to build and launch its own satellite.

Despite major technical problems, Ren Xinmin believes that the development of a restartable cryogenic upper stage for the launcher is the best long-term strategy and gives it a major push. He wins the argument to go ahead with this development despite other alternatives that would be easier to build.

Starting from 1978 onwards, the establishment of the new Xi Chang launch complex, the establishment of the Tracking Telemetry and Command (TT&C) network as well the ground stations for using the satellite all make speedy

---

[7] Xi Chang has the same latitude (28.5 °) as Cape Canaveral. Whether this is coincidence or a result of how US thinking has influenced Chinese space engineers is a debatable point.

[8] In the spring of 1978, Deng convened a meeting of the Special Committee at which the communications satellite program was discussed. Deng opened the discussion by observing, "If we invite a good teacher to give a lecture in the Great Hall of the People only 10,000 people can hear it, but if the same teacher were to give that lecture on television, and everyone had the equipment to receive it, that's a classroom of unlimited size." Obviously, Deng saw the significant benefits offered by a satellite for providing distance education. See p. 16.

progress. However, there are delays in the development and testing of the cryogenic engine and stage.

The first successful test of the cryogenic stage for the CZ 3 Launcher takes place in May 1983. Preparations accelerate for the launch of the communications satellite in 1984. Three satellites and launchers are made ready in case of any contingencies.

The first launch of a Chinese communications satellite takes place on January 29, 1984. Though the satellite is successfully placed in Low Earth Orbit (LEO) the cryogenic stage fails to ignite to take the satellite from LEO to a Geostationary Transfer Orbit (GTO). Though the Chinese tried to salvage the mission by firing the Apogee Boost Motor on the satellite and try to use the satellite for experimental purposes the mission is considered a failure.

After understanding the reasons for the failure, the cryogenic stage is tested six times in March 1984 before it is cleared for flight.

On April 8, 1984, the second launch of the Communications satellite is successful. Soon afterwards the satellite was moved from GTO and reached its assigned position in the GSO on April 16, 1984. The satellite transponders were turned on the next day. Transponder tests showed that the satellite was performing normally both for communications and TV transmissions. With this demonstration of its indigenous capabilities, China had finally emerged as a space power in the making.

2.4 Building Capabilities for Meeting Domestic Needs

The Deng period (1977–1990) and the renewed focus on meeting economic and societal needs transformed the space programme in a fundamental way. China placed 29 satellites in orbit using 22 launchers during this period. These included launches to Low Earth Orbit (LEO), Geostationary Orbit (GSO) as well Sun Synchronous Orbit (SSO). Two polar orbiting weather satellites were put into Sun Synchronous Orbits by the newly developed CZ 4 launcher from a new launch center located at Taiyuan.

Apart from meeting military needs through recoverable satellite launches (10 satellites) as well as ELINT (6 satellites) the programme was catering to the TV & Communications (7 satellites) and weather (2 satellites) needs of the country. Two satellites (one dummy satellite launch for an Australian telecom company to demonstrate launcher reliability) and one small satellite for Pakistan marked the beginning of China's entry into the international launch services market.

The production capacities and infrastructure created earlier in different parts of China were also used to produce consumer goods as desired by the directives given to it.

Apart from the workhorse CZ 2C and CZ 3 launchers for LEO and GTO orbits, China also developed a new launcher the CZ 4[9] for placing remote sensing satellites in sun synchronous orbits. An improved CZ 2D vehicle that would provide greater flexibility in the payloads delivered to LEO and SSO also began development during this period. China also demonstrated the injection of multiple satellites into orbit using the same launcher. The restructuring exercise ended the production of the Feng Bao launcher by the Shanghai Group.

Case Study 3

Building a New Launcher for The Burgeoning Communications Satellite Market—The CZ 2E Story

China also took steps to make sure that its products for export were technically competitive with global market needs. The CZ 3 vehicle that they had developed could only place about 1200–1300 kg in GTO, whereas the market needs of that time needed a launcher capable of placing about 2500 kg in GTO. To capitalize on this opportunity and to increase its market share, the Chinese decided to develop a new launcher by improving the performance of its existing CZ 2 stages by adding four strap-on boosters to it. This would increase the payload that it could deliver. With a suitable upper stage, the CZ 2E could place a payload of up to 4800 kg in GTO.

This launcher was developed in a record time of 18 months and was first used to launch a dummy payload for an Australian Telecommunications company that had contracted Hughes for the supply of the satellite.

The first launch that took place in June 1990 successfully placed the dummy satellite in LEO. A solid propellant perigee kick motor (the third stage of the launcher) was to then take it from LEO to GTO. However, this motor fired in the reverse direction preventing the satellite from reaching GTO.

This maiden launch did place a small Pakistani Badr satellite in Low Earth Orbit (LEO).

The failure of the Chinese made Perigee Kick Motor (PKM) during the first flight of the CZ 2E, resulted in a decision to replace it with a US made Star 63 F Perigee Kick Motor (PKM).

The changes made in the satellite launcher interface for accommodating the US Star motor, affected the reliability of the CZ 2E. It was launched seven times between 1990 and 1995 with three failures. Two Australian Optus satellites, one Asiasat satellite and one US built Echostar satellite reached GTO. In addition

---

[9] Stretched CZ 2C first and second stages along with a new liquid upper stage provided this new variant.

to the first failure with the dummy payload, one Australian Optus satellite as well as one Apstar satellite did not reach GTO because of launch failures.

The Apstar launch that took place in January 1995 from Xi Chang created a major problem. About 50 s into the flight, there was a huge explosion which seemed to have emanated from the top of the rocket rather than from the firing stage. The explosion killed 6 villagers and injured 23 others. The incompatibility between the US built Star 63 F PKM and first stage/strap-on booster firings under maximum dynamic load conditions may have been the cause of the problem.

After a review, the launcher was flown successfully flown twice more. The CZ 2E was retired after these flights. The availability of an improved CZ 3A launcher with the needed payload capacity may have also influenced this decision.

The development of the CZ 2E launch vehicle was not without its beneficial aspects. The CZ 2E served as the basic launcher for building the CZ 2F launcher used in China's human space flight programme. The strap on boosters were also used to improve the payload capacity of the CZ 3A launcher in its CZ 3B and CZ 3C variants[10] (Harvey 2013).

The joint investigation of the failure of the CZ 2E between US satellite manufactures and China's launch vehicle developers would also have far-reaching consequences on US-China relations in the space domain. These are addressed in the next chapter.

Just as Xi Chang was set up as a launch center for optimizing the payload delivered to GTO, a new launch center Taiyuan, was also created for optimizing payload delivery to Sun Synchronous Orbits (SSO). The three launch centers, Jiaquan, Taiyuan, and Xi Chang together would provide China with a significant augmentation in its launch capacity for its future programmes.

China also created domestic and international joint venture companies to cater to the growing domestic demand for communications satellite companies[11] (Harvey 2013). These companies were free to buy satellites from anywhere. Initially at least during the reform period many of them used lower cost Chinese launch services to reach GTO. After the US imposed export restrictions under ITAR in the second half of the 1990s, many of these companies were also free to procure launch services from the international marketplace.

---

[10] pp. 149–151.

[11] The three companies are Asiasat, APT (Asia Pacific Telecommunication which became APSTAR and after German investment Sinosat) as well as a domestic company China Telecommunications Broadcast Services (China Telecom). See p. 148.

2.5 Leveraging Capabilities for Promoting Exports

Though the export arm of the Seventh Ministry the CGWIC had been created in 1979, China had still to prove to the international community that it could provide reliable products and services. By the time it launched its first communications satellite in April 1984, it had established its credentials for placing satellites in LEO and recovering them as well. The obvious focus of such efforts was photo reconnaissance. The availability of the CZ 2C launcher, a recovery capsule that could be used to retrieve experiments and the newly developed and flight proven CZ 3 opened new international marketing opportunities.

In March 1985, China exhibited its CZ 3 launcher and the DFH 2 telecom satellite at the Zubo Exhibition in Japan. It also announced that its Long March (CZ) launchers would be available to interested international customers at a space technology conference in Geneva in May 1985[12] (Chen 1999).

On January 28, 1986, the space shuttle Challenger exploded 73 s after lift-off. A Titan 34D also exploded immediately after take-off in April 1986. This was preceded by another Titan failure in August 1985. A US Delta 3914 exploded in May. These failures also grounded the Atlas rocket that had design similarities with these launchers[13] (Smith 2006). The European Ariane 2 launcher also failed in May 1986 for the second time in four launches[14] (Holland 1986). Chinese launchers with their lower prices now seemed to be a quick and attractive way for satellite manufacturers to get their satellites into orbit. These global catastrophes provided an opportunity for China to export launch services. The CZ 3A launcher with a cryogenic upper stage was already under way. The accelerated development of the CZ 2E was also initiated to take advantage of the favorable conditions in the global market for launch services.

In 1986, China signed several launch service agreements. These included Sweden as well as US communications satellite companies such as Westar. In October 1988, the GWIC signed an agreement with the US McDonnell Douglas Company for the use of their PAM (Payload Assist Module) upper stage with Chinese Long March vehicles. In November 1988, China signed an agreement with Brazil for the launch and use of two remote sensing satellites. This was a joint project to be carried out by the two countries. China also signed a 5-year Agreement of Cooperation in Space Technology with Australia.

A French experimental capsule was flown on a recoverable satellite launch in August 1987 followed by another German experiment in August 1988. As mentioned, Badr, a Pakistani satellite was put into orbit in a test flight of the CZ 2E launcher. On the April 7, 1990, a CZ 3 launcher successfully placed a Hughes built 1250 kg

---

[12] p. 115.

[13] See p. CRS -2.

[14] https://www.upi.com/Archives/1986/07/08/Ariane-failure-investigated-as-sabotage/287352117 9200/.

satellite into GTO. China had begun its journey for realizing the Deng vision of societally relevant space programme[15] (Chen 1999).

Under pressure from US launcher companies, the US government tried to protect the interests of its domestic launcher companies by imposing conditions on China for launching US-built satellites. It enforced a quota system that limited the number of launches that China could provide during a certain period. It also specified pricing rules for launch services.[16] (Harvey 2013). Another condition imposed on China was that it should become a party to several international and space treaties. With the intervention of the prime minister, the treaties were ratified within 10 days[17] (Chen 1993).

Despite these restrictions, US satellite makers were keen to use Chinese launch services. As mentioned earlier, the failure of US launchers as well Europe's Ariane 2 had reduced supply. Chinese launchers were available and cheaper. Though problematic, US-China relations had not yet reached a stage of total breakdown. Even after the Tiananmen incident President Bush cleared the launch of three US satellites by Chinese launchers in December 1989. Towards the end of Deng's tenure, one more organizational change was made. In 1988, the Ministry of Astronautics Industry was merged with the Ministry of Aeronautics Industry to create the Ministry of Aerospace Industry (MASI). Even though this can be thought of as a merger, the two programmes appeared to be distinct and separate for the most part. Thus, despite this change, the space and missile programme continued to function in the same way as before[18] (Chen 1993).

2.6 The Impact of the 863 Programme on China's Space Efforts

The reforms initiated by Deng with a focus on immediate applications of technology created certain concerns among some of the top scientists in the country. They were worried that too much focus on economic and social relevance would significantly erode China's science and technology capabilities over the long term. A group of four technologists associated with China's nuclear weapons programme headed by Wang Gancheng proposed a Plan to upgrade China's Science and Technology capabilities to match those of the more advanced countries in the world[19] (Feigenbaum 1999). Their proposal popularly called the 863 programme, caught the attention of Deng and

---

[15] pp. 121–122.

[16] The quota for 1989–1994 was 9 satellites, while the quota for 1995–2001 was 11 satellites. See p. 148.

[17] See p. 51.

[18] pp. 45–46.

[19] Wang Deheng was an optical physicist. The other three members were Wang Ganchang a nuclear physicist, Yang Jiaxi an electrical engineer and Chen Fangyun an electronics and radio communications expert. See pp. 95–126.

the Prime Minister Zhao Ziyang. They asked the science and technology community to identify priorities and goals for the programme. Seven areas were identified as being critical for ensuring China's long-term security and economic competitiveness. These were automation, biotechnology, energy, information technology, lasers, new materials, and space technology.

In a meeting held in August 1986 under the Chairmanship of the Prime Minister Zhao Ziyang, space emerged as one of the most important priority areas for support under the 863 programme.

The long-term goals of the 863 programme for space were identified as:

- Building a space station;
- Developing a heavy launch vehicle;
- Developing a space transportation system.

As a result of this initiative, the space programme, which had been relegated to the background, once again became a privileged programme with direct access to the country's political elite. The current developments in China's space effort can be directly linked to the strategic direction provided by the 863 programme. Zhao Ziyang believed that a strong science and technology base was necessary to safeguard China's long-term security and economic interests. When he became the designated heir to Deng in 1987, the prime minister's position was passed on to Vice-Premier Li Peng who had an engineering background.[20] Li had studied in the Soviet Union and had been impressed with the Soviet space efforts. He was also an adopted son of Zhou Enlai and was influenced by his approach and style. When he assumed the office of the prime minister, he became the champion of the space programme at the highest level in the government. This support, reinforced after the Tiananmen incident, has ensured that the space programme continues to be supported by China's top leadership[21] (Chen 1999). Oversight responsibility for the implementation of the 863 projects including those running in the leading universities of the country was assigned to COSTIND.

Figure 2.1 depicts the connections between the space programme and the Political elite during the 1986–1990 period.

The organization structure at the end of the Deng era (~1990) is shown in Fig. 2.2.

2.7   A Critical Appraisal of the Deng Period

As mentioned earlier, the economic reforms introduced by Deng made the space programme more accountable for realizing tangible benefits. Domestic and international market needs became important drivers for the space effort. The organizational restructuring, the rationalization of work between competing entities and the closure

---

[20] Zhao backed the rebels in the Tiananmen incident and was removed from power. Li Peng backed action against the Tiananmen rebels and seemed to have become closer to Deng.

[21] Reference 4 pp. 117–119.

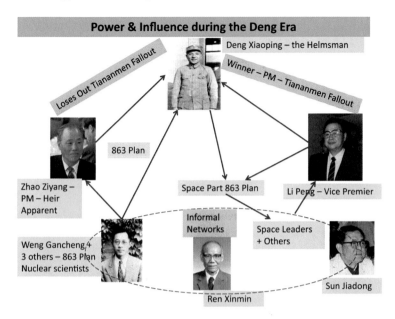

Fig. 2.1   Power and influence during the Deng Era

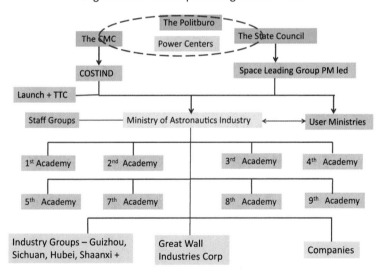

Fig. 2.2   The organization of the space programme–1990

of parallel launcher programmes, were aimed at improving efficiencies. In some cases, restructuring brought in additional resources and capabilities. The use of the space infrastructure for producing goods and services needed by domestic users and the one third system introduced by Deng were novel and unique approaches for making the space programme more accountable and domestically relevant. At an overall level, the restructuring reduced the power of COSTIND (created out of the merger of the preceding COSTND) and gave more power to the State Council and the user ministries.

The 29 satellites launched during this period also indicate a shift towards applications. Apart from satellites for communications, weather, remote sensing, and navigation satellites made their appearance. Two new launchers the CZ 3 and the CZ 4 were launched from two new launch locations set up at Xi Chang and Taiyuan. These were established to cater to Geostationary and Sun Synchronous Orbit payloads. Another remarkable development, that reflected the market driven approach of the space programme, was the development of an entirely new launch vehicle the CZ 2E for launching the heavier communications satellite into GTO. The development of this launcher (which was not a part of the original plan) was taken up on a crash basis by the First Academy (CALT). The launcher was developed and launched within a very short span of about 18 months.

Several commercial contracts to launch payloads and satellites for other countries were signed during this period. Experimental payloads for various European countries were launched on the recoverable satellite missions and a small Pakistani satellite Badr was placed in LEO during the first launch of the CZ 2E.

In the early part of this period, the space programme did not benefit from the privileged access to top political leaders that it had enjoyed earlier. It was treated like any other programme. However, after the advent of the 863 programme, that identified space as a priority area, it once again became a privileged programme with easy and ready access to top leaders. (See Sect. 2.6).

Though personal connections between the scientists and political leaders are still important, one sees the beginning of a transition from a personality driven, top-down programme towards a more technology driven, and development led agenda for space. The Deng reforms signal the beginnings of a more user based broader ecosystem emerging from the chaos of the Great Leap Forward and the Cultural Revolution. These developments set the stage for the space programme's subsequent development and growth over the next two decades.

References

Chen Y (1993) China's space commercialization effort organization, policy and strategies. Space policy

Chen Y (1999) China's space activities policy and organization 1956–1986. Dissertation, Faculty of the Columbian School of Arts and Sciences, George Washington University, ProQuest Dissertation and Theses

Feigenbaum EA (1999) Who is behind china's high technology revolution? How bomb makers remade Beijing's priorities, policies and institutions. Int Sec 24(1)

Harvey B (2013) China in space: the great leap forward. Springer Praxis Books, New York

Holland S (1986) Ariane failure investigated as sabotage. https://www.upi.com/Archives/1986/07/08/Ariane-failure-investigated-as-sabotage/2873521179200/

Kulacki G, Lewis JG (2009) A place for one's mat: China's space program 1956–2003. American Academy of Arts & Sciences

Smith MS (2006) CRS issue brief for congress, space launch vehicles: government activities, commercial competition, and satellite exports, congressional research service

Chapter 3
The End of the Cold War and the New Challenges to the Space Programme—1991–2000

3.1 Consolidation and Restructuring of China's Aerospace Ecosystem

Globally, the decade of the 1990s was one of the most turbulent periods in recent times. The breakup of the Soviet Union and the first Gulf War fundamentally altered the global power structure. The end of the Cold War and the dominance of the US in world affairs, had a direct impact on China's space programme. As the rationale for the alliance between the US and China directed against the Soviet Union started eroding, the US once more became concerned about the rise of China and the consequent implications for the world order. The Tiananmen incident that happened at the end of the Deng period provided fresh ammunition to the anti-China lobbies within the US national security establishments.

In the first few years, these developments did not immediately impact China's new thrust towards economic relevance. Though there were irritants to China's global commercialization efforts, especially with respect to the export of launch services, the Chinese space establishment seemed to be relatively unaffected and went about re-organizing the space effort to make it commercially and societally more relevant. As a part of this drive towards commercialization, they created a new corporate identity for the space programme. The idea may have been to provide an adequate resource base and a structure that would enable them to operate like Boeing or Lockheed.

In 1993, the Ministry of Aerospace Industry (MASI) was broken up into two broad entities that once again separated aviation and space activities. All the industrial entities and assets under MASI were reorganized into two corporations—The China Aviation Industry Corporation (AVIC) and the China Aerospace Corporation (CAC). Along with the creation of the CAC, China also created a new entity called the China National Space Agency (CNSA). The idea behind this restructuring appears to separate out the government related administration and coordination function from the supply of industrial products.

Unlike NASA in the US which is responsible both for the management of the space programme as well as for its interface with the government and the outside

© National Institute of Advanced Studies 2022
S. Chandrashekar, *China's Space Programme*,
https://doi.org/10.1007/978-981-19-1504-8_3

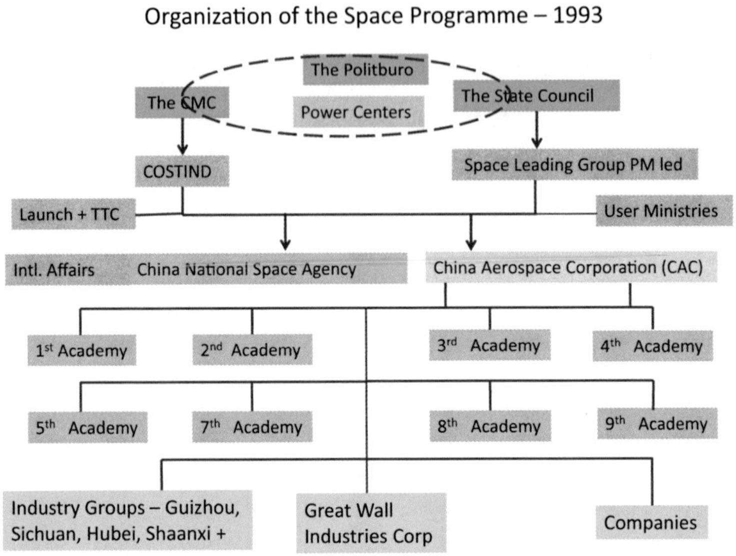

Organization of the Space Programme – 1993

Fig. 3.1 Organization of the Space Programme 1993

world, CNSA appears to be an agency that largely represents the international face of China's space programme.

The CAC was directly responsible for all other functions including the interfacing with COSTIND, the State Council and State Leading Groups. The CNSA plays a coordinating role with the real power vesting with the CAC. Policy, direction, and strategy come from the State Council and COSTIND probably operating through State Leading Groups on identified projects and programmes.

All the official white papers that China has put out on its space policies are issued by the State Council and not by CNSA[1] (Aliberti 2015).

Figure 3.1 provides the structure of the space Programme as of 1993.

3.2 A Go Ahead for the Human Space Flight Programme

The other major initiative during this period was a renewed focus on the human space flight programme. As mentioned in the previous chapter dealing with the Deng reforms, the 863-programme reinforced China's political commitment to establish a space station. Such a programme would also need the development of a suitable rocket to lift the various modules needed into space. It would also involve the development of manned spacecraft that could rendezvous with the station.

---

[1] pp. 11–12.

Case Study 1

Major Events in the Evolution of the Human Space Flight Programme

The origins of the human space flight programme go back to 1964 and Quian Xuesen who called for a Chinese programme for putting a human in space following the success of the Yuri Gagarin flight in 1962.

During the period 1964–1966, under the initiative of the CAS and the Shanghai Academy sounding rockets were developed for the testing of animals as precursors to human space flight.

The first T7A S sounding rocket carrying a cargo of white mice, white rats and fruit flies was flown on July 19, 1964. Two further missions were carried out in June 1965.

A modified T7AS rocket with a larger sealed cabin was also developed so that experiments could be performed on dogs. On July 15, 1966, a dog Xiabao, was recovered after a successful launch. This was followed by another flight in which Shanshan a female dog was also successfully recovered. The instrumentation needed for monitoring the animals during the flight as well as for monitoring parameters such as radiation levels were part of all flights. Cameras to record the behavior of the animals during the flight, especially under weightlessness conditions were also included in the package[2] (Harvey 2004).

The 1966 Space conference laid down guidelines for the artificial satellite project and the recoverable satellite project. Along with the main conference, a separate session on the human flight programme also took place. Later, a three-man COSTND committee also linked the human flight programme to the recoverable satellite programme.

An Institute of Space Medicine was set up in April 1968. Quian Xuesen was appointed as the Assistant Director. The Institute created all the facilities needed for a human flight programme that included accelerators, centrifuges, and pressure chambers. Though China went through many political upheavals, none of them seemed to have affected the functioning of this Institute in any significant way

The first astronauts were recruited in 1970 and Project 714 started in April 1971 after being approved by Mao himself. By now, the Institute of Space Medicine had become a full-fledged Training Institute. The astronaut trainees even went through weightlessness simulation exercises on British Trident aircraft.

However, the Lin Bao coup, the lack of resources and the secret nature of the project effectively killed it. The project was officially shelved in 1972. The fighter pilot recruits were all sent back to their home bases.

---

[2] pp. 44–45.

The first Chinese recoverable satellite flew in 1975. In principle, this would have made a human flight possible using a capsule like the one flown by John Glenn.

Throughout the cultural revolution and the Deng Reform period, there appears to have been unanimity within the political system that China did need a human-in-space programme. It is a continuing strand that has remained relevant, even under conditions of scarce resources, inadequate capabilities, and great political turmoil.

Immediately after Deng assumed power, 12 astronauts were recruited in 1979. However, resource constraints and other priorities dictated a deferral of the human in space programme to better times.

After the 863 programme became a reality in 1986, an expert group was set up in February 1987 to establish long-term goals for the space programme. The notion that a space station would give China a major political presence on the world stage and confer a "great power status" on it came out of these deliberations.

Two sub-committees were constituted under the 863 programme to move the matter further. One subcommittee dealt with the spacecraft and the launcher, while the other dealt with the space station. A competition for the best design was organized by the Ministry of Aerospace Industry (MASI).

Six designs came out of this competition. The evaluation of the designs by an expert team narrowed the choice to two alternatives. The China Academy of Launch Vehicle Technology (CALT) design which was also seen to be risky came out just ahead of the CAST design which was seen to be less risky. A second evaluation team then picked the CAST design as the one to be pursued.

The period 1988–1991 saw several ups and downs as far as the space station project was concerned. The debate centered around the need for a human space flight programme and its importance to development priorities. Deng had to come out of retirement to get the project approved. The architect of economic reform obviously saw no contradiction between the human space programme and the larger development agenda of China.

The project had one more hurdle to cross. CAST came up with three possible designs for the space station. One option was to have the reentry module on the top, the orbital module in the middle and the service module along with the propulsion units at the end. The second option was to have only a manned module and a service module without any orbital module. The third option was to have the orbital module on top, the manned module in the middle with the service module at the end. The third option which was very similar to the Russian Soyuz design was finally chosen.

On September 21, 1992, the Politburo gave the formal go ahead for the project. Wang Yongzhi was chosen as the Chief Designer. A new Human Flight

Office was created that reported to the State Council through the General Armament Department (GAD) of the PLA. The name Shenzhou (Divine Heavenly Vessel) was chosen for the manned module that would ferry the astronauts to the Space Station.

Along with the go-ahead, China also initiated actions for collaboration and procurement of key components from Russia[3] (Harvey 2013).They also sent two of their instructors to Star City for training. Recruitment for the Chinese Astronaut Corps began in 1996. Fourteen candidates were selected out of a larger group of about 1000 fighter pilots. Their training took place mainly in China with some weightlessness training taking place on IL-76 flights in Russia.

A huge expansion of the space infrastructure was also needed. A Taikonaut Training Centre was set up at Haidian, in the northwestern suburbs of Beijing. A new Mission Control Centre called the Beijing Aerospace Command and Control Centre (BACCC) was set up in the Yenshan District 40 km north of Beijing. High-speed communications links connected the BACCC to the National Data Centre (NDC) at Xian as well as to the Yuan Wang TT&C ships. A new Launch Complex at Jiuquan had also to be set up for[4] (Harvey 2013).Though the CZ 2E launcher was chosen as the basic configuration for the CZ 2F launcher, it had to undergo several modifications and an extensive series of tests to assure mission integrity.

CAST was chosen as the main system contractor for the Shenzhou with SAST assuming responsibility for the propulsion module. Chief Designers were appointed for the different systems required for the human space flight mission.

Though the human space flight programme had received support at the highest level, the programme did have some critics within the system. The CZ 2F rocket was ready for launch in 1999 though there were delays in the delivery of the Shenzhou spacecraft. A proposal to use the launcher capabilities for a scientific mission to the moon was made at that time. Supporters of the human space flight effort felt that any further delays could have a major impact on the future of the programme. A work around plan to use a refurbished engineering model of the reentry capsule to meet the 1999 deadline was adopted. Delays in the delivery of the propulsion module by SAST were also resolved through an accelerated testing programme. With these fixes, the programme was able to meet the 1999 launch target for the first Shenzhou mission[5] (Kulacki 2009).

On November 19, 1999, the CZ 2F rocket lifted off with the Shenzhou spaceship. After a 21-h 11-min flight, the Shenzhou capsule safely landed in the flat steppe grasslands about 415 km east of the Jiuquan launch site. Though China had carried out capsule recovery from orbit nearly 25 years earlier, this was the first time that a capsule compatible with sustaining the life of a human being had gone into orbit and

---

[3] The shopping list was very selective. Most of the development seems to be Chinese. Details are available in Chap. 11 on the human space flight programme. See pp. 257–270.

[4] pp. 257–270.

[5] pp. 19–29.

recovered safely. Though the first space flight with a human on board was still a few years away, China was on its way to becoming the third country to launch a human being into space and get him or her back safely.

The first four flights were all unmanned flights. They were used to progressively test the various components of the space station. The later manned flights also show an incremental approach. These illustrate China's conservative approach to its most ambitious and politically relevant programme.

3.3 Catering to Domestic and International Markets (1991–2000)

During this decade, China launched a total of 52 satellites using 42 launchers. Thirty-nine of these launchers were Chinese, 2 of them were procured launches from Russia, and 1 Tsinghua microsatellite was launched as a piggyback payload on a Russian Kosmos 3 M launcher.

Satellites

The decade from 1991 to 2000 saw a major expansion in China's unmanned space effort. During this period, China built and launched its second generation of communication satellites. In parallel it also bought state-of-the-art communications satellites and placed them in orbit with its own rockets. Because of China's up and down relationship with the US, the end of this decade also saw China buying communications satellites from international vendors and launch them with procured launchers. A dedicated communications satellite to meet military needs was built and launched during this period. Fifteen communications satellites were launched to cater to Chinese domestic needs. Two of these were procured launches because of US imposed restrictions on satellite exports to China.

In keeping with the Deng mandate of commercialization, this period saw a major expansion in China's efforts to export its space capabilities. A total of six communications satellites were launched into GTO for various international customers. China also launched 12 Iridium satellites into LEO for creation of the US Iridium constellation of orbiting communications satellites and one satellite Freja (a co-passenger along with a recoverable satellite) for Sweden.[6]

The recoverable satellite programme continued with the launch of five more capsules. One commercial payload for Japan was flown along with other equipment on the last recoverable satellite flight of the decade that took place in 1996.

The weather satellite programme continued with the launch of a third polar orbiting weather satellite in 1999. This was preceded by the launch of China's first GSO-based weather satellite in 1997 followed by another one in 2000.

---

[6] To demonstrate the performance of the launcher to place two Iridium satellites in orbit at the same time, China also launched a rocket with a dummy payload to demonstrate performance of the Smart Dispenser needed for launching two satellites on the same rocket.

The end of this decade also witnessed Chinese satellite launches for meeting both remote sensing and navigation needs. The China-Brazil Earth Resources Satellite (CBERS) built in collaboration with Brazil was launched into a 730 km Sun Synchronous Orbit in 1999.[7] This was followed by another similar Ziyuan satellite launch into a lower orbit in 2000 to cater to military needs. In the year 2000, China also launched two Beidou satellites into GSO for providing regional navigation services.[8]

Towards the end of the 1990s, the space programme also began work on small satellites. Tsinghua University's partnership with the UK Surrey Space Technologies Ltd (SSTL) resulted in the launch of a 50 kg microsatellite carrying a camera as a piggyback payload on a Russian Kosmos 3 M launcher. This was the first step for the creation of a major programme on small satellites that would evolve over the next 15 years. The Russian launch may have been needed to bypass US International Traffic Arms Regulations (ITAR).

Two experimental Shijian satellites were also launched during this decade. The Shijian 4 was launched as co-passenger along with a DFH 3 test satellite into GTO. It had a radiation monitoring payload. The Shijian 5 was a prototype test of the future Haiyang Ocean satellites. It was launched as a co-passenger along with the Fengyun 1C polar weather satellite.

Table 3.1 shows the breakup of the 52 satellites launched during this period.

Table 3.1 Chinese satellites 1991–2000

Satellite type	Market	Nos
GSO Communications (Built + Bought)	Domestic	15
GSO Communications (Launch services)	International	6
Orbiting Communications (LEO Iridium)	International	13
Recoverable Satellites	Domestic	5
GSO Weather Satellite (Fengyun)	Domestic	2
Orbiting (SSO) Weather Satellite	Domestic	1
Remote Sensing Satellites (CBERS + Ziyuan)	Domestic + International	2
GSO Navigation Satellites (Beidou)	Domestic	2
Shenzhou (Precursor Human Flight)	Domestic	1
Test Satellites (Co-passengers)	Domestic	2
Small Satellite (Test International partners)	Domestic	1
Piggyback	International	2
Total		**52**

[7] The CBERS satellite had another Brazilian Scientific Applications Satellite as a piggyback payload.

[8] Unlike the US GPS system that uses a 24-satellite constellation in different inclination 20,000 km orbits, the original Chinese system proposed the use of satellites in GSO for providing position fixes on the ground. Japan and India too are experimenting and implementing such systems that use satellites in GSO and Geosynchronous orbits.

Launchers

Forty-two launchers were used by China to place 51 satellites in orbit. One small Tsinghua satellite built in collaboration with SSTL of the UK was launched by a Russian Kosmos launcher as a piggyback.

The CZ 2E launcher, that had been developed in the earlier period, was used to launch GSO Communications satellites for domestic and international customers. Two of the six launches carried out during this period failed. When the improved CZ 3A launcher with a bigger cryogenic third stage became available the Chinese decided to use that and effectively retired the CZ 2E. However, when the human space flight programme was approved, the CZ 2E got a new lease of life. Since it had the largest payload capability of all available Chinese launchers at that time, it became the logical choice for boosting the Shenzhou spacecraft into orbit. A major programme for human rating the CZ 2E was initiated. This resulted in the CZ 2F launcher which has been used for all of China's human flight missions so far with a 100% reliability record.

The CZ 3A which had a larger cryogenic third stage and which could place a 2300 kg satellite in GTO was flown successfully during this period. Another variant of the CZ 3A called the CZ 3B with four liquid strap-on boosters also made its debut in 1996. The CZ 3B could place over 5000 kg in GTO and was expected to be China's largest and most capable rocket. It was meant to address the launch services requirement for heavy satellites that was emerging as the high growth segment of the communications satellite market. On its first flight carrying an Intelsat satellite in 1996, the rocket veered off its trajectory immediately after launch and exploded. Many people in villages around the launch center were killed. This incident coupled with the earlier failures of CZ 2E launches, became a trigger for a major downturn in US-China relations. The CZ 3B was later used to launch the Agila satellite for the Philippines the following year.

Though the CZ 2C launcher was to be replaced by the CZ 2D, it got a new lease of life through a major contract for launching Iridium satellites. China demonstrated its capability for these missions, through a test demonstration in which it used a newly developed Smart Dispenser to place a pair of satellites into the required LEO. As mentioned earlier 12 Iridium satellites were placed in orbit using this version of the CZ 2C launcher.

This period also witnessed the emergence of a new launcher variant for placing satellites in near polar Sun Synchronous Orbits (SSO). The CZ 4 which had been developed earlier for placing the polar weather satellites in SSO was improved further to provide the operational variant CZ 4B. In its original avatar, it was developed as an alternative to the CZ 3 for placing satellites in GTO using a conventional instead of a cryogenic upper stage. This was developed by the Shanghai Academy of Space Technology (SAST) (Eighth Academy). It became the preferred choice along with the variants CZ 4B and CZ 4C for getting to Sun Synchronous Orbits from the Taiyuan Launch Centre.

3.4 US Imposes Sanctions on China

The Tianmen Square incident in June 1989, in which many protestors wanting democratic reforms were killed, evoked a sharp reaction from the western countries including the US. The US imposed sanctions on China the following year. The breakup of the Soviet Union in 1991, also led to a re-evaluation of the close US-China ties that that had characterized much of the latter half of the Cold War. These developments had a spillover effect on US-China space cooperation as well.

The entry of cheaper Chinese launchers into the international launch services market resulted in a series of actions instigated by major aerospace companies who were directly affected by this Chinese move. Though US satellite companies liked the lower prices offered by China, they were not powerful enough to counter the influence of the aerospace majors, who had a major stake in the global launch services business.

Several failures of the CZ 2E between 1990 and 1995, indicated some incompatibility between the Chinese rocket and the US-supplied Star F Perigee Kick Motor. In January 1995, a CZ 2E rocket carrying a Hughes built Apstar satellite crashed after lift-off. Six people were killed, and many others were injured. The Perigee Kick Motor is used to transfer the orbit of a satellite from a near earth orbit to a geostationary transfer orbit. Technically, there is a coupling between the launch vehicle and the satellite that requires close interface management between them. To resolve problems that arise in such situations, joint investigations between the satellite maker, the perigee motor provider and the launcher agency are needed. These investigations into the causes of the failures were undertaken by US satellite manufacturers and their Chinese launcher counterparts. The results of these investigations by US-based satellite makers were provided to the launcher agencies in China.

In February 1996, a CZ 3B launcher carrying a Loral-built Intelsat 708 Communication satellite also malfunctioned immediately after lift-off. Six people were killed and 57 injured. In this case too, there appeared to be some incompatibility between the launch vehicle and the satellite. A joint investigation team that included both Hughes and Loral, was established to understand the nature of the problem. The results were provided to the Chinese.

This transfer of information evoked concerns within the US defence and strategic community. These related to the possibility that the transferred information would enable the Chinese to improve their ballistic missiles[9] (Smith 2006).

The coupling between the launcher and the satellite for a commercial geostationary orbit launch is very similar to that of a ballistic missile carrying Multiple Independent Re-entry Vehicles (MIRV). A special committee headed by Congressman Christopher Cox was set up to investigate the allegations that sensitive information had been provided by US satellite manufacturers to Chinese launch vehicle entities[10] (Select Committee US House of Representatives 1999). Based on the findings of

---

[9] See pp. CRS 11–CRS13.

[10] The Cox Committee Report also included commercial intelligence theft, thermo-nuclear warheads, high-performance computing missiles and space. Only the non-sensitive part of the report has been made public.

this Committee, sanctions were imposed on the concerned companies. The US also amended its export control laws to treat satellites as munitions requiring special licenses before they can be exported. In practice, these amendments made it difficult for US satellites to be launched on Chinese launch vehicles. Since the US has a major share of the global market for communications satellites, this effectively excluded Chinese launchers from the biggest segment of the launch services market. In a tightening of these laws to eliminate certain loopholes, any satellite that carried a US component would also need the clearance of the US State Department before it could be launched by another country. Since at that time, it was very difficult to build a satellite without components from the US, these laws ensured that other satellite manufacturers, especially from Europe were effectively prevented from using Chinese launchers for their satellite missions. From the year 2000 onwards, these actions by the US, prevented China from becoming a bigger player in the lucrative global launch services market[11] (Smith 2005).

Though over time, China has taken steps to mitigate the effects of these measures, the US imposed sanctions continue to be in place. They serve as a significant barrier for China's entry into the global market for launch services.

3.5 Overview (1991–2000)

The end of the Cold War, the breakup of the Soviet Union and the US-led invasion of Iraq in the first Gulf War created a new set of challenges for the Chinese Space Programme.

The major shift that took place inside China during this period was the decision to establish a space station. Though not a priority area from a development perspective, there seemed to have been near unanimity in the political circles within China, that this was necessary for China to establish its credentials as an emerging global power.

China's approach to its space station programme appears to be cautious and conservative with a minimal risk approach. Though the overall development effort is substantial, it builds upon China's capabilities in different areas. It has not been averse to seeking help in certain critical areas. China is fully self-sufficient in all aspects of the human space flight effort.

A major part of the Chinese effort (see Table 3.1) was devoted to the provision of domestic satellite communications in accordance with the Deng mandate. The focus on importing state-of-the-art satellites to provide value-added services is in tune with the development agenda put forth during the Deng era. However, it is also clear that Chinese decision-makers were aware that they need to catch up technologically, for being able to compete in the global marketplace.

The extension of weather observations into GSO was the other major application thrust. Navigation and remote sensing applications become operational towards the

---

[11] p CRS-13.

close of the decade. These developments are also directly related to the Deng mandate. They also have dual use implications that China would exercise in the future.

The developments in launcher capabilities including the CZ 2E clearly indicate that the various Chinese establishments were responding and adapting to the needs of the international marketplace.

Though the first Gulf War revealed the value of space assets in fighting and winning conventional wars, most of the focus during this period is on dual use civilian applications. However, some effort to develop dedicated military systems for communications, remote sensing, and navigation are visible towards the end of the decade. Of the 52 satellites launched during this period, 21 satellites are for various international customers. Of the remaining 31 satellites, only 9 of them seem to have a direct military function. Thus, there is a significant move towards societal and economic relevance in China's space efforts during this period.

The decision to bring the human space flight programme directly under the State Council, with the operational responsibility effectively being given to the PLA, seems to be a harbinger of change that would manifest itself in the next phase of the programme's evolution.

References

Aliberti M (2015) When China goes to the Moon, European Space Policy Studies Institute (ESPI), Studies in Space Policy, vol 11. Springer

Harvey B (2004) China's space programme from conception to manned space flight. Springer Praxis

Harvey B (2013) China in space: the great leap forward. Springer Praxis Books, New York

Kulacki G, Lewis JG (2009) A place for one's mat: China's space program 1956–2003, American Academy of Arts & Sciences

Select Committee US House of Representatives (1999) US National Security and Military/Commercial Concerns with the People's Republic of China. https://www.govinfo.gov/content/pkg/GPO-CRPT-105hrpt851/pdf/GPO-CRPT-105hrpt851.pdf

Smith SM (2006) CRS issue brief for Congress, space launch vehicles: government activities, commercial competition, and satellite exports, Congressional Research Service

Chapter 4
The Growth Phase of China's Space Programme (2001–2020)

4.1 Restructuring to Cope with New Challenges—2001–2008

The dawn of the twenty-first century saw a re-orientation of China's space programme. While this was driven in part by increasing national capabilities, a major part of the changes came about because of developments in the geo-political arena. Though the seeds were sown in the early 1990s (the First Gulf War, the Break-up of the Soviet Union, US anti-China sentiments and embargos) it was only toward the close of the twentieth century that China initiated major actions to deal with these changes.

Since its emergence as a nation state in 1949, the Taiwan issue has been a continuing irritant in the US-China relationship. Apart from the Korean War, there have been three major crises between the US and China over Taiwan. The first two crises took place in 1954–55 and 1958. In both these cases, US naval presence through its aircraft carriers deterred China from taking aggressive actions on Taiwan. The most recent of these confrontations took place in 1995–96 when PLA forces started shelling various Taiwanese targets. This was a response to the granting of a visa to South Korean President Lee for visiting the US. In response to these provocations, the US dispatched two aircraft carriers to secure Taiwan's interests. The US Naval presence in the South China Sea deterred further aggressive actions by China.

Unlike the other crises, the third Taiwan crisis took place at a time when China was on an ascending trajectory toward becoming a major power. In the aftermath of the crisis, Chinese strategists and military planners became concerned about how they could prevent or deter the US from intervening in a future Taiwan conflict.

© National Institute of Advanced Studies 2022
S. Chandrashekar, *China's Space Programme*,
https://doi.org/10.1007/978-981-19-1504-8_4

A re-orientation in approach that involved nuclear weapons, conventional weapons, missiles, and space assets that would deter both nuclear and conventional war over Taiwan became the central tenet of China's strategy.[1] The Anti-Access Area Denial (A2AD) capabilities that China sought to create involved a significant space component. These not only included major enhancements in C4ISR capabilities,[2] but also conventional weapon systems such as the Anti-Ship Ballistic Missile (ASBM) as well as Space Situational Awareness (SSA), Anti-Satellite (ASAT), and Ballistic Missile Defence (BMD) components.

This re-orientation, in which military needs became more important, resulted in a major re-organization effort in 1999. The PLA created four new organizational entities under itself. These were the General Armaments Department (GAD), General Staff Department (GSD), General Political Department (GPD), and the General Logistics Department (GLD). Under the new dispensation, the GAD and to a lesser extent the GSD were made responsible for the military component of the space programme (Stokes 2015).

COSTIND whose power had already been reduced in the re-organization of 1993, saw its power being reduced even further. It now had oversight only over the Nuclear Power Programme and administrative responsibilities for the management of defence firms.

This re-organization also directly affected the structure of the space industry. The China Aerospace Corporation (CAC) which had been created in 1993 by merging all space capabilities into one organization, was broken up into two separate corporations—the China Aerospace Science & Technology Corporation (CASC) and the China Aerospace Science & Industry Corporation (CASIC). While there is a fair degree of overlap in the activities of these two space conglomerates, the purpose of this restructuring appeared to be to separate the weapon-oriented activities from the more benign dual use activities. Though this separation is far from perfect given the interconnectedness between technologies, this appears to have been the purpose of this re-organization[3] (Stokes 2003). The various Academies and the industry groups that were all under the CAC were divided among these two new corporations[4] (Stokes 2003). Figure 4.1 provides an overview of the restructuring exercise carried out in 1999.

Individual functions relating to the civilian domain such as communications, remote sensing, weather and maybe even navigation come notionally under the State Council. The Leading Groups under the State Council that involve representation

---

[1] The 1995–96 Taiwan Strait Crisis was one of the more obvious reasons for China to worry about US intervention in a Taiwan scenario. The ASBM System and the Anti-Access Area Denial (A2AD) strategy involving a significant Space component can be directly linked to this crisis. The War in Kosovo and the bombing of the Chinese embassy in 1999 may have also influenced the re-organization and restructuring exercise.

[2] Command, Control, Communications, Computers, Intelligence, Surveillance, Reconnaissance.

[3] pp 214–218.

[4] pp 217–218.

Re-organization of the Space & Missile Programme – 1999

Fig. 4.1 Re-organization of the space and missile programme 1999

from user ministries, the relevant space corporations, maybe the CAS and of course the General Armaments Division provide the forum for the formulation of strategy and its implementation.

Similar Leading Groups under the CMC with the help of the GAD and GSD may have the responsibility for the formulation and implementation of the military requirements. These programmes are probably knit into an overall strategy by an Oversight Leading Group, that reports jointly to the CMC and the State Council. This Leading Group may be controlled by the GAD which would allot funds and resources based on the relative importance of the various projects and programmes. The human space flight programme that began in 1992, and the Lunar exploration programme initiated after 2000 have leading groups. These leading groups may provide the link between the organizations responsible for product development and delivery, with the higher-level organizations involved in the formulation of strategy. The crucial entity that appears to have a dominant say is the GAD.

4.2   The Human Space Flight Programme

This period saw the near completion of the first phase of China's human space flight programme that was to have as its grand finale in the establishment of a permanent space station. Though the programme has been delayed because of the failure of the heavy lift launcher the CZ 5, the recent return to flight of this rocket suggests that the space station will soon be a reality. There are however still some issues related to the availability of the CZ 7 that is needed for ferrying supplies to the space station. Given China's track record in resolving such technical problems, the space station is likely to become operational in the next 2–3 years.

During this period, China became the third country in the world to put a human into space and bring them back to earth. China also demonstrated several other capabilities including launching a satellite from the space capsule, space walks by astronauts and putting the first Chinese woman astronaut into space. In most of the launches a laboratory with many experiments on-board continued to orbit after the human crew had been safely returned to earth.

In addition to these relatively short-lived orbital laboratories, two successive launches spaced apart by a couple of days first placed the Tiangong mini space station in orbit followed by the docking of the Shenzhou modules with the station and its occupation by astronauts for performing a variety of tasks. Table 4.1 provides an overview of these landmark events that took place during this period.

China's ability to operate a larger space station saw a gradual and planned increase during the period 2001–2020. These actions indicate that China will continue this as a showpiece programme for sustaining a continued human presence in space. The failure of the second launch of the large CZ 5 launcher in July 2017 has delayed the establishment of the space station.

After a detailed review, the CZ 5 was successfully launched at the end of 2019 carrying a payload to the GSO. Three successful flights followed in 2020. The first tested a new crew capsule along with a novel heat shield for reentry. The two other

Table 4.1 Human space flight missions

Mission	Date	Mission duration	Orbital module life
Shenzhou 1	19-11-1999	21 h	12 days
Shenzhou 2	09-01-2001	7 days	226 days
Shenzhou 3	25-03-2002	7 days	232 days
Shenzhou 4	29-12-2002	7 days	247 days
Shenzhou 5	15-10-2003	1 day	227 days
Shenzhou 6	12-10-2005	6 days	532 days
Shenzhou 7	25-09-2008	3 days	466 days
Tiangong 1	29-09-2011	Years	Space Laboratory Mini-Station
Shenzhou 8	31-10-2011	18 days	137 days. Unmanned docking Tiangong
Shenzhou 9	16-06-2012	13 days	Visit to Tiangong 1. Woman Astronaut
Shenzhou 10	11-06-2013	15 days	Visit to Tiangong 1
Tiangong 2	15-09 2016	Years	Replacement for Tiangong 1
Shenzhou 11	16-10-2016	30 days	Two-member crew 30 day stay Tiangong 2
Tianzhou 1	20-04-2017	Supply ship	Docked cargo vessel with Tiangong 2 First flight of the CZ 7 medium lift launcher
XJY 6 Test	16-03-2020	Supply Ship	CZ 7 Test Launch—failure
New Crew Capsule	07-09-2020	Test	CZ 2F used to launch a new crew capsule

missions were the launch of the sample recovery and return from the moon and a Mars Lander and Rover to explore the Red Planet. These successful launches of the CZ 5 make it possible for China to accelerate its space station programme. It also gives it the capability to undertake more sophisticated missions to the moon and other parts of the solar system.

In April 2017, the newly developed CZ 7 launcher put the unmanned Tianzhou cargo carrier into orbit for a rendezvous with the Tiangong 2. A second launch of the CZ 7 in March 2020, failed. Since the CZ 7 is the ferry for carrying supplies to the planned space station this is likely to create some delays in the establishment of China's Space Station. Both the CZ 5 and the CZ 7 launchers are needed for the sustained operations of the Space Station. In view of the importance of the human space flight programme and the space station, an early return to flight for the CZ 7 can be expected.

Table 4.1 provides details about these Chinese activities related to the human space flight programme.

A more detailed assessment of the human space flight programme is available in Chap. 11 of the book. The programme has also seen major shifts related to the oversight and management functions. In the initial phase, control was taken away from COSTIND and handed over to the PLA's GAD.

The abolition of the GAD and the re-organization of the PLA along with the creation of a Strategic Support Force (SSF) with oversight responsibilities for military space in 2015 may also see changes in the way the Human Space Flight Programme will be managed.

4.3 The Lunar Exploration Programme

The successes achieved by the human space flight programme, brought China a lot of prestige associated with such high visibility ventures. This may have the reasons behind China's start of a major exploration programme of the Moon. As a part of a well-crafted strategy, it has since launched six Chang'e missions to the Moon. The first two phases of the programme—orbit around the moon and landing on the moon with the release of a rover—were all completed by 2017.

The organization structure that China has created for the Lunar Exploration Programme is very similar to the one it has established for the Human Space Flight Programme. The GAD was the focal point at the very top of the management line of the programme. The operational aspect comes under a new entity the State Administration for Science Technology and Industry for National Defence (SASTIND) with a three-member troika responsible for overall management, design, and science parts of the programme. An elaboration of the genesis, the approval, and the organization structure of the Programme is provided in Chap. 12 of this book.

The failure of the second launch of the CZ 5 in 2017 delayed the launch of the lunar sample return mission. The re-organization of the PLA military space programme in 2015 may have also resulted in some organizational rearrangements.

To retain the momentum of the programme, China launched another lunar mission the Chang'e 4 in December 2018 using its well-proven 3B launcher. Chang'e 4 landed on the dark side of the moon in early January 2019. This was the first time that a lunar probe had landed on the dark side of the moon. It then deployed a moon rover to explore the surface around the landing area. Data from the dark side are being relayed to the earth via another Chang'e satellite located at the earth-moon Lagrange point. Several novel experiments were also carried out.

After the return to flight of the CZ 5 in December 2019, China completed the last leg of its current Lunar Exploration Programme. On November 23, 2020, the Chang'e 5 was launched on a CZ 5 rocket from Wenchang. Lunar orbit was realized on November 28, 2020. The Lander-Ascent vehicle composite landed on the surface of the Moon on the 1st of December 2020. After collecting lunar samples, the ascent vehicle lifted off from the lunar surface for a rendezvous with the orbiter on December 3, 2020. The samples were then transferred to the return capsule which undocked from the orbiter and set course for a return rendezvous with the Earth. The return capsule with the lunar samples made a safe reentry on December 15, 2020.

With this successful mission, China has successfully achieved all the objectives of the first phase of its Lunar Exploration Programme. It is now poised to look at continuation programmes that may involve human colonies on the moon and other planets.

4.4   China's Mars Mission

In 2020 China achieved another major landmark in its planetary exploration efforts. The Tianwen 1 Mars probe lifted off from Wenchang sitting atop a CZ 5 on July 23, 2020. After entering Martian orbit in February 2021, China successfully placed the Lander Rover combination on the Martian surface in May. The Rover was deployed on the 22nd of May. China has put out several images of the Martian surface and reported on the scientific findings of the mission. It became the third country after the US and the Soviet Union to successfully land on the surface of Mars. It is only the second country to deploy a Rover on the Martian surface after the US.

4.5   Consolidation and Restructuring—2008

Another major re-organization that has a direct impact on the space programme took place in early 2008. COSTIND whose powers had been whittled down in the earlier re-organization was abolished. A new super ministry called the Ministry of Industry and Information Technology (MIIT) was created. All COSTIND functions were absorbed into departments under the new super ministry. Operations and entities under the earlier Ministry of Information Technology were also absorbed by the MIIT.

For managing the critical areas related to national defence, a new entity called the State Administration for Science Technology and Industry for National Defence (SASTIND) was established under this ministry. SASTIND would ostensibly exercise oversight and regulatory responsibilities for all components of the space programme. The restructuring exercise also created a new department under the MIIT called the Civil Military Integration Promotion Department (CMIPD) with the job of better integrating civil and military activities.

The creation of a new super ministry added more layers between the top decision-making bodies and the individual ministries. The ostensible purpose may be the better integration of civil and military activities. The abolition of COSTIND, which enjoyed a similar status to the GAD in the earlier arrangements and the transfer of most of its responsibilities to departments under the new super ministry, strengthened the role of the GAD as the entity that directly links with the decision-making elite. This process of consolidation could help in the integration of activities across industries. The CMIPD may also promote better integration of civilian and military needs.

This restructuring exercise at the top provides further avenues for the GAD to leverage its position and extend its spheres of influence into the various arms of government. The creation of SASTIND in a sense replicates the earlier coordination and oversight function that COSTIND had over the space programme. However, unlike COSTIND which directly linked the space programme with the country's political elite, SASTIND remains at a much lower level in the hierarchy of power. Its creation would certainly help improve the coordination between the space corporations, CNSA as well as other major users such as the CAS, the Weather Services, Agriculture, Oceans, etc.

Figure 4.2 provides an overview of the new structure of the space programme in 2008. As we can see from the figure the coordination and oversight functions have become much more complex. Such a trend is consistent with the challenges of using space for a variety of tasks typical of a rising space power.

4.6   Major PLA Reform and Military Space 2015

A major organizational restructuring exercise of the PLA with a special focus on military space activities was carried out at the end of 2015. Unlike the earlier changes carried out in 1999 and 2008 these shifts in responsibilities suggest a radical change in the internal power structure of China's armed forces. The new transformed organization structure indicates a military posture that is aligned with the strategic imperative of "fighting and winning local wars under conditions of informationization."

The shift away from the notion of "Military Regions" toward theatre commands that control all the resources needed to fight and win local wars, transfers combat responsibilities away from the Army, Navy, Air Force, and the Rocket Forces toward the theatre commands.

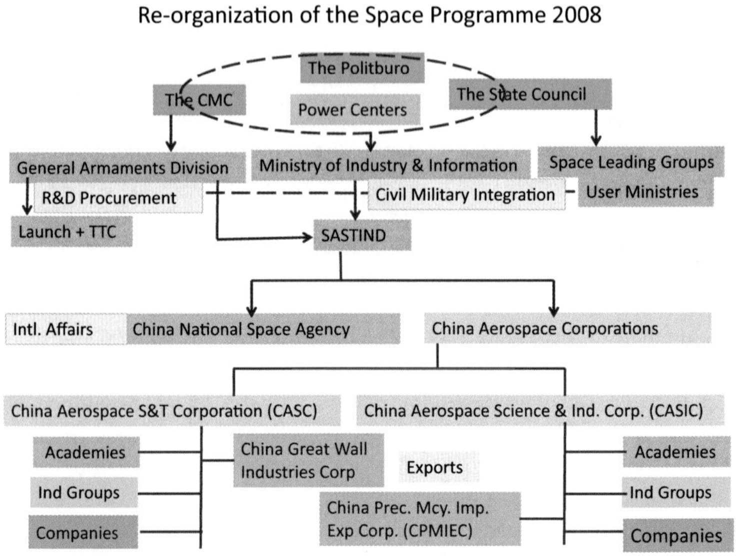

Re-organization of the Space Programme 2008

Fig. 4.2 Re-organization of the space programme 2008

The creation of a new Strategic Support Force (SSF) directly reporting to the Central Military Commission (CMC) that is responsible for space, cyber and Electronic Warfare reinforces the critical role of information dominance during conflicts.

The abolition of the various General Staff Departments that acted as intermediaries between the CMC and the combat forces and their regrouping into smaller entities directly reporting to the CMC suggest a more direct control of the CMC over the conduct of integrated combat operations by the theatre commands. These shortened commands remove power from the middle layers and make theatre commands come directly under the CMC. The overall efficiency is likely to improve because of these changes.

The restructuring exercise will also be somewhat specific to the theatre though mobility, flexibility and rapid redeployment seem to be inbuilt into the proposed changes (Rachel and Stokes 2018). The power of the Army (though a new Army HQ has been created) has been diluted while those of the other three services strengthened.

All the services have lost power and status to some degree. Their functions have been changed from direct roles in combat operations to suppliers of equipment and trained personnel to the theatre commands. The theatre commands will decide on combat operations under the direct control of the CMC.

These far-reaching organizational changes in which space capabilities play a major role lend teeth to China's proclaimed strategy of "fighting and winning local wars under conditions of informationization"[5] (Chandrashekar and Ramani 2018).

---

[5] pp 21–23.

How these changes will affect the Human Space Flight and Lunar Exploration Programmes both of which came under the GAD is still to be understood. These may be transferred to more direct oversight by agencies dealing with the civilian aspects of the programme. The restructuring of the military space activities and their implications are covered in greater detail in Chap. 13 of this book.

4.7 Major Achievements and Strategic Directions (2001–2020)

During the period 2001–2020 China launched 585 satellites using 310 launchers. Apart from the 17 Human Flight related missions, China also completed the first phase of its Lunar Exploration programme. The Chang'e 5 was launched from Wenchang on November 23, 2020. It successfully brought back samples from the moon and returned to earth on December 6, 2020. The six successful lunar missions are a major scientific triumph that establishes China as a major space power.

China also launched several application satellites during this period. Many of them had a direct military or dual use function. These larger numbers suggest that China is now entering the growth phase of its space programme with satellite products covering the gamut from human spaceflight to small satellites. The scaling up has significantly increased. There have also been periodic restructuring of the programme that aligns it more closely with space and national power aspirations. The evidence suggests that apart from diversifying into several applications, a major thrust toward the military use of space for both defensive and offensive purposes is under way.

The pace of activity has significantly picked up during the last three years (2018–2020). In 2018 China launched 106 satellites using 39 launchers. The US had only 34 launches. In 2019, China launched 82 satellites using 34 launchers. The US had only 21 launches. In 2020 however China was pushed into second place with 39 launches as compared to 44 launches by the US.

Based on the satellite catalogue Fig. 4.3 provides an overview of the thrust of China's programme for the 2001–2020 period.

One of the major contributions of the space programme to China's national security has been the creation of a constellation of operational Yaogan satellites consisting of Electronic Intelligence (ELINT), Synthetic Aperture Radar (SAR), and Electro-Optical (EO) satellites operating at different altitudes.

The 68 satellites that have been launched under the Yaogan name perform the vital C4ISR function needed for China's Anti-Access Area Denial strategy.[6] The ELINT Capabilities of this series have been significantly enhanced through the launch of a series of coplanar constellations starting in 2017 and continuing into 2020.

China conducted an ASAT test on a defunct weather satellite in 2007 to showcase its kinetic kill capabilities for destroying critical assets in space. Demonstration of

---

[6] C4ISR stands for Command, Control, Communications, Computer, Intelligence, Surveillance and Reconnaissance.

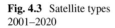

Fig. 4.3 Satellite types 2001–2020

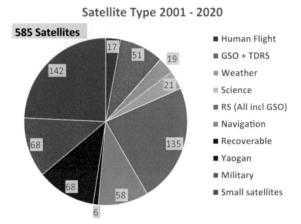

its ability to do this is a key component of its A2AD strategy for deterring the US from getting involved in a Taiwan crisis scenario.

The Yaogan satellites were preceded by several ELINT precursors. Satellites under the Shijian, Chuangxin, Shiyan, Gaofen and several other names have also been tested for a variety of military operations that include maneuvers, proximity approach and rendezvous, early warning, and space situational awareness functions. These suggest a careful well thought out strategy involving several types of space-based assets that can blind, jam, damage or destroy critical US satellites that are important for the US projection of power around the world.

51 Communications and Tracking and Data Relay Satellites (TDRS)[7] were launched during this period. These included a few bought-out satellites for domestic use launched from abroad to bypass ITAR[8] regulations. This also includes several ITAR free satellites that China delivered in orbit for several countries. In 2020, the Palapa launch that was contracted for delivery of an Indonesian satellite in-orbit failed to reach orbit due to the malfunctioning of the CZ 3B rocket.

In 2018 China began test launches of future constellations of orbiting communications satellites. These have continued in a testing mode during 2019 and 2020 as well. The constellations are likely to be both for civilian and military use and will be in orbits of 500–1100 km with several inclinations. Some recent launches of experimental constellations also suggest integration of both the remote sensing and communication aspects within one constellation. Satellites in different orbits and altitudes may be performing these functions in tandem.

The first phase of the GSO-based navigation system initiated during 1999 was transformed into a full-fledged independent navigation satellite system consisting of satellites in MEO, GSO and Inclined GSO. A 35-satellite navigation constellation has become operational in 2020. China will use this capability to make its Beidou

---

[7] These satellites in Geostationary Orbit relay data collected by orbiting satellites to specially located ground stations. They facilitate the global collection of data by orbiting satellites.

[8] ITAR is a US law and stands for International Traffic in Arms Regulation.

navigation system a global standard that could compete with other standards like the GPS.

Thirteen operational weather satellites consisting of both orbiting and GSO components were launched for meeting both domestic civilian and military operational needs. In addition, China launched test satellites to develop new capabilities and instruments for improving weather and climate predictions. About 6 such flight tests of new instruments have been carried out in the period 2018–2020. A China France Joint experiment satellite CFOSAT for oceanographic studies was launched in 2018.

Six recoverable satellite missions carrying application and research payloads including microgravity experiments were launched under both the FSW and Shijian labels. In the last three to four years there have been no recoverable satellite launches.

Apart from the Yaogan series China also launched several civilian/dual use remote sensing satellites for a variety of applications.

These include 11 CBERS Ziyuan satellites (Brazil cooperation), 5 Huanjing satellites for the use of the Asia Pacific Space Cooperation Organization (APSCO) countries, Haiyang Ocean satellites (6), Tansuo, Tianhui and other satellites for meeting various user needs.

Remote sensing satellite constellations operated by competing companies using smaller satellites (DMC constellation, Jilin constellation) promise to become more prominent in the domestic and global markets for remote sensing data. The Gaofen civilian remote sensing satellite system consisting of satellites in LEO, SSO and GSO also got a major fillip in 2019 and 2020. The constellation has become operational.

China also provided launch services for a Turkish Remote sensing satellite and built and delivered two remote sensing satellites in SSO for Venezuela during this period. Two Saudi Arabian satellites were launched with Chinese rockets. A fifteen small satellite constellation for an Argentinian company was successfully placed in orbit. Satellites for Spain, Sudan, Ethiopia and test satellites for Germany and France were launched in a piggyback mode. Two remote sensing satellites were launched for Pakistan in 2018. One was built and delivered in orbit by China while the other was developed by Pakistan as a test satellite and launched on the CZ 2C together as a pair.

The period also saw the launch of several piggyback satellites many of which may have involved testing of constellations for both military and civilian applications.

On the launcher side the period witnessed the development of new solid fueled launchers based on the DF 21, the DF 31 and other solid propellant missiles. The Kaituozhe launcher and Kuaizhou launcher were flight tested during this period with mixed results. The Kuaizhou launcher failed in two consecutive launch attempts in 2020. There have not been any launches of the Kaituozhe after 2017.

In 2015 and 2016 four other new launchers the CZ 5, the CZ 6, the CZ 7, and the CZ 11 were also flight tested. The new large heavy launcher the CZ 5 needed both for the human space flight programme and for the lunar sample return mission was launched from the new Wenchang site in 2016.[9]

These launchers will replace the earlier fleet of CZ 2, 3, 4 series of launchers in a phased manner. The new launchers use a more environmentally friendly cryogenic (liquid oxygen, liquid hydrogen) or semi-cryogenic (liquid oxygen, kerosene) propellants.

The second launch of the CZ 5 failed during the early part of the lift-off in 2017. This had affected both its human space flight programme and the Lunar Exploration Programme.

The CZ 5 returned to flight in 2019. Three further flights took place in 2020. After the successful placing of a future crewed capsule in Low Earth Orbit, it launched a mission to Mars. It was then used for the final Chang'e sample return mission from the moon that was completed in December 2020.

In 2019 and 2020 China launched the CZ 11 from offshore platforms on the Yellow Sea. It has also demonstrated its abilities to launch satellites from different launch sites on the same day. In 2019 it also launched two satellites on the same day from mobile platforms located within the same launch site.

The CZ 7 that will ferry supplies to the space station was flight tested from the Wenchang site for the first time in 2016. The Tianzhou Cargo vessel was launched using the CZ 7 in 2017 and successfully docked with the Tiangong 2 Space Station. The third launch of the CZ 7 failed in 2020. This could have an impact on the schedule for the planned space station.

Several private players are also entering the launch services business. Most of them use solid rocket stages possibly derived from China's missile programmes. Launches in 2018 and 2019 have mostly failed though one Hyperbola rocket did place four small satellites in LEO.

China continued to support several science missions in addition to its major Lunar Exploration Programme. In 2020 it successfully sent a spacecraft into a Mars orbit. It released a lander followed by a rover to explore the Martian surface. These are covered in greater detail in Chap. 12 of this book.

From the detailed scrutiny of the satellite record **229 of the total 585 satellites** launched during this period can be linked to the military. This provides clear evidence of the increasing importance that China attaches to the role of space in the pursuit of its national security objectives.

---

[9] The first launch of the CZ 5 took place on November 3 2016 from the Wenchang Launch Complex.

References

Burton Rachel, Stokes Mark (2018), The Peoples Liberation Army Theater Command Leadership: The Eastern Theater Command, Project 2049 Institute, Asia Eye Blog, August 14 2018 at https://project2049.net/2018/08/13/the-peoples-liberation-army-theater-command-leadership-the-eastern-theater-command/

Chandrashekar S, Ramani N., China's Space Power & Military Strategy – The role of the Yaogan Satellites (2018), ISSSP Report No. 02–2018. Bangalore: International Strategic and Security Studies Programme, National Institute of Advanced Studies, July 2018, http://isssp.in/wp-con tent/uploads/2018/07/ISSSP-Report-July-2018.pdf

Stokes A Mark (2003), The People's Liberation Army and China's Space and Missile Development: Lessons from the Past and Prospects for the Future, Chapter 6 in Lessons of History: The Chinese People's Liberation Army at 75, Burkitt Laurie, Scobel Andrew, Nortzel, M Larry(Editors), Strategic Studies Institute July 2003

Stokes A Mark (2015), Prepared Statement of Mark A Stokes Executive Director 2049 Institute before the US-China Economic & Security Review Commission Hearings on China's Space and Counterspace Programmes, Washington DC February 18, 2015.

Chapter 5
Strategic Trends and Future Directions

5.1 The Satellite Record

China launched a total of 678 satellites on 396 launchers from the first satellite launch of 1970 till the end of 2020. Figure 5.1 provides a year wise timeline of the satellites. Most of these satellites are Chinese satellites built for meeting domestic needs. Some of the satellites are satellites launched for other countries. These numbers also include "Chinese owned" bought satellites launched using foreign launchers.

After an initial period of lowkey activity up to the early 1990s, there is a steady growth till about 2009. The last decade has seen a significant increase in the pace of space activities. In 2018, the number of satellites launched annually had increased from about 30 to 40 per year to 106 satellites. There was a reduction to 82 satellites in 2019 with an increase to 89 satellites in 2020. The evidence indicates a doubling in China's capacity to build and launch satellites. Many of the launchings in the last 3–4 years are multiple satellites launched on the same rocket. Though there have been failures some of which have been documented, there is not much information available on satellite failures. Most reported failures are launcher failures. Of the 678 satellites, 32 satellites have failed for an overall reliability of 95%. Figure 5.2 provides a function-wise breakup of the various satellites launched by China. Figure 5.3 provides the same data in percentage terms.

Though the Human Flight programme accounts for only a small number of spacecraft (18), it has taken up a significant share of resources since 1992.

Launch services for foreign satellites as well as satellites built by China and delivered in orbit make up a total of 73 satellites. These include communications and remote sensing satellites built and delivered in orbit by China, other launches of foreign-built remote sensing satellites, smaller satellites, and piggyback payloads. This accounts for 11% of all satellites launched by China. As mentioned in Sect. 3.4, the US amended its export control laws to bring satellites under the munitions list. This resulted in a situation where US built satellites as well as satellites built elsewhere but carrying US parts had to get an export license from the US State Department for launch by a Chinese rocket. These restrictions prevented China from gaining a

© National Institute of Advanced Studies 2022
S. Chandrashekar, *China's Space Programme*,
https://doi.org/10.1007/978-981-19-1504-8_5

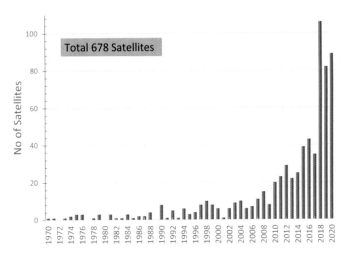

Fig. 5.1   Satellite trends China 1970–2020

Fig. 5.2   Satellite types nos.
China 1970–2020

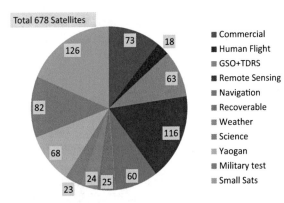

Fig. 5.3   Satellite types
percentages China
1970–2020

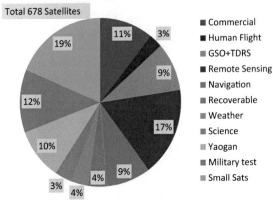

larger share of the launch services market. Except for a small interlude in the 1990s, China has been unable to export launch services. More recently, it has succeeded in delivering satellites in orbit which do not contain any US components. They are therefore not subject to sanctions under from the International Traffic in Arms Regulations (ITAR) (see Chap. 16 for details).

China has launched a total of 63 satellites (9%) for meeting its domestic communications needs. These include some satellites built elsewhere, a few of which have also been launched by foreign launchers. Fourteen of these satellites are dedicated for military use including 5 Tianlian satellites that were launched for catering to the tracking and data relay (TDRS) needs of the human space flight effort (see Chap. 7 for details).

The Yaogan series of Intelligence Surveillance and Reconnaissance (ISR) satellites (68 satellites, 10%) is dedicated for military use and constitutes a vital link in China's Anti-Ship Ballistic Missile (ASBM) as well as its Anti-Access/Area Denial (A2AD) strategy. China has launched 116 civilian or dual use remote sensing satellites. These account for 17% of the total satellites. The last 3 years have seen many of the state sponsored remote sensing satellites becoming operational. There are also many companies establishing small satellite remote sensing constellations for various applications (see Chap. 9 for details).

Weather satellites have remained a high priority area for providing operational services. China has launched 24 satellites (4%) into both polar orbits and GSO for meeting these operational needs (see Chap. 8 for details).

China has launched 60 navigation satellites. This represents 9% of all satellites launched. In 2020, China's state-of-art Beidou navigation satellite system became operational. The Beidou system has satellites in Geostationary, Inclined Geostationary and Medium Earth Orbits. The operational constellation consists of 35 satellites. Though the primary driver of this effort has been a military need, China is taking steps to make it available to all users across the world just like the US GPS navigation system. Beidou is compatible with the GPS and the European Galileo systems. China has plans to compete with the US GPS as well as Europe's Galileo for meeting civilian timing, navigation, and position fixing needs (see Chap. 10 for details).

The Recoverable Satellite Programme's original purpose was to take photographs from orbit and return the exposed film to the ground for use by the military. It has also been used as a test platform for developing and testing instruments and payloads for operational use. Equipment carried on these flights includes test setups for performing physical, chemical, materials science, and biology experiments. Twenty-five of these satellites (5%) have been launched so far. Some of them have carried payloads for other countries. A dedicated recovery satellite under the Shijian military label (Shijian 10) was launched in 2016 (see Chap. 6 for details).

Apart from the Yaogan series, several ELINT, BMD as well as ASAT test satellites (Shijian Shiyan Chuangxin and Gaofen) have been launched by China. While a detailed look at the military function is provided later, China launched about 301 satellites (under different labels) directly related to military functions that include C4ISR, Navigation, ELINT, BMD, ASAT, and the Shenzhou manned spacecraft.

There are also five recent TJS and Goafen satellite launches to the GSO that could be precursor versions of an early warning and Signal Intelligence (SIGINT) capabilities. Taking out the 73 commercial satellites launched for other countries, nearly 50% of the total number of Chinese satellites are for military use.

A scrutiny of the military satellite record suggests that China has in the past largely tried to catch-up with the US. Most of its early activities in the space realm had to do with C4ISR, infrastructure and Space Situational Awareness (SSA) functions. These can be construed to be predominantly defensive in orientation. As the international space order has evolved, space has also been transformed into a realm that directly supports military operations as seen in several conflicts including the Gulf War of 1991. Recognizing these global trends, Chinese strategists decided that space could be used advantageously to counter US projection of power in its neighbourhood. These developments led to a major change in China's approach to the military uses of space. Starting from the late 1990s, it has moved towards building operational capabilities not only in the traditional domains of space use such as C4ISR, navigation, weather services, SSA, and infrastructure but also in space war. It has formulated and implemented an Anti-Access Area Denial (A2AD) strategy that has a significant space component including ASAT and BMD weapons.

As the importance of the role of space in future wars and deterrence strategies increases, this aspect of the uses of space will become more prominent (see Chap. 13 for details of the military part of the programme). The restructuring of the space activities since 1999 had resulted in an increase in the power of the PLA's GAD. Both the human space flight space programme and the Lunar Exploration Programme come under the overall control of the GAD. **In 2015, the entire military space effort and the overall management of the space programme saw a major shift.** The GAD was eliminated with the CMC directly dealing with the military space effort through the setting up of a Strategic Support Force (SSF) responsible for Space, Cyber and Electronic Warfare (see Chaps. 11, 12 and 13 for details).

Though China has launched only 23 satellites (3%) dedicated to science, the Chinese Academy of Sciences (CAS) has been closely involved with the space programme from the very beginning. Both Recoverable Satellites as well as the Shenzhou human space flight missions have been used as platforms for performing basic science and applications experiments. Chinese institutions under the CAS have independent capabilities to design and build satellites too. As China moves towards an "innovation driven space ecosystem" the focus on breakthrough research using space is becoming an area of national importance. This is evident in its Lunar Exploration Research Initiative. The launch of the world's first Quantum Experimental Satellite in 2016, an X-ray pulsar navigation research satellite, and a Hard X-ray Modulation Telescope Satellite in 2018 suggests that China is willing to make the investments to be at the cutting edge of research in key science areas. In 2020, China landed a spacecraft to Mars and then deployed a rover to study the Martian landscape. It also completed the first phase of its Lunar Exploration Programme with the return of samples collected from the moon to earth. Though the number of science missions are small, they need advanced capabilities and complex mission planning. They also require large amounts of money and the deployment of significant human resources.

These achievements showcase China's space prowess in a very favourable way and enhance China's image on the world stage. Such missions are likely to become more important in China's future space endeavors. These aspects are covered in detail in Chap. 12 of this book.

China is also investing a lot of resources into developing the capabilities to build and launch small satellites that can perform most if not all the functions performed by their larger and more customized cousins. The launch of the Jilin and several other commercial constellations in the 2017–2020 period as well as China's purchase of the entire capacity of the three-satellite constellation of the UK's Surrey Small Satellite Company are indicative of what is likely to follow (see Chaps. 9 and 16).

5.2 The Launcher Record

The development of launch capability is a key requirement for any emerging space power. Chinese leaders have always supported indigenous development of capabilities in key strategic areas. The evolution of China's launch capabilities is dealt with in Chap. 14. Over time, China has progressively raised its capacity for placing larger payloads into various orbits of interest as well as for catering to the needs of the human space flight and solar system exploration programmes. Figure 5.4 provides a timeline of the total launches carried out by China.

China used a total of 396 launchers to place 678 satellites into various orbits during the period 1970–2020. Eleven launches were procured launches not involving any Chinese launcher. Out of these 396 launches, there were 29 failures providing an overall reliability of 93%. Many of the failures were during the early phases of the launcher development effort. The launch record since 1995 has been better.

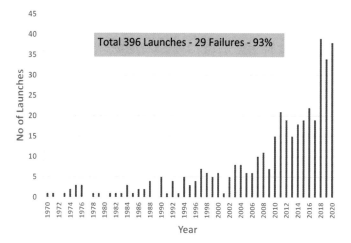

Fig. 5.4 China's space launches 1970–2020

There were 352 launches from 1995 to 2020 with 20 failures giving an overall launch reliability of 94%. This is comparable with the reliability record of other spacefaring nations. The number of launches doubled from 19 in 2017 to 39 in 2018. In 2019, there were 34 launchings which increased to 38 in 2020.

Of the 678 satellites put into orbit by China, 570 satellites were launched using 330 launch vehicles belonging to the CZ 2, CZ 3, and CZ 4 families.

China has formally retired several other launchers used earlier. These include the CZ 1 used for launching the first two Chinese satellites, the Feng Bao launcher developed by the Shanghai Group during the Cultural Revolution, as well as the CZ 2E launch vehicle developed and used in the 1990s for commercial launchings into the GSO.

Figure 5.5 provides an overview of the various CZ 2, CZ 3, and CZ 4 series of launch vehicles used by China.

Figure 5.6 presents the same data as in Fig. 5.5 in terms of percentages.

Fig. 5.5 CZ 2, 3, 4 launches 1970–2020

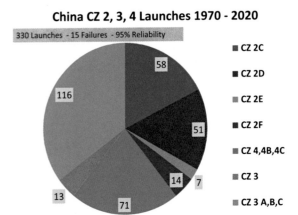

Fig. 5.6 CZ 2, 3, 4 launches 1970–2020 percentages

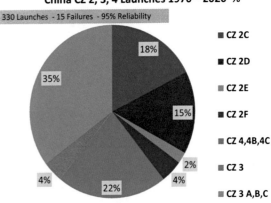

The CZ 2C and the CZ 3 were the earlier generation vehicles. The CZ 2D and the CZ 4 series were later developments. Further improvements through the addition of strap-on boosters resulted in the extension of the 2C/2D series into the CZ 2E series. With the addition of a third solid stage, this configuration tried to address the needs of the burgeoning market for launch services into Geostationary Transfer Orbit (GTO) in the 1990s. The CZ 2E in turn with improvements became the CZ 2F launcher for the Shenzhou human space flight spacecraft.

An improved cryogenic upper stage to the CZ 3 led to the creation of the CZ 3A. With the additions of strap-on boosters, more powerful launchers such as the CZ 3B, 3BE and CZ 3C launcher families have been created. These not only cater to the different communications satellite market segments but also provide the launchers for its lunar exploration programme.

The CZ 4 series (originally a non-cryogenic route for meeting the launch services requirements for communications satellites) has become the workhorse vehicle for launch into polar and sun synchronous orbits.

The CZ 2E, CZ 3 and the CZ 3A, 3B, 3BE, and 3C launchers are mainly used for placing communications satellites in GSO. They account for 41% of the launches. The CZ 2C, CZ 2D, CZ 4, CZ 4B, and CZ 4C that are used for LEO and SSO missions account for 55% of the launchings. The CZ 2F launcher used for the manned Shenzhou missions account for the remaining 4%.

Along with the Human Space Flight and Lunar Exploration Programmes, China also initiated several projects to upgrade its fleet of launch vehicles. Apart from the functional requirement for heavier payloads, these efforts address the need for several kinds of launch vehicles to cater to the different segments of the emerging satellite market. The new launchers are greener. They do not use the toxic and noxious hydrazine nitrous oxide propellant combinations but the more benign and less polluting cryogenic and semi-cryogenic rocket engines. China is also moving towards using many of its solid rocket motors and stages from its missile programme to configure launchers for the growing small satellite market.

The 2015–2020 period saw the fruition of many of these plans. The heavy lift launcher to be used for launching heavy and very heavy satellites to the GSO as well for the space station and the sample return lunar mission, was first flown successfully in 2016. However, it failed during the second launch. After an investigation, it was launched successfully in 2019. In 2020, it was successfully tested again following which it was used for a mission to Mars and for the sample return mission from the moon.

The medium lift launcher the CZ 7, needed as a cargo ferry for the Space Station as well for medium-sized communications satellites was launched in 2016 and again in 2017. The rocket however failed during its third launch in 2020. This may have some impact on China's plans to establish a space station in the next 2–3 years.

The small lift launcher the CZ 6 has been successfully launched four times starting with its inaugural flight in 2015.

The last launcher belonging to this new generation of liquid propellant launchers that use either cryogenic (liquid oxygen, liquid hydrogen) or semi-cryogenic (liquid oxygen, kerosene) engines, the CZ 8 was also successfully launched for the first time in 2020.

The DF 31 missile-derived CZ 11 small satellite launcher has also had eleven successful launches over the last 5 years. It can be considered operational.

There have also been several launchings of small launchers such as the Kaitouzhe and Kuaizhou. These are derived from the solid propellant rocket motors used in Chinese missiles. Companies in China have tried to configure new rockets using solid rocket motors from the missile programme. The pace of these launchings has seen an increase in the last 3 years. Several of these launches have failed. It may take some time for some or all of them to be able to provide a commercial launch service.

The transition from the older fleet to the "newer" "greener" fleet of launchers is still ongoing. As of 2020, most launchings have used the proven, well-established, fleet of CZ 2, CZ 3, and CZ 4 launchers. The move towards the complete replacement of the older fleet may still take another 2–3 years to complete.

5.3   Space Infrastructure

China also has in place the required infrastructure to cater to the nation's evolving needs for space products and services. Apart from the three launch bases at Jiuquan, Taiyuan, and Xi Chang, a new launch base for Geostationary and other deep space missions has become operational at Wenchang. The heavy CZ 5 launcher and the CZ 7 needed for the establishment of the space station, the launch of multiple heavy satellites into GTO as well as for the sample recovery mission from the moon, have been flown from this site from 2016 onwards.

In 2019, China launched two separate space missions from two different launch sites on the same day. It also launched two different satellites using mobile platforms from its Jiuquan space port on the same day.

The first offshore launchings were carried out from sea-based platforms in the Yellow Sea Area in 2019 with the successful launching of a CZ 11 rocket carrying seven satellites into a Low Earth Orbit (LEO).

These suggest that China is creating a flexible, launch-on-need, response capability to any military challenge it may face in the space domain.

China has a network of TT&C stations including seven ships for meeting its mission tracking needs. It has also launched five Tianlian Tracking & Data Relay Satellites (TDRS) to augment its land-based stations. Dedicated communications, weather and remote sensing data reception stations are also available for controlling the satellite and for collecting the data (see Chap. 15 for details).

China has created a substantial human resource base and a network of organizations and institutions that cover the spectrum of space activities from basic science to satellites, launchers, and applications. It has created a strong industry infrastructure for production, launch, and mission operations. Several user agencies and companies span the gamut of applications to add value to the services offered by the large constellation of satellites (see Chap. 15 for more details).

5.4 Overall Assessment

As China moves from a position of catching up towards becoming a leader in meeting civilian and military space needs, there will be significant growth. This will not only be in terms of product and services diversity but also of scale. China will also look to leverage its position in space for becoming a major power in the international geopolitical arena.

Though China became a space faring nation in 1970, it is only from around 2010 onwards that it has become a major player. It can now match both Russia and the US in terms of the number of launches as well in the number of satellites. In 2018, China had a total of 39 launches as compared to 34 launches of the US and 20 by Russia. In 2019, China was once again ahead of the US with 34 launches. The US had 27 with Russia coming third with 25 launches. Year 2020 saw a reversal with the US having launched 44 times. China was second with 38 launches while Russia was behind with 17 launches. These suggest that the China US competition will drive the global space agenda in the short and medium terms. In early 2021, China has about 210 military satellites that are currently operational. This is not too far away from the US number of about 216 (UCS 2021).

China has deployed the entire spectrum of capabilities for performing space based C4ISR functions. Five recent launches to the GSO indicate that it is not too far behind in the early warning function needed for Ballistic Missile Defence (BMD). It has also indicated that it can develop and deploy both ASAT and BMD capabilities.

Space-based C4ISR as well as deployed capabilities like the ASBM and ASAT suggest that the space component is well integrated with its Anti-Access and Area Denial Strategy. The number of operational Yaogan satellites coupled with the completion of the Beidou Navigation System have placed China almost on par with the US on the military uses of space (see Chap. 13 for details on Military Satellites).

China has turned to Russia for help in critical areas for its Human Space Flight Programme. However, these appear to be back up options for validation and benchmarking indigenous capabilities. China has created all the facilities and infrastructure within the country to cater to the increasingly sophisticated needs of the human space flight effort (see Chap. 11 for details).

Though originally behind in the performance of its indigenous communications satellites, China has now almost caught up with the more advanced western space powers including the USA. After the significant erosion of its relations with the US, China has increasingly turned to Europe for help on the technology front. However, as China has become more aggressive in its pursuit of great power status, it faces a backlash from many Western countries. How these will affect its space programme is still an open question.

Strong bilateral links with France and French Companies, the involvement of German Companies for sourcing critical technologies for communications satellites and the collaboration with the UK on the small satellite front are the means through which China has made major headway in catching up with the more advanced countries (see Chap. 7 for details).

These bilateral initiatives have been complemented through a major collaboration effort with ESA on the Space Science front via the Cluster Doublestar programme for studies on the Sun and its impact on the Earth's environment. A research satellite "China France Oceanography Satellite" (CFOSAT) has also been launched in 2018 as a part of a joint research programme.

China's Lunar Exploration Programme has grown from strength to strength. Despite the failure of the heavy lift launcher the CZ 5, China has successfully completed all goals with the successful return of samples from the moon in 2020. In 2018, it launched a Chang'e rover that landed on the dark side of the moon. This was the first satellite to do so.

Recent launches of a Quantum communication satellite, an X-ray Pulsar time measurement satellite, a hard X-ray Modulation Telescope, and other state-of-the-art science missions suggest that it wants to catch up and even out-perform some of the more advanced science powers. The Mars mission completed in 2020 has brought in additional laurels and political prestige to China (see Chap. 12 for more details on the Science programme).

Its remote sensing programme has always had a very strong component of indigenous development. It has also shown through its deployment of the Yaogan constellation that it has the industrial capacity to scale up its space operations.

However, a closer look at the satellite record for remote sensing does suggest a significant amount of duplication. Internal issues related to the coordination of the needs of various user groups could be one reason for some of this. It is also possible that the Chinese are intentionally creating independent capabilities at multiple locations to foster competition and innovation (see Chap. 9 for details).

In both navigation and meteorology, China has already deployed operational systems. These are well on their way to becoming more robust and efficient.

China took a 5% stake in the European Galileo system when it was first proposed in Europe in the 1990s. However, the collaborative efforts have not fructified. There has been speculation that China did benefit significantly from this involvement and got the technology needed for building space quality atomic clocks for use in navigation satellites. After Europe made the Galileo project a 100% government supported effort, this cooperation has come to a standstill.

Both Europe and China are well on their way towards establishing competing navigation systems. They will compete with the US GPS (see Chaps. 8 and 10 for details on Weather Satellites and Navigation Satellites).

China has also invested adequately in the emerging technologies related to small satellites that may fundamentally alter the structure of the space industry. Multiple organizations spread across the entire space ecosystem including Universities are engaged in building, launching, and using small satellites for a variety of applications. Many of them may become commercial ventures catering to the various user segments that require space services (see Chap. 15).

A new stable of advanced launchers is also becoming available for meeting civilian and military needs. A privately developed small launcher was launched for the first time in 2018. Other company launches took place in 2019 as well. Though what is meant by private investment in China is not very clear, many more companies will develop new launchers as China attempts to mimic the US ecosystem (see Chap. 14 for details of China's Launch Vehicles).

China can also use its assets in space for improving its standing among the nations of the world. Regional initiatives such as Asia Pacific Space Cooperation Organization (APSCO) and global initiatives like its tie-up with UN Office for Outer Space Affairs (UNOOSA) provides opportunities for countries to fly their experiments on China's Space station (see Chap. 16 for details).

5.5 China as a Space Power

China's major effort to establish and operate a space station as well as its investments in the Lunar and Mars exploration efforts suggest that its global ambitions and aspirations are closely linked with its space capabilities. The role of these major initiatives in the transformation of China's S & T capabilities from that of a follower country to that of an innovative leader has long been recognized by the political leadership. The elite scientists and technologists of China seem to have become a more influential voice in the realization of these grand ambitions.

The Union of Concerned Scientists (UCS) provides periodic country-wise updates on the current operational satellites. The latest release is dated April 30, 2021. Figure 5.7 provides a snapshot of the space capabilities of the major space powers in 2021 that is based on this data set (UCS 2021).

China has the second largest number of operational satellites with 431. The US has 2505 satellites and is ahead. However, 1441 satellites out of the US total of 2505

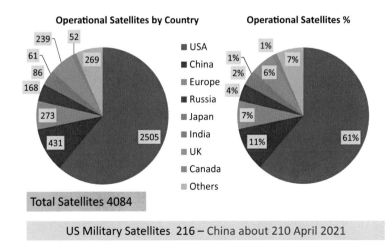

Fig. 5.7 The 2021 space order—the China US dynamic

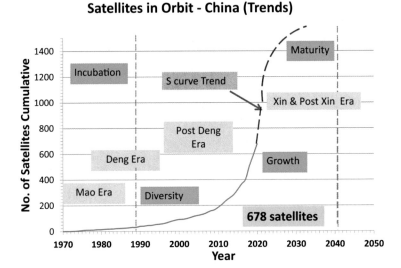

Fig. 5.8 Satellites in orbit China trends

represents the small satellite Starlink Constellation pioneered by Elon Musk through his company SpaceX. In the last 5 years, US companies have been at the forefront of a major transformation of the space industry. This would ensure that the US will continue to be the dominant player in space.

If the Starlink satellites are taken out to provide a more balanced comparison across the spectrum of other space activities, China has 431 operational satellites as compared to the US count of 1004. The US is still ahead but China is the only other contender that is close to matching it.

In terms of military satellites however, the two countries are almost on par. China has 210 military satellites as compared to 216 military satellites for the US (UCS 2021).

It is clear from the above that China is well on its way to becoming a major player in the global space order. As of now it appears to be the only power that can even remotely threaten the US domination of the high ground of space. Figure 5.8 plots the cumulative no of satellites launched by China up to and including the year 2020. The figure shows that the cumulative no of satellites is growing exponentially as of 2020. An S curve can be fitted to the trend line that is observed.[1]

---

[1] Ideally, one must plot the total number of operational satellites at any given point in time to fit an S curve. Since such a number is not readily available, the cumulative number is used as a surrogate variable.

From the S curve, it appears that the pace of satellite launches is likely to continue to grow till at least 2040.[2] The cumulative satellite number by that time is likely to cross the 1500-satellite mark. The advent of small satellites has created a new growth opportunity in space.

The various phases of the S curve can be linked to the periods identified in the earlier part of this narrative. The incubation phase of the programme took place during the Mao period starting from around 1958 and going on to 1976 (see Chap. 1).

The diversity phase during which the programme started to make several types of satellites for various applications lasted from 1977 to about 1990 when Deng formally relinquished power (see Chap. 2).

The post-Deng era saw the programme get into its growth phase. The move towards human space flight as well as the extension of capabilities for going to the moon took place in the 1990s and the early post-2000 period, respectively. This period also saw a renewed shift towards the military uses of space with the GAD of the PLA becoming a major focal point for many of the programmes (see Chap. 3).

The advent of Xi around 2010 resulted in a major increase in the scope and scale of China's space activities. A more aggressive China that is willing to create the space assets needed for countering US actions in its neighborhood is evident from the satellite and launch record. A major re-organization of the PLA and the military space programme was initiated in 2015 (see Chap. 4).

From the trends seen so far, it is likely that China will remain on a growth trajectory, with some leveling off in the period after 2040. The US is still ahead though China is catching up. In terms of military space, it is almost on par with the US on numbers but may still be behind in capabilities. However, unlike the US, China is only worried about its region and not about projecting power across the world. This will tilt the balance towards China if there is a US China confrontation in the South or East China seas. Figure 5.9 provides the relative competitive positions of the major space powers on the global Space Industry evolutionary S curve.

As stated earlier, the world space industry is still in the growth phase and continues to be US-dominated. Though China started its space programme almost immediately after the launch of Sputnik, it lost precious time during the Chairmanship of Mao. Despite this, it has managed to catch up and broaden its use of space to forge ahead of both Europe and Russia. It has now emerged as the major competitor to the US in the realm of space. The US dominance of both the civilian and military uses of space is now under threat.

---

[2] Though one can fit a proper S curve to the data, only an approximate curve has been used. A detailed forecasting exercise is not needed since only broad trends are being studied.

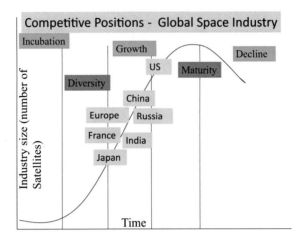

Fig. 5.9 Competitive positions global space industry

The breakup of the Soviet Union has led to a major setback of Russia's space industry. Though the situation has improved somewhat, it is likely to have difficulties in going ahead. Developments along its borders with Europe, especially in Ukraine, are also likely to constrain its growth and its competitive position in the global space industry.

Historically, Europe has been a major space power with France, the UK, Germany, and Italy being major players. The exit of the UK from the European Union (EU) is likely to erode this position. France and Germany have been the pillars of the EU. The collective capabilities of the EU still make it a powerful space power that has benefited from cooperation with China in the past. The strong EU ties with the US that had eroded during the Trump Administration are likely to strengthen during the Biden Presidency. These developments are likely to have an impact on the China-EU relationship. How this will playout is still an open and context-dependent question. On the whole, Europe may be somewhat more guarded in its cooperative endeavors with China.

India too has a buoyant largely civilian space industry dominated by one major player. Structural issues within the Indian space ecosystem can hinder growth for India to catch up with the other major powers. Its competitive position in key functional areas is not as strong as those of the other powers though it has a comprehensive space effort that covers the entire value chain and the spectrum of applications. The absence of a clear national strategy on the military uses of space is also likely to come in the way of its emergence as a significant space power.

Japan in recent years has moved away from a largely peaceful orientation in the uses of space towards a more militaristic position. Problems with a rising China and China's North Korean allies have forced it into adopting a more militaristic stance.

Though there are strong internal pressures not to move towards a more aggressive posture, the problems of history, especially in relation to some of its neighbors, may make it move in a direction of increased military use.

From the above assessment, it appears that the US-China space dynamic will drive the global space ecosystem. Europe and Russia have chosen to cooperate rather than confront China and have worked with China to sell products, services, and technology. Though Europe may still choose to side with the US in crunch situations, they recognize China's emergence as a space power and may be prepared to work with them for mutual benefit. While under the Obama Administration, it was largely a China-Russia alliance against the other space powers with Europe being selectively neutral, the advent of Trump had changed these equations. The US China relations took a nosedive under the Trump administration. There is no reason to believe that Biden wants to improve US-China relations. Biden unlike Trump will also try to build an alliance against China. The divisions within the Western powers during the Trump regime may have worked in China's favor despite sanctions affecting its economy. This may not be possible under the new US regime. Emerging powers like India, which now has a major border problem with China, are also likely to become a more active partner in a US-led anti-China alliance. An increasingly hostile US-China relationship will directly affect the global space industry. Unlike in the past, China is today in a much stronger position to directly compete with the US both in the civilian and military uses of Space. One can see the emergence of two competing networks one led by the US and the other led by China. The US-led network under Biden is likely to become stronger. How this dynamic will play out in terms of the longer term relationships between major space powers and the other countries of the world is still an ongoing work in progress.

From this overview of the global space system, it is also clear that the only challenger to the US dominance of space is China. Increasingly space will be one of the major routes through which China will channel its responses to various US moves for its continued dominance of the world including the Indo Pacific region.

Annexure 5.1: List of Satellite Launchings 1970–2020

No	Satellite Name	Launch Date	Launcher	Launch site	Perigee Km	Apogee Km	Inclination Degree	Mass Kg	Comments
1	Dong Fang Hong	4/24/1970	Long March 1	Jiuquan	440	2386	68.50	173	Technology Military
2	Shi Jian 1	3/3/1971	Long March 1	Jiuquan	267	1830	69.90	221	Technology Military
3	JSSW	9/18/1973	FengBao 1	Jiuquan	Failure	Failure	NA	1100	Technology Military
4	JSSW	7/14/1974	FengBao 1	Jiuquan	Failure	Failure	NA	1100	Technology Military
5	FSW	11/5/1974	Long March 2A	Jiuquan	Failure	Failure	NA	1800	Reconnaissance Military
6	JSSW 1	7/26/1975	FengBao 1	Jiuquan	183	460	69.90	1107	ELINT Military
7	FSW 0-1	11/26/1975	Long March 2A	Jiuquan	179	479	63.00	1790	Reconnaissance Military
8	JSSW 2	12/16/1975	FengBao 1	Jiuquan	186	387	69.00	1109	ELINT Military
9	JSSW 3	8/30/1976	FengBao 1	Jiuquan	198	2145	69.10	1108	ELINT Military
10	JSSW 6	11/10/1976	FengBao 1	Jiuquan	Failure	Failure	NA	1100	ELINT Military
11	FSW 0-2	12/7/1976	Long March 2A	Jiuquan	171	480	59.10	1790	Reconnaissance Military
12	FSW 0-3	1/26/1978	Long March 2A	Jiuquan	161	479	57.00	1810	Reconnaissance Military
13	Shijian-1	7/28/1979	FengBao 1	Jiuquan	Failure	Failure	NA	663	ELINT Military?
14	Shijian-1A	7/28/1979	FengBao 1	Jiuquan	Failure	Failure	NA	663	ELINT Military?
15	Shijian-1B	7/28/1979	FengBao 1	Jiuquan	Failure	Failure	NA	663	ELINT Military?
16	Shi Jian 2A	9/19/1981	FengBao 1	Jiuquan	390	1600	59.50	257	ELINT Military
17	Shi Jian 2	9/19/1981	FengBao 1	Jiuquan	235	1645	59.50	483	ELINT Military
18	Shi Jian 2B	9/19/1981	FengBao 1	Jiuquan	232	1595	59.50	28	ELINT Military
19	FSW 0-4	9/9/1982	Long March 2C	Jiuquan	174	393	62.90	1780	Reconnaissance
20	FSW 0-5	8/19/1983	Long March 2C	Jiuquan	172	389	90.10	1840	Reconnaissance
21	Shiyan Weixing	1/29/1984	Long March 3	Xichang	290	460	31.00	461	GSO Comsat first mission Failed
22	STTW	4/8/1984	Long March 3	Xichang	35521	36383	0.70	461	GSO Comsat - first success
23	FSW 0-6	9/12/1984	Long March 2C	Jiuquan	174	400	67.90	1810	Reconnaissance
24	FSW 0-7	10/21/1985	Long March 2C	Jiuquan	171	393	62.90	1810	Reconnaissance
25	STTW 1	2/1/1986	Long March 3	Xichang	35895	36225	0.2.	433	GSO Comsat

No	Satellite Name	Launch Date	Launcher	Launch site	Perigee Km	Apogee Km	Inclination Degree	Mass Kg	Comments
26	FSW 0-8	10/6/1986	Long March 2C	Jiuquan	173	385	57.00	1770	Reconnaissance Military
27	FSW 0-9	8/5/1987	Long March 2C	Jiuquan	172	400	62.90	1810	Reconnaissance Military
28	FSW 1-1	9/9/1987	Long March 2C	Jiuquan	206	310	63.00	2070	Reconnaissance Military
29	Chinasat 1/STTW 2	3/7/1988	Long March 3	Xichang	35784	36612	0.50	441	GSO Comsat
30	FSW 1-2	8/5/1988	Long March 2C	Jiuquan	206	310	63.00	2130	Reconnaissance Military
31	Feng Yun 1-1	9/6/1988	Long March 4	Taiyuan	881	995	99.10	750	1st Orbiting Weather satellite
32	Chinasat 2 / STTW 3	12/22/1988	Long March 3	Xichang	35756	37318	0.50	441	GSO Comsat
33	Chinasat 3 / STTW 4	2/4/1990	Long March 3	Xichang	35780	37199	0.50	441	GSO Comsat
34	Asiasat 1	4/7/1990	Long March 3	Xichang	35791	36744	0.50	1250	GTO Mass. GSO Comsat
35	Aussat demo	7/16/1990	Long March 2E	Xichang	Failure	Failure	NA	7353	Launch demo for Australia
36	Badr-1	7/16/1990	Long March 2E	Xichang	205	986	28.50	52	Commercial Launch Pakistan
37	Qi Weixing 1	9/3/1990	Long March 4	Taiyuan	884	900	98.90	NA	Balloons for Weather
38	Qi Weixing 2	9/3/1990	Long March 4	Taiyuan	862	902	98.90	NA	Balloons for Weather
39	Feng Yun 1-2	9/3/1990	Long March 4	Taiyuan	885	900	98.90	880	2nd Orbiting Weather satellite
40	FSW 1-3	10/5/1990	Long March 2C	Jiuquan	208	311	57.00	2080	Reconnaissance Military
41	Chinasat 4 / STTW 5	12/28/1991	Long March 3	Xichang	219	2451	31.10	1024	GTO Mass – GSO Comsat
42	FSW 2-1	8/9/1992	Long March 2D	Jiuquan	172	330	63.17	2590	Reconnaissance Military
43	Optus B-1	8/13/1992	Long March 2E	Xichang	35657	37330	0.33	1582	Australian Commercial COMSAT
44	Freja	10/6/1992	Long March 2C	Jiuquan	596	1763	63.00	259	Swedish Satellite - commercial
45	FSW-1 4	10/6/1992	Long March 2C	Jiuquan	214	312	63.00	2060	Reconnaissance Military
46	Optus B-2	12/21/1992	Long March 2E	Xichang	Failure	Failure	NA	7650	Australian Commercial COMSAT
47	FSW-1-5	10/8/1993	Long March 2C	Jiuquan	209	300	57.00	2100	Recon Military Satellite Failed
48	DFH 3 Test / Kuafu	2/8/1994	Long March 3A	Xichang	210	36054	28.50	1600	GTO Mass Generation 2 Comsat
49	Shi Jian 4	2/8/1994	Long March 3A	Xichang	209	39118	28.50	400	Technology - Military
50	FSW 2-2	7/3/1994	Long March 2D	Jiuquan	174	343	63.00	2760	Reconnaissance Military
51	Apstar 1	7/21/1994	Long March 3	Xichang	35269	41820	3.94	1,383	Domestic GSO COMSAT
52	Optus B-3	8/27/1994	Long March 2E	Xichang	30778	38978	1.60	1700	Australian Commercial COMSAT
53	DFH3/Zhongxing 6A	11/29/1994	Long March 3A	Xichang	35181	35993	0.30	2232	Domestic GSO COMSAT
54	Apstar 2	1/25/1995	Long March 2E	Xichang	Failure	Failure	NA	7650	Domestic GSO COMSAT

No	Satellite Name	Launch Date	Launcher	Launch site	Perigee Km	Apogee Km	Inclination Degree	Mass Kg	Comments
55	Asiasat 2	11/28/1995	Long March 2E	Xichang	34375	3932	0.60	3485	Domestic GSO COMSAT
56	Echostar 1	12/28/1995	Long March 2E	Xichang	17572	35072	5.00	3288	US GSO COMSAT Commercial
57	Intelsat 708	2/14/1996	Long March 3B	Xichang	Failure	Failure	NA	4576	US GSO COMSAT Commercial
58	Apstar 1A	7/3/1996	Long March 3	Xichang	34089	42010	3.24	1,383	Domestic GSO COMSAT
59	Chinasat7/Zhongxing7	8/18/1996	Long March 3	Xichang	200	17229	27.00	1200	Domestic GSO COMSAT
60	FSW 2-3	10/20/1996	Long March 2D	Jiuquan	171	342	63.00	2970	Reconnaissance Military
61	Chinasat6/Zhongxing6B	5/11/1997	Long March 3A	Xichang	35778	35788	0.30	2232	Domestic GSO COMSAT
62	Feng Yun 2-1	6/10/1997	Long March 3	Xichang	35786	35792km	0.30	1200	1st GSO Weather Satellite
63	Agila	8/19/1997	Long March 3B	Xichang	35777	35791	0.30	3770	Philippines GSO COMSAT
64	Iridium Dummy SD test	9/1/1997	Long March 2C	Taiyuan	623	632	87.00	650	US orbiting Comsat demo
65	Apstar 2R	10/16/1997	Long March 3B	Xichang	NA	NA	0.03	3,747	Domestic GSO COMSAT
66	Iridium 42	12/8/1997	Long March 2C	Taiyuan	623	632	86.30	650	US orbiting Comsat - commercial
67	Iridium 44	12/8/1997	Long March 2C	Taiyuan	623	633	86.30	650	US orbiting Comsat - commercial
68	Asiasat 3	12/24/1997	Proton-K	Baikonur	Failure	Failure	NA	3400	Bought COMSAT Russian Launch
69	Iridium 51	3/25/1998	Long March 2C	Taiyuan	624	641	86.60	650	US orbiting Comsat - commercial
70	Iridium 61	3/25/1998	Long March 2C	Taiyuan	624	641	86.60	650	US orbiting Comsat - commercial
71	Iridium 69	5/2/1998	Long March 2C	Taiyuan	625	641	86.40	650	US orbiting Comsat - commercial
72	Iridium 71	5/2/1998	Long March 2C	Taiyuan	626	641	86.40	650	US orbiting Comsat - commercial
73	Chinastar 1	5/30/1998	Long March 3B	Xichang	35778	35795	0.02	2,984	Domestic GSO COMSAT
74	Sinosat 1	7/18/1998	Long March 3B	Xichang	36074	36130	1.80	2,840	Domestic GSO COMSAT
75	Iridium 76	8/19/1998	Long March 2C	Taiyuan	617	630	86.40	657	US orbiting Comsat - commercial
76	Iridium 78	8/19/1998	Long March 2C	Taiyuan	616	630	86.40	657	US orbiting Comsat - commercial
77	Iridium 88	12/19/1998	Long March 2C	Taiyuan	623	624	86.40	657	US orbiting Comsat - commercial
78	Iridium 89	12/19/1998	Long March 2C	Taiyuan	623	624	86.40	657	US orbiting Comsat - commercial
79	Asiasat 3S	3/21/1999	Proton-K	Baikonur	GSO	GSO	0.02	4,137	Bought COMSAT-Russian Launch
80	Fengyun 1-3	5/10/1999	Long March 4B	Taiyuan	849	868	98.80	954	3rd Orbiting Weather Satellite
81	Shi Jian 5	5/10/1999	Long March 4B	Taiyuan	946	865	98.80	298	Technology Military
82	Iridium 93	6/11/1999	Long March 2C	Taiyuan	746	752	96.50	657	US orbiting Comsat - commercial
83	Iridium 92	6/11/1999	Long March 2C	Taiyuan	626	647	86.40	657	US orbiting Comsat - commercial

No	Satellite Name	Launch Date	Launcher	Launch site	Perigee Km	Apogee Km	Inclination Degree	Mass Kg	Comments
84	SAC 1	10/14/1999	Long March 4B	Taiyuan	732	747	98.60	60	Brazil Piggyback
85	CBERS 1	10/14/1999	Long March 4B	Taiyuan	728	745	98.50	1450	China Brazil Remote sensing
86	Shenzhou 1	11/19/1999	Long March 2F	Jiuquan	197	325	42.60	7600	Manned Mission – first test
87	Feng Huo 1	1/25/2000	Long March 3A	Xichang	35782	35789	0.75	2310	Military GSO COMSAT
88	Fengyun 2-2	6/25/2000	Long March 3	Xichang	35786	35864	1.07	1200	2nd GSO Weather Satellite
89	Tsinghua 1	6/28/2000	Kosmos 3M	Plasetsk	677	703	98.70	50	Smallsat-SSTL UK collaboration
90	Zi Yuan 2A	9/1/2000	Long March 4B	Taiyuan	468	493	97.40	1500	Domestic RS recon satellite
91	Beidou 1-1	10/30/2000	Long March 3A	Xichang	36093	36252	0.00	2300	1st GSO Navigation Satellite
92	Beidou 1-2	12/20/2000	Long March 3A	Xichang	35979	36166	1.52	2300	2nd GSO Navigation Satellite
93	Shenzhou 2	1/9/2001	Long March 2F	Jiuquan	197	336	42.60	7600	Manned mission 2nd test
94	Shenzhou 3	3/25/2002	Long March 2F	Jiuquan	195	336	41.40	8600	Manned mission 3rd test
95	Feng Yun 1-4	5/15/2002	Long March 4B	Taiyuan	851	873	98.80	954	Orbiting Weather Satellite
96	Haiyang 1A	5/15/2002	Long March 4B	Taiyuan	792	795	98.80	367	First Domestic Ocean Satellite
97	HTSTL-1	9/15/2002	Kaituozhe 1	Taiyuan	Failure	Failure	NA	50	Technology Military
98	Zi Yuan 2B	10/27/2002	Long March 4B	Taiyuan	473	490	97.40	1500	Domestic RS recon satellite
99	Shenzhou 4	12/29/2002	Long March 2F	Jiuquan	331	337	42.40	7600	Manned mission 4th test
100	AsiaSat 4	4/12/2003	Atlas III B	CCAS	NA	NA	0.00	3,760	Bought COMSAT-Russian Launch
101	Beidou 1-3	5/24/2003	Long March 3A	Xichang	35884	35886	0.14	2300	3rd GSO Navigation Satellite
102	Shenzhou 5	10/15/2003	Long March 2F	Jiuquan	197	328	42.40	7970	Yan Liwei first Taikonaut
103	CBERS 2	10/21/2003	Long March 4B	Taiyuan	731	750	98.50	1450	China Brazil civilian RS Satellite
104	Chuangxin 1-01	10/21/2003	Long March 4B	Taiyuan	732	750	98.50	88	Technology Military
105	FSW 3-1	11/3/2003	Long March 2D	Jiuquan	194	325	63.00	3000	Reconnaissance / Science
106	Zhongxing 20	11/14/2003	Long March 3A	Xichang	NA	NA	0.31	2300	Domestic GSO COMSAT
107	Double Star 1	12/29/2003	Long March 2C	Xichang	570	78948	28.20	350	Solar Mission-ESA Collaboration
108	Naxing 1	4/18/2004	Long March 2C	Xichang	600	615	97.60	25	Technology Military
109	Tansuo 1	4/18/2004	Long March 2C	Xichang	600	615	97.60	204	Earth Remote Sensing Mapping
110	Double Star 2	7/25/2004	Long March 2C	Taiyuan	560	38568	90.00	330	Solar Mission-ESA Collaboration
111	FSW 3-2	8/29/2004	Long March 2C	Jiuquan	199	491	63.00	3800	Reconnaissance
112	Shi Jian 6-1A	9/8/2004	Long March 4B	Taiyuan	588	604	97.70	375	Probably Electronic Intelligence

No	Satellite Name	Launch Date	Launcher	Launch site	Perigee Km	Apogee Km	Inclination Degree	Mass Kg	Comments
113	Shi Jian 6-1B	9/8/2004	Long March 4B	Taiyuan	593	602	97.70	975	Probably Electronic Intelligence
114	FSW 3-3	9/27/2004	Long March 2C	Jiuquan	205	297	63.00	2600	Reconnaissance / Science
115	Fengyun 2-3	10/19/2004	Long March 3A	Xichang	35786	GSO	NA	1380	GSO Weather Satellite
116	Zi Yuan 2C	11/6/2004	Long March 4B	Taiyuan	473	483	97.30	2500	Remote Sensing Reconnaissance
117	Tansuo 2	11/18/2004	Long March 2C	Xichang	397	581	98.10	300	Earth Remote Sensing Mapping
118	Apstar 6	4/12/2005	Long March 3B	Xichang	35781	35791	0.06	4680	Domestic GSO COMSAT
119	Shijian 7	7/5/2005	Long March 2D	Jiuquan	547	570	97.60	400	BMD Military
120	FSW 3-4	8/2/2005	Long March 2C	Jiuquan	166	494	63.00	3800	Reconnaissance / Science
121	FSW 3-5	8/29/2005	Long March 2D	Jiuquan	203	298	63.00	3600	Reconnaissance / Science
122	Shenzhou 6	10/12/2005	Long March 2F	Jiuquan	193	337	42.40	8100	2nd Manned Flight
123	Beijing 1/DMC 4	10/27/2005	Kosmos 3M	Plasetsk	683	707	98.18	166	Smallsat RS SSTL Russian Launch
124	Yaogan 1	4/26/2006	Long March 4C	Taiyuan	603	626	97.80	2721	ISR SAR satellite Military
125	Shi Jian 8	9/9/2006	Long March 2C	Jiuquan	178	449	63.00	3600	Reconnaissance / Science
126	Zhongxing-22A	9/13/2006	Long March 3A	Xichang	35767	35807	0.07	2300	Domestic GSO COMSAT
127	Shijian 6-2A	10/23/2006	Long March 4B	Taiyuan	594	600	97.70	375	ELINT Military
128	Shijian 6-2B	10/23/2006	Long March 4B	Taiyuan	596	601	97.70	975	ELINT Military
129	Sinosat 2	10/28/2006	Long March 3B	Xichang	Failure	Failure	NA	5100	Domestic GSO COMSAT
130	Fengyun 2-4	12/8/2006	Long March 3A	Xichang	35777	35791	NA	1380	GSO Weather Satellite
131	Beidou 2-1 G-1	2/2/2007	Long March 3A	Xichang	35647	36428	NA	2300	GSO Navigation Satellite
132	Haiyang 1B	4/11/2007	Long March 2C	Taiyuan	783	814	98.60	442	Ocean RS Satellite
133	Compass M1	4/13/2007	Long March 3A	Xichang	21519	21545	55.30	2300	MEO Navigation Satellite
134	NIGCOMSAT-1	5/13/2007	Long March 3BE	Xichang	35782	35789	42.0	5155	Nigeria COMSAT-Commercial
135	Picosat	5/25/2007	Long March 2C	Jiuquan	631	655	97.90	1	Small satellite Technology
136	Yaogan-2	5/25/2007	Long March 2C	Jiuquan	631	655	97.90	600	EO Reconnaissance Military
137	Sinosat 3	5/31/2007	Long March 3A	Xichang	NA	NA	0.08	2300	Domestic GSO COMSAT
138	Chinasat 6	7/5/2007	Long March 3B	Xichang	35771	35801	0.02	4600	Domestic GSO COMSAT
139	CBERS 2B	9/19/2007	Long March 4B	Taiyuan	773	775	98.60	1450	China Brazil Remote sensing
140	Chang e 1	10/24/2007	Long March 3A	Xichang	Lunar	Lunar	64.00	2350	Planetary Probe (Lunar)
141	Yaogan-3	11/11/2007	Long March 4C	Taiyuan	613	624	97.80	2730	Military SAR Satellite

No	Satellite Name	Launch Date	Launcher	Launch site	Perigee Km	Apogee Km	Inclination Degree	Mass Kg	Comments
142	Tian Lian 1	4/25/2008	Long March 3C	Xichang	35638	35949	0.06	3000	TDRS Date Relay Satellite
143	Feng Yun 3-1	5/27/2008	Long March 4C	Taiyuan	804	811	98.80	2298	Orbiting Weather Satellite
144	Zhongxing 9	6/9/2008	Long March 3B	Xichang	NA	NA	0.03	4600	Domestic GSO COMSAT
145	Huanjing 1A	9/6/2008	Long March 2C	Taiyuan	626	660	98.00	470	APSCO Remote Sensing
146	Huanjing 1B	9/6/2008	Long March 2C	Taiyuan	627	672	98.00	470	APSCO Remote Sensing
147	Shenzhou 7	9/25/2008	Long March 2F	Jiuquan	200	337	42.40	7890	3-member manned flight -walk
148	Banxing	9/27/2008	Shenzhou 7	Jiuquan	280	296	42.40	40	Satellite released from Shenzhou
149	Shi-Jian 6-3A	10/25/2008	Long March 4B	Taiyuan	580	605	96.50	375	ELINT Military
150	Shi Jian 6-3B	10/25/2008	Long March 4B	Taiyuan	582	605	96.50	975	ELINT Military
151	Simon Bolivar	10/29/2008	Long March 3BE	Xichang	GSO	GSO	GSO	5100	Venezuela GSO Comsat
152	Chuangxin 1-02	11/5/2008	Long March 2D	Jiuquan	786	804	98.50	88	Technology Military
153	Tansuo 3	11/5/2008	Long March 2D	Jiuquan	786	804	98.50	204	Remote Sensing Mapping
154	Yaogan 4	12/1/2008	Long March 2D	Jiuquan	634	652	97.90	600	EO Reconnaissance Military
155	Yaogan 5	12/15/2008	Long March 4B	Taiyuan	488	494	97.40	2500	EO Reconnaissance Military
156	Feng Yun 2-5	12/23/2008	Long March 3A	Xichang	35786	na	2.61	1380	GSO Weather Satellite
157	Compass 2 G2	4/14/2009	Long March 3A	Xichang	35775	35798	0.95	2200	GSO Navigation Satellite
158	Yaogan 6	4/22/2009	Long March 2C	Taiyuan	486	521	97.60	1000	ISR SAR satellite Military
159	Asiasat 5	8/11/2009	Zenit 3SLB	Baikonur	35788	35798	0.00	3,700	Bought COMSAT-Russian Launch
160	Palapa D1	8/31/2009	Long March 3B	Xichang	35783	35803	98.10	4100	Indonesia GSO COMSAT
161	Shijian 11-1	11/12/2009	Long March 2C	Jiuquan	699	703	98.30	300	BMD Early Warning
162	Yaogan 7	12/9/2009	Long March 2C	Jiuquan	623	659	97.80	500	EO Reconnaissance Military
163	Xiwang	12/15/2009	Long March 4C	Taiyuan	1192	1204	100.50	50	Technology Military
164	Yaogan 8	12/15/2009	Long March 4C	Taiyuan	1192	1204	100.50	1040	Medium EO Recon Satellite
165	Compass 2 G1	1/16/2010	Long March 3C	Xichang	35776	35799	1.78	3000	GSO Navigation Satellite
166	Yaogan 9A	3/5/2010	Long March 4C	Jiuquan	1083	1100	63.40	1000	ELINT Constellation
167	Yaogan 9B	3/5/2010	Long March 4C	Jiuquan	1083	1100	63.40	200	ELINT Constellation
168	Yaogan 9C	3/5/2010	Long March 4C	Jiuquan	1083	1100	63.40	200	ELINT Constellation
169	Compass 2 G3	6/2/2010	Long March 3C	Xichang	35772	35799	1.83	3000	GSO Navigation Satellite
170	Shi Jian 12	6/15/2010	Long March 2D	Jiuquan	575	597	97.70	300	ASAT / SSA Military

No	Satellite Name	Launch Date	Launcher	Launch site	Perigee Km	Apogee Km	Inclination Degree	Mass Kg	Comments
171	Compass 2 IG1	7/31/2010	Long March 3A	Xichang	35667	35895	55.00	2200	IGSO Navigation Satellite
172	Yaogan 10	8/9/2010	Long March 4C	Taiyuan	607	621	98.70	2700	ISR SAR satellite Military
173	Tianhui 1	8/24/2010	Long March 2D	Jiuquan	488	504	97.30	2500	Remote Sensing Mapping
174	Sinosat 6	9/4/2010	Long March 3B	Xichang	35791	35800	0.37	5100	Domestic GSO COMSAT
175	Yaogan 11	9/22/2010	Long March 2D	Jiuquan	624	657	98.00	600	EO Reconnaissance Military
176	Zheda 1A-1	9/22/2010	Long March 2D	Jiuquan	624	657	98.00	2.5	Small satellite
177	Zheda 1A2	9/22/2010	Long March 2D	Jiuquan	624	657	98.00	3.5	Small Satellite
178	Chang e 2	10/1/2010	Long March 3C	Xichang	Lunar	Lunar	NA	2500	Planetary Probe (Lunar)
179	Shi Jian 6-4B	10/6/2010	Long March 4B	Taiyuan	566	604	97.80	975	ELINT Military
180	Shi Jian 6-4A	10/6/2010	Long March 4B	Taiyuan	588	604	97.70	375	ELINT Military
181	Compass-2 G4	10/31/2010	Long March 3C	Xichang	36769	35804	1.78	6000	GSO Navigation Satellite
182	Feng Yun 3-2	11/4/2010	Long March 4C	Taiyuan	827	847	98.70	2298	Orbiting Weather Satellite
183	Shentong 1	11/24/2010	Long March 3A	Xichang	35772	35798	0.55	2300	GSO Military COMSAT
184	Compass-2 IG2	12/17/2010	Long March 3A	Xichang	35717	35856	55.40	2230	Inclined GSO Navigation Satellite
185	Compass-2 IG3	4/9/2011	Long March 3A	Xichang	35694	35871	55.29	2300	Inclined GSO Navigation Satellite
186	Zhonxing 10	6/20/2011	Long March 3BE	Xichang	35771	35814	0.04	5100	Domestic GSO COMSAT
187	Shijian 11-3	7/6/2011	Long March 2C	Jiuquan	690	704	98.20	300	BMD Early Warning
188	Tian Lian 1-2	7/11/2011	Long March 3C	Xichang	GSO	GSO	1.00	2500	TDRS Date Relay Satellite
189	Compass-2 IG4	7/26/2011	Long March 3A	Xichang	35699	35871	55.00	4200	Inclined GSO Navigation Satellite
190	Shijian 11-2	7/29/2011	Long March 2C	Jiuquan	689	704	98.10	300	BMD Early Warning
191	Paksat 1R	8/11/2011	Long March 3BE	Xichang	35778	35805	NA	5120	Pakistan Comsat Commercial
192	Haiyang 2A	8/15/2011	Long March 4B	Taiyuan	906	916	99.40	1500	Ocean RS Satellite
193	Shijian 11-04	8/18/2011	Long March 2C	Jiuquan	Failure	Failure	NA	NA	BMD Early Warning
194	Zhongxing 1A	9/18/2011	Long March 3BE	Xichang	35778	35796	0.30	5320	Domestic GSO COMSAT
195	Tiangong-1	9/29/2011	Long March 2F	Jiuquan	261	314	42.80	8500	Space Station (test module)
196	Eutelsat W3C	10/7/2011	Long March 3BE	Xichang	35779	35795	0.20	5370	Europe COMSAT Commercial
197	Shenzhou 8	10/31/2011	Long March 2FH	Jiuquan	261	314	42.80	8082	Shenzhou Tiangong Docking
198	Yinghuo-1	11/8/2011	Zenit 3F	Baikonur	206	341	51.40	115	Mars Mission Ukraine launcher
199	Tianxun 1	11/9/2011	Long March 4B	Taiyuan	491	497	97.40	58	Technology Military

No	Satellite Name	Launch Date	Launcher	Launch site	Perigee Km	Apogee Km	Inclination Degree	Mass Kg	Comments
200	Yaogan 12	11/9/2011	Long March 4B	Taiyuan	491	497	97.40	2500	EO Reconnaissance Military
201	Chuangxin 1-03	11/20/2011	Long March 2D	Jiuquan	783	804	98.50	88	Technology Military
202	Tansuo 4 / Shiyan 4	11/20/2011	Long March 2D	Jiuquan	783	804	98.50	400	Remote Sensing Mapping
203	AsiaSat 7	11/25/2011	Proton M	Baikonur	35782	35793	0.02	4,500	Domestic Comsat Russian launch
204	Yaogan 13	11/29/2011	Long March 2C	Taiyuan	505	510	97.10	1000	ISR SAR satelite Military
205	Compass-2 IG5	12/1/2011	Long March 3B	Xichang	35704	35866	55.00	4200	Inclined GSO Navigation Satellite
206	NIGCOMSAT-1R	12/19/2011	Long March 3BE	Xichang	35793	35795	0.20	5150	Nigeria Comsat Commercial
207	Zi Yuan 1-02C	12/22/2011	Long March 4B	Taiyuan	773	774	98.50	1540	Domestic RS Satellite
208	Vesselsat 2	1/9/2012	Long March 4B	Taiyuan	472	484	97.43	2650	Orbcom (US) AIS Commercial
209	Zi Yuan 3 / 3A	1/9/2012	Long March 4B	Taiyuan	498	506	97.00	2636	Domestic RS Recon Satellite
210	Feng Yun 2-6	1/13/2012	Long March 3A	Xichang	35786	35816	2.19	1380	GSO Weather Satellite
211	Compass-2 G5	2/24/2012	Long March 3C	Xichang	35766	35799	1.80	4200	GSO Navigation Satellite
212	Apstar 7	3/31/2012	Long March 3BE	Xichang	35784	35802	0.05	5054	Domestic GSO Communications
213	Beidou M3 DW12	4/29/2012	Long March 3BE	Xichang	21460	21595	55.00	2200	MEO Navigation Satellite
214	Beidou M4 DW13	4/29/2012	Long March 3BE	Xichang	21452	21603	55.00	2200	MEO Navigation Satellite
215	Tianhui 1-02	5/6/2012	Long March 2D	Jiuquan	490	506	97.40	1000	Remote Sensing Mapping
216	Tiantuo 1	5/10/2012	Long March 4B	Taiyuan	468	471	97.20	9	Small Satellite AIS
217	Yaogan 14	5/10/2012	Long March 4B	Taiyuan	473	479	97.20	2700	EO Reconnaissance Military
218	Zhongxing 2A	5/26/2012	Long March 3BE	Xichang	35781	35792	0.19	5200	Domestic GSO Communications
219	Yaogan 15	5/29/2012	Long March 4C	Taiyuan	1201	1200	100.10	1040	Medium EO Recon Satellite
220	Shenzhou 9	6/16/2012	Long March 2F	Jiuquan	261	315	42.80	8000	Woman astronaut-Space Station
221	Tianlian-1C	7/25/2012	Long March 3C	Xichang	35768	35802	2.04	2462	TDRS Date Relay Satellite
222	Beidou M5	9/18/2012	Long March 3BE	Xichang	21477	21574	55.90	4000	MEO Navigation Satellite
223	Beidou M2	9/18/2012	Long March 3BE	Xichang	21462	21591	55.00	4000	MEO Navigation Satellite
224	VRSS-1	9/29/2012	Long March 2D	Jiuquan	622	654	98.04	880	Venezuela RS Commercial
225	Shijian 9A	10/14/2012	Long March 2C	Taiyuan	623	650	98.00	790	Technology Military
226	Shijian 9B	10/14/2012	Long March 2C	Taiyuan	623	650	98.00	260	Technology Military
227	Beidou G6	10/25/2012	Long March 3C	Xichang	35775	35799	1.84	3800	GSO Navigation Satellite
228	FN-1A	11/18/2012	Long March 2C	Taiyuan	487	503	97.40	130	Small Satellites

No	Satellite Name	Launch Date	Launcher	Launch site	Perigee Km	Apogee Km	Inclination Degree	Mass Kg	Comments
229	FN-1B	11/18/2012	Long March 2C	Taiyuan	487	503	97.40	30	Small Satellites
230	HJ-1C	11/18/2012	Long March 2C	Taiyuan	487	503	97.40	890	APSCO SAR Satellite
231	XY-1	11/18/2012	Long March 2C	Taiyuan	487	503	97.40	140	Small Satellites
232	Yaogan-16 A	11/25/2012	Long March 4C	Jiuquan	1079	1089	63.40	NA	ELINT Constellation
233	Yaogan-16 B	11/25/2012	Long March 4C	Jiuquan	1079	1089	63.40	NA	ELINT Constellation
234	Yaogan-16 C	11/25/2012	Long March 4C	Jiuquan	1032	1081	63.40	NA	ELINT Constellation
235	Zhonxing-12	11/27/2012	Long March 3BE	Xichang	35759	35807	0.04	5000	Domestic GSO Communications
236	Gokturk-2	12/18/2012	Long March 2D	Jiuquan	669	690	98.06	450	Turkey RS Satellite Commercial
237	CubeBug-1	4/26/2013	Long March 2D	Jiuquan	934	660	98.10	2	Argentina Small Satellite
238	NEE 01pogos	4/26/2013	Long March 2D	Jiuquan	634	660	98.10	1	Ecuador Small Satellite
239	TurkSat-3USat	4/26/2013	Long March 2D	Jiuquan	634	660	98.00	4	Turkey Smallsat commercial
240	GF-1	4/26/2013	Long March 2D	Jiuquan	830	853	90.00	1000	Technology Military?
241	Zhonxing-11	5/1/2013	Long March 3BE	Xichang	35762	35825	0.30	5000	Domestic GSO Communications
242	Shenzhou-10	6/11/2013	Long March 2F	Jiuquan	260	313	NA	7700	3-member docking Tiangong
243	Shijian 11-05	7/15/2013	Long March 2C	Jiuquan	680	703	98.10	NA	BMD Early Warning
244	Chuangxin 3	7/19/2013	Long March 4C	Taiyuan	787	801	98.10	200	Technology Military
245	Shiyan 7	7/19/2013	Long March 4C	Taiyuan	669	680	98.10	NA	Technology Military
246	Shijian 15	7/19/2013	Long March 4C	Taiyuan	584	632	97.70	NA	Technology Military
247	Yaogan-17 A	9/1/2013	Long March 4C	Jiuquan	1089	1116	63.40	NA	ELINT Constellation
248	Yaogan-17 B	9/1/2013	Long March 4C	Jiuquan	1085	1110	63.40	NA	ELINT Constellation
249	Yaogan-17 C	9/1/2013	Long March 4C	Jiuquan	1084	1111	63.40	NA	ELINT Constellation
250	Fengyun-3C	9/23/2013	Long March 4C	Taiyuan	801	815	98.80	2300	Orbiting Weather Satellite
251	Kuaizhou-1	9/25/2013	Kuaizhou	Jiuquan	299	322	96.60	NA	Technology New Launcher
252	Shijian 16	10/25/2013	Long March 4B	Jiuquan	600	616	75.00	NA	BMD Early Warning
253	Yaogan-18	10/29/2013	Long March 2C	Taiyuan	492	510	97.60	NA	ISR SAR satellite Military
254	Yaogan-19	11/20/2013	Long March 4C	Taiyuan	1201	1207	100.50	1100	Medium EO Recon Satellite
255	Shiyan-5	11/25/2013	Long March 2D	Jiuquan	739	754	97.80	204	Technology Military
256	Chang'e 3	12/1/2013	Long March 3B	Xichang	210	389109	NA	1340	Planetary Probe (Moon)
257	CBERS-3	12/9/2013	Long March 4B	Taiyuan	1252	1252	98.50	1980	China Brazil Remote Sensing

No	Satellite Name	Launch Date	Launcher	Launch site	Perigee Km	Apogee Km	Inclination Degree	Mass Kg	Comments
258	Tupac Katari 1	12/20/2013	Long March 3B	Xichang	35782	35806	0.20	5100	Bolivia Comsat Commercial
259	Shijian 11-06	3/31/2014	Long March 2C	Jiuquan	687	704	98.30	1000	BMD Early Warning
260	Asiasat 8	8/5/2014	Falcon 9	CCAS	35785	35799	0.07	4535	Bought Comsat-Russian Launch
261	Yaogan-20 A	8/9/2014	Long March 4C	Jiuquan	1086	1103	63.40	NA	ELINT Constellation
262	Yaogan-20 B	8/9/2014	Long March 4C	Jiuquan	1086	1103	63.40	NA	ELINT Constellation
263	Yaogan-20 C	8/9/2014	Long March 4C	Jiuquan	1086	1103	63.40	NA	ELINT Constellation
264	BLITE PL	8/19/2014	Long March 4B	Taiyuan	608	631	98.00	10	Poland Piggyback Commercial
265	Gaofen-2	8/19/2014	Long March 4B	Taiyuan	608	631	98.00	2000	Technology Military
266	CX-1 04	9/4/2014	Long March 2D	Jiuquan	778	809	98.50	90	Communication Tech. Military
267	Ling Qiao	9/4/2014	Long March 2D	Jiuquan	778	809	98.50	135	Communication Tech Military
268	Asiasat 6	9/7/2014	Falcon 9	CCAS	35791	35795	0.07	3,813	Bought Comsat-Russian Launch
269	Tiantuo-2	9/8/2014	Long March 4B	Taiyuan	475	488	97.40	67	Technology Military
270	Yaogan-21	9/8/2014	Long March 4B	Taiyuan	481	492	97.40	NA	EO Reconnaissance Military
271	Shijian 11-07	9/28/2014	Long March 2C	Jiuquan	686	705	98.10	1000	BMD Early Warning
272	Yaogan-22	10/20/2014	Long March 4C	Taiyuan	1196	1207	100.30	1040	Medium EO Recon Satellite
273	Chang'e 5 T1	10/24/2014	Long March 3C	Xichang	Lunar	Lunar	NA	NA	Planetary Probe (Lunar)
274	Shijian 11-08	10/27/2014	Long March 2C	Jiuquan	688	703	NA	1000	BMD Early Warning
275	Yaogan-23	11/14/2014	Long March 2C	Taiyuan	510	514	97.30	1000	SAR Reconnaissance Military
276	Yaogan-24	11/20/2014	Long March 2D	Jiuquan	629	654	97.90	1000	EO Reconnaissance Military
277	Kuaizhou-2	11/21/2014	Kuaizhou	Jiuquan	283	303	96.55	NA	Technology Military?
278	CBERS-4	12/7/2014	Long March 4B	Taiyuan	742	751	98.50	1980	China Brazil Remote sensing
279	Yaogan-25 A	12/10/2014	Long March 4C	Jiuquan	1091	1097	63.40	NA	ELINT Constellation
280	Yaogan-25 B	12/10/2014	Long March 4C	Jiuquan	1090	1098	63.40	NA	ELINT Constellation
281	Yaogan-25 C	12/10/2014	Long March 4C	Jiuquan	1089	1097	63.40	NA	ELINT Constellation
282	Yaogan-26	12/27/2014	Long March 4B	Taiyuan	482	488	97.40	2000	EO Reconnaissance Military
283	Fengyun-2G	12/31/2014	Long March 3A	Xichang	35786	GSO	2.27	1380	GSO Weather Satellite
284	Beidou 3 I1	3/30/2015	Long March 3C	Xichang	35652	35959	55.00	830	IGSO Navigation Satellite
285	GF 8	6/26/2015	Long March 4B	Taiyuan	471	481	97.30	NA	Technology Military
286	Beidou 3 M1	7/25/2015	Long March 3B	Xichang	21520	21551	55.00	2028	MEO Navigation Satellite

No	Satellite Name	Launch Date	Launcher	Launch site	Perigee Km	Apogee Km	Inclination Degree	Mass Kg	Comments
287	Beidou 3 M2	7/25/2015	Long March 3B	Xichang	21520	21551	55.00	2028	MEO Navigation Satellite
288	Yaogan-27	8/27/2015	Long March 4C	Jiuquan	1194	1206	100.46	1040	Medium EO Recon Satellite
289	TJSSW 1	9/12/2015	Long March 3BE	Xichang	195	35823	0.10	NA	GSO Early Warning?
290	Gaofen 9	9/14/2015	Long March 2D	Jiuquan	625.2	671.7	98.00	NA	Technology Military
291	LilacSat 2	9/19/2015	Long March 6	Taiyuan	521	542	97.47	11	Small Satellites New Launcher
292	Luliang 1	9/19/2015	Long March 6	Taiyuan	516	536	97.46	NA	Small Satellites New Launcher
293	Zheda Pixing-2A	9/19/2015	Long March 6	Taiyuan	521.4	540.2	97.46	12	Small Satellites New Launcher
294	Xiwang-2A	9/19/2015	Long March 6	Taiyuan	499	533	97.46	25	Small Satellites New Launcher
295	Kaituo 1A	9/19/2015	Long March 6	Taiyuan	519	537	97.46	130	Small Satellites New Launcher
296	Xiwang-2C	9/19/2015	Long March 6	Taiyuan	520	540	97.46	10	Small Satellites New Launcher
297	Xiwang-2D	9/19/2015	Long March 6	Taiyuan	550	540	97.46	10	Small Satellites New Launcher
298	XW-2E	9/19/2015	Long March 6	Taiyuan	519	539	97.46	1.5	Small Satellites New Launcher
299	Xiwang-2B	9/19/2015	Long March 6	Taiyuan	520	540	97.46	10	Small Satellites New Launcher
300	Kaituo 1B	9/19/2015	Long March 6	Taiyuan	541	536	97.46	1	Small Satellites New Launcher
301	NUDT-PHONESAT	9/19/2015	Long March 6	Taiyuan	518.3	537.9	97.45	1	Small Satellites New Launcher
302	Zheda Pixing-2B	9/19/2015	Long March 6	Taiyuan	521.8	540.7	97.45	12	Small Satellites New Launcher
303	Naxing 2	9/19/2015	Long March 6	Taiyuan	520	541	97.45	NA	Small Satellites New Launcher
304	XW-2F	9/19/2015	Long March 6	Taiyuan	520	541	97.45	1.5	Small Satellites New Launcher
305	TW-1 / Tianwang 1B	9/25/2015	Long March 11	Jiuquan	466	485	97.32	NA	Small Satellites New Launcher
306	Pujian-1	9/25/2015	Long March 11	Jiuquan	469	486	97.31	NA	Small Satellites New Launcher
307	TW-1	9/25/2015	Long March 11	Jiuquan	466	487	97.31	NA	Small Satellites New Launcher
308	TW-1	9/25/2015	Long March 11	Jiuquan	467	487	97.31	NA	Small Satellites New Launcher
309	Beidou 12 S	9/29/2015	Long March 3B	Xichang	192	35828	55.02	4200	Inclined GSO Navigation Satellite
310	Jilin-1 Technical	10/7/2015	Long March 2D	Jiuquan	640	664	98.04	55	Commercial RS Constellation
311	Jilin-1 Video	10/7/2015	Long March 2D	Jiuquan	640	664	98.04	95	Commercial RS Constellation
312	Jilin-1 Video	10/7/2015	Long March 2D	Jiuquan	639	664	98.04	95	Commercial RS Constellation
313	Jilin 1	10/7/2015	Long March 2D	Jiuquan	639	664	98.04	664	Commercial RS Constellation
314	Apstar 9	10/16/2015	Long March 3BE	Xichang	188	41774	0.08	4000	Domestic GSO Communications
315	Tianhui-1C	10/26/2015	Long March 2D	Jiuquan	490	501	97.35	3500	Remote Sensing Mapping

No	Satellite Name	Launch Date	Launcher	Launch site	Perigee Km	Apogee Km	Inclination Degree	Mass Kg	Comments
316	ChinaSat-2C	11/3/2015	Long March 3B	Xichang	35779	35793	0.56	5170	Domestic GSO Communications
317	Yaogan-28	11/8/2015	Long March 4B	Taiyuan	460	482	97.24	3000	EO Reconnaissance Military
318	LaoSat 1	11/20/2015	Long March 3B	Xichang	35781	35794	0.11	4600	Laos GSO COMSAT Commercial
319	Yaogan-29	11/26/2015	Long March 4C	Taiyuan	615	619	97.84	3000	ISR SAR satellite Military
320	Chinasat 1C	12/9/2015	Long March 3BE	Xichang	177	35816	0.11	5200	Domestic GSO Communications
321	Wukong (DAMPE)	12/17/2015	Long March 2D	Jiuquan	488	505	97.30	1900	Science Dark Matter Explorer
322	Gaofen 4	12/28/2015	Long March 3BE	Xichang	205	35807	0.58	4600	GSO Early Warning?
323	Belintersat 1	15/1/2016	Long March 3B	Xichang	197	41781	26.38	5223	Commercial, Republic of Belarus
324	Beidou-3 M3	02/01/16	Long March 3C	Xichang	21513	21982	55	1014	3rd MEO Navigation Satellite
325	Beidou-2 I3-S	03/29/16	Long March 3A	Xichang	35676	35890	54.97	2000	2nd IGSO Navigation Satellite
326	Shijian-10	04/05/16	Long March 2D	Jiuquan	220	482	63	3600	Microgravity Research
327	Yaogan 30	05/15/16	Long March 2D	Jiuquan	628	656	98.1	1300	EO Recon Military
328	Ziyuan III-02	05/30/16	Long March 4B	Taiyuan	507.4	510.9	97.5	2640	Domestic RS satellite
329	ÑuSat-1	05/30/16	Long March 4B	Taiyuan	485	503	97.42	35	Commercial, Argentina
330	ÑuSat-2	05/30/16	Long March 4B	Taiyuan	485.7	503.1	97.42	35	Commercial, Argentina
331	Beidou-2 G7	06/12/16	Long March 3C	Xichang	35676	35890	19.28	3800	2nd GSO Navigation Satellite
332	Cargo to Station	06/25/16	Long March 7	Wenchang	288	382	40.81	2800	Demonstration, New Launcher
333	Star of Aoxiang	06/25/16	Long March 7	Wenchang	287	380	40.8	18	Small satellite, New Launcher
334	Aolong-1	06/25/16	Long March 7	Wenchang	200	375	40.8	NA	Small satellite, New Launcher
335	Tiange-1	06/25/16	Long March 7	Wenchang	200	394	40.8	NA	Small satellite, New Launcher
336	Tiange-2	06/25/16	Long March 7	Wenchang	200	394	40.8	NA	Small satellite, New Launcher
337	Shiyan Zhuangzhi	06/25/16	Long March 7	Wenchang				NA	Small satellite, New Launcher
338	Shijian 16-02	06/29/16	Long March 4B	Jiuquan	595	615	75	2800	LEO, Military, Recon
339	Tiantong-1-01	08/05/16	Long March 3BE	Xichang	35767.9	35819.4	4.8	5300	GSO, Mobile Comsat
340	Gaofen 3	08/09/16	Long March 4C	Taiyuan	757.9	758.8	98.4	2950	RS Civilian
341	QUESS	08/15/16	Long March 2D	Jiuquan	492.4	510	97.4	640	Quantum Communication
342	3CAT-2	08/15/16	Long March 2D	Jiuquan	485	503	97.4	7	Commercial, Spain
343	Lixing 1	08/15/16	Long March 2D	Jiuquan	124	140	97.4	110	Technology, Military
344	Gaofen 10	08/31/16	Long March 4C	Taiyuan			Failure	NA	Failure

No	Satellite Name	Launch Date	Launcher	Launch site	Perigee Km	Apogee Km	Inclination Degree	Mass Kg	Comments
345	Tiangong-2	09/15/16	Long March 2F	Jiuquan	378.6	388.1	42.8	8600	Human flight
346	BANXING-2	09/15/16	Long March-2F	Jiuquan	378.1	386.9	42.8	47	Small satellite
347	Shenzhou 11	10/16/16	Long March 2F	Jiuquan	379.5	394.9	42.8	8082	Human flight
348	Shijian 17	11/03/16	Long March 5	Wenchang	35787.8	35820.5	0.8	3800	GSO Domestic Comsat
349	XPNAV-1	11/09/16	Long March 11	Jiuquan	497.9	521.9	97.4	240	Mini Satellite, Pulsar Navigation
350	Xiaoxiang 1 (XX 1)	11/09/16	Long March 11	Jiuquan	496.8	520.6	97.4	7.5	Small satellite
351	Lishui 1	11/09/16	Long March 11	Jiuquan	495.2	521.3	97.4	NA	Small satellite
352	Pina-2 01	11/09/16	Long March 11	Jiuquan	496.1	520.4	97.4	NA	Small satellite
353	Pina-2 02	11/09/16	Long March 11	Jiuquan	510.7	1035.9	97.4	NA	Small satellite
354	CAS 2T	11/09/16	Long March 11	Jiuquan	511.5	1040.8	98.8	NA	Small satellite
355	KS 1Q	11/09/16	Long March 11	Jiuquan	511.6	996.1	98.8	NA	Small satellite
356	YUNHAI 1	11/11/16	Long March 2D	Jiuquan	760	787	98.5	1300	SSO, Weather satellite
357	TIANLIAN 1-04	11/22/16	Long March-3C	Xichang	201	41776	3	2100	GSO, Communication Satellite
358	FENGYUN 4A	12/10/16	Long March-3B	Xichang	191.8	35812.4	28.5	5300	GSO, Weather Satellite
359	TanSat	12/21/16	Long March 2D	Jiuquan	698.0	726.3	98.2	620	SSO, Weather Satellite
360	Spark-01	12/21/16	Long March 2D	Jiuquan	698.0	726.3	98.2	43	Small satellite
361	Spark-02	12/21/16	Long March 2D	Jiuquan	698.0	726.3	98.2	43	Small satellite
362	YIJIAN	12/21/16	Long March 2D	Jiuquan	696.8	728.4	98.2	50	Small satellite
363	SuperView 1 - 01	12/28/16	Long March 2D	Taiyuan	214	524	97.6	560	SSO, EO, Commercial
364	SuperView 1 - 02	12/28/16	Long March 2D	Taiyuan	214	524	97.6	560	SSO, EO, Commercial
365	BY70-1	12/28/16	Long March 2D	Taiyuan	214	524	97.6	2	Small satellite
366	TJS 2	1/5/2017	Long March 3B	Xichang	210.3	35,809.20	27.5		EW satellite in GSO Military
367	Xingyun Shiyan 1	1/9/2017	Kuaizhou	Jiuquan	537.5	553.9	97.5	2.79	Cubesat University
368	Lingqiao Shipin 1-03	1/9/2017	Kuaizhou	Jiuquan	538.1	552.6	97.54	165	Commercial RS Jilin Constellation
369	Kaidun 1	1/9/2017	Kuaizhou	Jiuquan	536.8	548.9	97.54	2.72	University small satellite
370	Tiankun-1	3/2/2017	Kaituozhe-2	Jiuquan	374	404	96.9	350	RS Test satellite
371	Shijian-13	4/12/2017	Long March 3B	Xichang	35,776	35,812	0.1	4,600	GSO Comsat civilian
372	Tianzhou-1	4/20/2017	Long March 7	Wenchang	387.50	395.7	42.80	12910	CZ 7 Ferry for space station test
373	SilkRoad-1/ Silu1	4/20/2017	Long March 7	Wenchang	387.00	395.00	42.8	4.5	Smallsat RS surveying mapping

No	Satellite Name	Launch Date	Launcher	Launch site	Perigee Km	Apogee Km	Inclination Degree	Mass Kg	Comments
374	HXMT	6/15/2017	Long March 4B	Jiuquan	543.3	554.6	43.0	2700	Hard X-ray Science Satellite
375	ÑuSat3	6/15/2017	Long March 4B	Jiuquan	541.6	551.8	43	37	Argentina RS satellite
376	CAS-4A	6/15/2017	Long March 4B	Jiuquan	540.0	551.7	43.00	55	Test RS satellite
377	CAS-4B	6/15/2017	Long March 4B	Jiuquan	540.5	551.6	43.0	55	Test RS satellite
378	Zhongxing-9A	6/18/2017	Long March 3BE	Xichang	GSO	GSO		~5200	Comsat in GSO civilian
379	NUDTSat	6/23/2017	PSLV XL	SDLC SHAR	497.00	515.00	97.45	2	Small Satellite PSLV launch
380	Shijian 18	7/2/2017	Long March 5	Wenchang	Failure	Failure	Failure	7600	2nd test flight of CZ 5 Failed
381	Yaogan 30-01	9/29/2017	Long March 2C	Xichang	604.9	608.8	35.0		Tactical ELINT constellation
382	Yaogan 30-02	9/29/2017	Long March 2C	Xichang	604.5	609.2	35.0		Tactical ELINT constellation
383	Yaogan 30-03	9/29/2017	Long March 2C	Xichang	605.0	608.7	35.0		Tactical ELINT constellation
384	VRSS-2	10/9/2017	Long March 2D	Jiuquan	635.5	662.9	98	1000	RS Orbit delivery Venezuela
385	BeiDou-3M1	11/5/2017	Long March 3B	Xichang	21,513.5	21556.60	55.00	1014	MEO Navigation Satellite
386	BeiDou-3M2	11/5/2017	Long March 3B	Xichang	21,548.5	22201.00	55.00	1014	MEO Navigation Satellite
387	Feng Yun 3D	11/14/2017	Long March 4C	Taiyaun	804.8	819.3	98.7		SSO Orbiting weather satellite
388	HEAD-1	11/14/2017	Long March 4C	Taiyaun	803.7	816.6	98.7	45	AID satellite for tracking ships
389	Jilin-1-04	11/21/2017	Long March 6	Taiyuan	539.4	555.9	97.5	95	Jilin commercial constellation
390	Jilin-1-05	11/21/2017	Long March 6	Taiyuan	530	542	97.5	95	Jilin commercial constellation
391	Jilin-1-06	11/21/2017	Long March 6	Taiyuan	530	542	97.5	95	Jilin commercial constellation
392	YAOGAN-30 D	11/24/2017	Long March 2C	Xichang	607.3	616.2	35.0		Tactical ELINT constellation
393	YAOGAN-30 E	11/24/2017	Long March 2C	Xichang	597.0	609.9	35.0		Tactical ELINT constellation
394	YAOGAN-30 F	11/24/2017	Long March 2C	Xichang	592.2	598.2	35.0		Tactical ELINT constellation
395	LKW-1	12/4/2017	Long March 2D	Jiuquan	495.8	510.4	97.46	3000	Military RS satellite Yaogan type
396	ALCOMSAT-1	12/10/2017	Long March 3B	Xichang	187.2	41,802.50	26.4	5225	Orbit delivery Comsat Algeria
397	LKW-2	12/23/2017	Long March 2D	Jiuquan	496.1	509.8	97.5	1000	Military RS satellite Yaogan type
398	Yaogan-30 G	12/25/2017	Long March 2C	Xichang	592.5	602.8	35.0		Tactical ELINT constellation
399	Yaogan-30 H	12/25/2017	Long March 2C	Xichang	601.3	610.3	35.0		Tactical ELINT constellation
400	Yaogan-30 J	12/25/2017	Long March 2C	Xichang	611.2	618	35.0		Tactical ELINT constellation
401	SUPERVIEW-1 03	1/9/2018	Long March 2D	Taiyuan	527.8	535.3	97.6	560	Civilian RS Constellation
402	SUPERVIEW-1 04	1/9/2018	Long March 2D	Taiyuan	523.1	539.8	97.6	560	Civilian RS Constellation

No	Satellite Name	Launch Date	Launcher	Launch site	Perigee Km	Apogee Km	Inclination Degree	Mass Kg	Comments
403	BEIDOU 3M7	1/11/2018	Long March 3B	Xichang	21545.8	22200.9	55.0	1014	MEO Navigation Satellite
404	BEIDOU 3M8	1/11/2018	Long March 3B	Xichang	21542.9	21552.7	55.0	1014	MEO Navigation Satellite
405	LKW-3	1/13/2018	Long March 2D	Jiuquan	491.4	508.5	97.3	3100	Yaogan Type military satellite
406	Xiaoxiang 2	1/19/2018	Long March 11	Jiuquan	530.6	554.0	97.6	8	Small satellite
407	Xiaoxiang 6	1/19/2018	Long March 11	Jiuquan	534.4	555.4	97.5	8	Small satellite
408	Jilin 1 - 07	1/19/2018	Long March 11	Jiuquan	525.0	1065.0	98.5	95	Civilian RS Constellation
409	Jilin 1 - 08	19/1/2018	Long March 11	Jiuquan	525.0	1065.0	98.5	95	Civilian RS Constellation
410	Zhou Enlai	19/1/2018	Long March 11	Jiuquan	525.0	1065.0	98.5	2.4	Small satellite
411	KIPP	19/1/2018	Long March 11	Jiuquan	525.0	1065.0	98.5	NA	Small satellite
412	WEINA 1A	25/1/2018	Long March 2C	Xichang	602.7	608.2	35.0	NA	Small satellite
413	YAOGAN-30 K	25/1/2018	Long March 2C	Xichang	602.1	611.4	35.0	NA	Tactical ELINT constellation
414	YAOGAN-30 L	25/1/2018	Long March 2C	Xichang	604.5	609.0	35.0	NA	Tactical ELINT constellation
415	YAOGAN-30 M	25/1/2018	Long March 2C	Xichang	604.2	609.2	35.0	NA	Tactical ELINT constellation
416	FENGMANIU 1	2/2/2018	Long March 2D	Jiuquan	495.5	517.0	97.3	3	China small satellite
417	ZHANGZHENG-1	2/2/2018	Long March 2D	Jiuquan	494.0	516.1	97.3	730	Ionosphere Science SSO satellite
418	NUSAT 4	2/2/2018	Long March 2D	Jiuquan	492.8	515.8	97.3	37	Argentina RS satellite
419	GOMX4-B	2/2/2018	Long March 2D	Jiuquan	490.0	514.4	97.3	6	Denmark small satellite
420	GOMX4-A	2/2/2018	Long March 2D	Jiuquan	490.1	515.0	97.3	6	Denmark small satellite
421	SHAONIAN XING	2/2/2018	Long March 2D	Jiuquan	491.9	514.5	97.3	2	China small satellite
422	NUSAT 5	2/2/2018	Long March 2D	Jiuquan	492.8	515.8	97.3	37	Argentina RS satellite
423	BEIDOU 3M3	12/2/2018	Long March 3B	Xichang	21,540.00	22,198.9	55.0	1014	MEO Navigation Satellite
424	BEIDOU 3M4	12/2/2018	Long March 3B	Xichang	21,550.60	22,198.6	55.0	1014	MEO Navigation Satellite
425	LKW-4	17/3/2018	Long March 2D	Jiuquan	496.6	509.6	97.3	3100	Yaogan Type military satellite
426	BD-3 M9	29/3/2018	Long March 3B	Xichang	21,529.40	21,545.50	55.0	1060	MEO Navigation Satellite
427	BD-3 M10	29/3/2018	Long March 3B	Xichang	21,549.6	22,199.20	55.0	1060	MEO Navigation Satellite
428	Gaofen-1-02	30/3/2018	Long March 4C	Taiyuan	638.00	642.00	98.0	805	Coplanar RS constellation
429	Gaofen-1-03	30/3/2018	Long March 4C	Taiyuan	638.00	642.00	98.0	805	Coplanar RS constellation
430	Gaofen-1-04	30/3/2018	Long March 4C	Taiyuan	638.00	642.00	98.0	805	Coplanar RS constellation
431	Yaogan 31A	10/4/2018	Long March 4C	Jiuquan	1,090	1,100	63.4	1000	Triangular Large Area ELINT

No	Satellite Name	Launch Date	Launcher	Launch site	Perigee Km	Apogee Km	Inclination Degree	Mass Kg	Comments
432	Yaogan 31B	10/4/2018	Long March 4C	Jiuquan	1,090	1,100	63.4	1000	Triangular Large Area ELINT
433	Yaogan 31C	10/4/2018	Long March 4C	Jiuquan	1,090	1,100	63.4	1000	Triangular Large Area ELINT
434	Weina 1B	10/4/2018	Long March 4C	Jiuquan	1,092.50	1,106.60	63.4	NA	Small nano orbiting comsat
435	Zhuhai - OVS 2 A	26/4/2018	Long March 11	Jiuquan	512.2	527.5	97.4	90	second generation co-orbital RS
436	Zhuhai - OHS 2 A	26/4/2018	Long March 11	Jiuquan	512.6	527	97.4	100	second generation co-orbital RS
437	Zhuhai - OHS 2 B	26/4/2018	Long March 11	Jiuquan	513.1	526.5	97.4	100	second generation co-orbital RS
438	Zhuhai - OHS 2 C	26/4/2018	Long March 11	Jiuquan	516.1	523.7	97.4	100	second generation co-orbital RS
439	Zhuhai - OHS 2 D	26/4/2018	Long March 11	Jiuquan	507.6	524.1	97.4	100	second generation co-orbital RS
440	APStar-6C	3/5/2018	Long March 3B	Xichang	35,792.6	35,796.70	0.1	5200	Domestic GSO Comsat
441	Gaofen 5 (GF 5)	8/5/2018	Long March 4C	Taiyuan	706.3	708.7	98.1	2700	GSO Weather Satellite test
442	Queqiao	20/5/2018	Long March 4C	Xichang	402	383,116.8	27.5	425	Relaysat-Dark side of moon
443	DSLWP A1	20/5/2018	Long March 4C	Xichang	402	383,116.80	27.5	45	Long Wavelength observation
444	DSLWP A2	20/5/2018	Long March 4C	Xichang	402	383,116.80	27.5	45	Long Wavelength observation
445	Gaofen 6 (GF 6)	2/6/2018	Long March 2D	Jiuquan	641.7	654.6	98.1	1,064	RS Satellite civilian
446	Luojia 1-01	2/6/2018	Long March 2D	Jiuquan	640.8	654.8	98.1	10	University small satellite
447	Feng-Yun - 2H	5/6/2018	Long March 3A	Xichang	202.8	35,912.70	24.0	1380	GSO Weather Satellite
448	XJSW A	27/6/2018	Long March 2C	Xichang	487.7	495.6	35.0	NA	35-degree ELINT Linked?
449	XJSW B	27/6/2018	Long March 2C	Xichang	495.6	494.2	35.0	NA	35-degree ELINT Linked?
450	PRSS-1	9/7/2018	Long March 2C	Jiuquan	596	631.4	98.1	NA	Pakistan RS orbit delivery
451	PakTES-1A	9/7/2018	Long March 2C	Jiuquan	601.6	635.4	98.1	285	Pakistan RS Test satellite
452	BeiDou-2-I7	9/7/2018	Long March 3A	Xichang	200.1	35,785.5	55.1	NA	Inclined GSO Navigation Satellite
453	BD-3 M5	29/7/2018	Long March 3B	Xichang	21,538.5	21,903.40	55.0	1014	MEO Navigation Satellite
454	BD-3 M6	29/7/2018	Long March 3B	Xichang	21,527	22,200.80	5.0	1014	MEO Navigation Satellite
455	Gaofen 11	31/7/2018	Long March 4B	Taiyuan	254.7	700.9	97.4	NA	Low perigee elliptical SSO orbit?
456	BEIDOU 3 M11	24/8/2018	Long March 3B	Xichang	21,546.6	22,204.70	55.0	1060	MEO Navigation Satellite
457	BEIDOU 3 M12	24/8/2018	Long March 3B	Xichang	21,525.6	21,544.6	55.0	1060	MEO Navigation Satellite
458	HAIYANG 1C	7/9/2018	Long March 2C	Taiyuan	777.0	793.2	98.6	442	Ocean RS Satellite in SSO
459	BD-3 M13	19/9/2018	Long March 3B	Xichang	21,540.3	22,200.3	55.0	1014	MEO Navigation Satellite
460	BD-3 M14	19/9/2018	Long March 3B	Xichang	21,545.4	22,200.4	55.0	1014	MEO Navigation Satellite

No	Satellite Name	Launch Date	Launcher	Launch site	Perigee Km	Apogee Km	Inclination Degree	Mass Kg	Comments
461	CentiSpace-1 S1,	29/9/2018	Kuaizhou 1A	Jiuquan	701.7	717.6	98.2	97	Navigation Test Satellite
462	Yaogan 32-01-01	9/10/2018	Long March 2C	Jiuquan	702.2	703.7	98.3	NA	Yaogan doublet odd orbit?
463	Yaogan 32-01-02	9/10/2018	Long March 2C	Jiuquan	702.1	703.8	98.3	NA	Yaogan doublet odd orbit?
464	BD-3 M15	15/10/2018	Long March 3B	Xichang	21,548.9	22,202.9	55.0	1060	MEO Navigation Satellite
465	BD-3 M16	15/10/2018	Long March 3B	Xichang	21,544.7	22,202.5	55.0	1060	MEO Navigation Satellite
466	Haiyang 2B	24/10/2018	Long March 4B	Taiyuan	972.3	974.7	99.4	1500	2nd generation Ocean Satellite
467	Tangguo Guan	24/10/2018	Long March 4B	Taiyuan	641	940	99.5	20	Store & forward Alibaba P/L
468	Weilai 1	27/10/2018	ZhuQue 1	Jiuquan	NA	NA	NA	40	New Zhuque launcher – failure
469	CFOSAT	29/10/2018	Long March 2C	Jiuquan	523.1	525.1	97.5	600	China France Ocean Satellite
470	Xiaoxiang 1-02	29/10/2018	Long March 2C	Jiuquan	516.9	530.7	97.5	8	China small satellite RS or Com
471	Zhaojin 1	29/10/2018	Long March 2C	Jiuquan	515.8	530.9	97.5	8	China small satellite RS or Com
472	Tiange 1	29/10/2018	Long March 2C	Jiuquan	514.0	531.3	97.5	8	China small satellite RS or Com
473	Tianfuguoxing 1,	29/10/2018	Long March 2C	Jiuquan	518.6	491.2	97.5	8	China small satellite RS or Com
474	Changshagaoxin	29/10/2018	Long March 2C	Jiuquan	517.1	530.3	97.5	8	China small satellite RS or Com
475	Cubebel 1	29/10/2018	Long March 2C	Jiuquan	512.1	531.3	97.5	2	Belarus small satellite
476	BEIDOU 3G1	1/11/2018	Long March 3B	Xichang	189.8	35,817.3	28.5	4600	IGSO navigation satellite
477	BD-3 M17	18/11/2018	Long March 3B	Xichang	21,530.4	22,201.4	55.0	1014	MEO Navigation Satellite
478	BD-3 M18	18/11/2018	Long March 3B	Xichang	21,539.20	22,079.2	55.0	1014	MEO Navigation Satellite
479	Shiyan-6	20/11/2018	Long March 2D	Jiuquan	494.5	511.8	97.4	NA	Military test mission ELINT Test
480	Jiading-1	20/11/2018	Long March 2D	Jiuquan	496.1	510.8	97.4	50	Small satellite comsat
481	Tianzhi-1	20/11/2018	Long March 2D	Jiuquan	495.3	512	97.4	27	Small satellite Technology
482	Tianping-1A	20/11/2018	Long March 2D	Jiuquan	493.8	511.7	97.4	NA	Small satellite Technology
483	Tianping-1B	20/11/2018	Long March 2D	Jiuquan	495.6	511.2	97.4	NA	Small satellite Technology
484	SaudiSat 5A	7/12/2018	Long March 2D	Jiuquan	541.1	559.0	97.6	425	Saudi RS satelite China launch
485	SaudiSat 5B	7/12/2018	Long March 2D	Jiuquan	539.6	558.4	97.6	425	Saudi RS satelite China launch
486	Sagittarius 01	7/12/2018	Long March 2D	Jiuquan	537.8	559.4	97.6	NA	Small RS Com satellites company
487	Ladybird 1	7/12/2018	Long March 2D	Jiuquan	537.8	558.7	97.6	NA	Small RS Com satellites company
488	Ladybird 2	7/12/2018	Long March 2D	Jiuquan	537.0	559.1	97.6	NA	Small RS Com satellites company
489	Ladybird 3	7/12/2018	Long March 2D	Jiuquan	536.6	559.1	97.6	NA	Small RS Com satellites company

No	Satellite Name	Launcher	Launch Date	Launch site	Perigee Km	Apogee Km	Inclination Degree	Mass Kg	Comments
490	Ladybird 5	Long March 2D	7/12/2018	Jiuquan	535.6	558.2	97.6	NA	Small RS Com satellites company
491	Ladybird 6	Long March 2D	7/12/2018	Jiuquan	534.3	559.5	97.6	NA	Small RS Com satellites company
492	Ladybird 7	Long March 2D	7/12/2018	Jiuquan	534.0	558.7	97.6	NA	Small RS Com satellites company
493	Tianyi 3-	Long March 2D	7/12/2018	Jiuquan	534.0	558.4	97.6	NA	Small RS Com satellites company
494	Tianyi y-0y	Long March 2D	7/12/2018	Jiuquan	533.6	558.2	97.6	NA	Small RS Com satellites company
495	Tianyi z-0z	Long March 2D	7/12/2018	Jiuquan	537.9	552.4	97.6	NA	Small RS Com satellites company
496	TFStar	Long March 2D	7/12/2018	Jiuquan	532.1	557.1	97.6	NA	Small RS Com satellites company
497	Chang'e 4	Long March 3B	7/12/2018	Xichang	270.6	283.9	28.9	3780	Lunar rover Dark Side of Moon
498	Hongyun1	Long March 11	21/12/2018	Jiuquan	1,068	1,085.7	99.9	NA	Test of Internet Constellation
499	TJS 3	Long March 3C	24/12/2018	Xichang	35,786	35,801.3	0.1	NA	EW GSO Military Satellite
500	Hongyan 1	Long March 2D	29/12/2018	Jiuquan	1,097.6	1,107.6	50.0	NA	Test of Internet Constellation
501	Yunhai-2 01	Long March 2D	29/12/2018	Jiuquan	523.4	532.0	50.0	NA	Test of Internet Constellation
502	Yunhai-2 02	Long March 2D	29/12/2018	Jiuquan	521.5	531.6	50.0	NA	Test of Internet Constellation
503	Yunhai-2 03	Long March 2D	29/12/2018	Jiuquan	519.2	531.5	50.0	NA	Test of Internet Constellation
504	Yunhai-2 04	Long March 2D	29/12/2018	Jiuquan	1,095.1	1,105.5	50.0	NA	Test of Internet Constellation
505	Yunhai-2 05	Long March 2D	29/12/2018	Jiuquan	803.0	811.0	50.0	NA	Test of Internet Constellation
506	Yunhai-2 06	Long March 2D	29/12/2018	Jiuquan	1,098.9	1,107	50.0	NA	Test of Internet Constellation
507	Shen Tong 2D	Long March 3B	1/10/2019	Xichang	35791.2	35798	0.1	5100	Shentong military Comsat
508	Jilin-1 - 09	Long March 11	1/21/2019	Jiuquan	527	549	97.5	NA	Commercial RS Jilin
509	Jilin-1 - 10	Long March 11	1/21/2019	Jiuquan	523	552	97.5	NA	Commercial RS Jilin
510	Lingque 1A	Long March 11	1/21/2019	Jiuquan	519	545	97.5	NA	RS small satellite constellation
511	Xiaoxiang-1 03	Long March 11	1/21/2019	Jiuquan	528	547	97.5	NA	Test satellite small
512	ChinaSat 6C	Long March 3B	3/9/2019	Xichang	35787	35799	0.1	NA	Commercial
513	Lingque 1B	Chongqing	3/27/2019	Jiuquan	Failure	Failure	Failure	205	New launcher solid missile based
514	Tian Lian 2 - 01	Long March 3B	3/31/2019	Xichang	35652	35938	2.8	5200	TDRS
515	Beidou 3I	Long March 3B	4/20/2019	Xichang	35722	35,862	55.0	4200	Navigation Beidou
516	Tianhui 2-01A	Long March 4B	4/29/2019	Taiyuan	522	523	97.5	NA	RS satellites continuity
517	Tianhui 2-01B	Long March 4B	4/29/2019	Taiyuan	522	523	97.5	NA	RS satellites continuity
518	Beidou 2G8	Long March 3C	5/17/2019	Xichang	35780	35808	1.7	4600	Navigation Beidou

No	Satellite Name	Launch Date	Launcher	Launch site	Perigee Km	Apogee Km	Inclination Degree	Mass Kg	Comments
519	Yaogan 33	5/22/2019	Long March 4C	Taiyuan	Failure	Failure	Failure	NA	SAR Satellite Failure
520	Jilin 1	6/5/2019	Long March 11	YSLA	568	586	45.0	42	RS Commercial
521	Bufeng 1A	6/5/2019	Long March 11	YSLA	563	583	45.0	NA	Meteorology
522	Bufeng 1B	6/5/2019	Long March 11	YSLA	562	582	45.0	NA	Meteorology
523	Xiaoxiang 1-04	6/5/2019	Long March 11	YSLA	562	583	45.0	NA	Communication
524	Tianqi 3	6/5/2019	Long March 11	YSLA	562	583	45.0	8	Communication
525	Tianxiang 1	6/5/2019	Long March 11	YSLA	558	583	45.0	NA	Communication
526	Tianxiang 2	6/5/2019	Long March 11	YSLA	565	590	45.0	NA	Communication
527	Beidou 46	6/24/2019	Long March 3B	Xichang	35739	35841	55.1	4200	Navigation Beidou
528	CAS 7 - BP 1B	12/20/2019	Hyperbola-1	Jiuquan	290	290	42.7	3	First Private sector launch test
529	Star Age 6	12/20/2019	Hyperbola-1	Jiuquan	290	290	42.7	NA	Test satellite small
530	Hangtian	12/20/2019	Hyperbola-1	Jiuquan	290	290	42.7	NA	Test satellite small
531	Technology	12/20/2019	Hyperbola-1	Jiuquan	290	290	42.7	NA	Test satellite small
532	Yaogan 30-05-01	7/26/2019	Long March 2C	Xichang	609	618	35.0	NA	ELINT Tactical
533	Yaogan 30-05-02	7/26/2019	Long March 2C	Xichang	598	611	35.0	NA	ELINT Tactical
534	Yaogan 30-05-03	7/26/2019	Long March 2C	Xichang	590	602	35.0	NA	ELINT Tactical
535	Qiancheng-1 01	8/17/2019	Jielong-1	Jiuquan	537	568	97.6	65	China Rocket Co CALT Co
536	Xingshidai-5	8/17/2019	Jielong-1	Jiuquan	534	568	97.6	10	small RS test
537	Tianqi-2	8/17/2019	Jielong-1	Jiuquan	534	569	97.6	8	small relay test
538	Zhongxing-18	8/19/2019	Long March 3B	Xichang	Failure	Failure	Failure	NA	Communications
539	Xiaoxiang 1-07	8/30/2019	Kuaizhou 1A	Jiuquan	599	617	97.8	8	Small satellite test
540	KX-09	8/30/2019	Kuaizhou 1A	Jiuquan	599	615	97.8	NA	Small satellite test
541	Ziyuan 1-2D	9/12/2019	Long March 4B	Taiyuan	781	781	98.6	1840	Continuation of Ziyuan EO series
542	BNU-1	9/12/2019	Long March 4B	Taiyuan	739	758	98.6	16	Research test small satellite
543	Taurus 1	9/12/2019	Long March 4B	Taiyuan	739	758	98.6	NA	Test satellite small
544	ZHUHAI-1 03A	9/19/2019	Long March 11	Jiuquan	510	530	97.4	90	RS Constellation commercial
545	ZHUHAI-1 03B	9/19/2019	Long March 11	Jiuquan	513	527	97.4	90	RS Constellation commercial
546	ZHUHAI-1 03C	9/19/2019	Long March 11	Jiuquan	514	526	97.4	90	RS Constellation commercial

No	Satellite Name	Launch Date	Launcher	Launch site	Perigee Km	Apogee Km	Inclination Degree	Mass Kg	Comments
547	ZHUHAI-1 03D	9/19/2019	Long March 11	Jiuquan	513	526	97.4	90	RS Constellation commercial
548	ZHUHAI-1 03E	9/19/2019	Long March 11	Jiuquan	498	519	97.4	90	RS Constellation commercial
549	Beidou 47	9/22/2019	Long March 3B	Xichang	21553	22116	55.0	1014	Navigation Beidou
550	Beidou 48	9/22/2019	Long March 3B	Xichang	21515	21515	55.0	1014	Navigation Beidou
551	Yunhai-1 02	9/25/2019	Long March 2D	Jiuquan	789	791	98.6		Ocean satellite 2nd launch
552	GAOFEN 10R	10/4/2019	Long March 4C	Taiyuan	630	635	97.8	NA	Replacement SAR satellite
553	TJS 4	10/17/2019	Long March 3B	Xichang	207	35817	27.0	2700	Early Warning satellite in GSO
554	Gaofen 7 (GF 7)	11/3/2019	Long March 4B	Taiyuan	502	517	97.5	2400	Stereo mapping RS satellite
555	Huangpu 1	11/3/2019	Long March 4B	Taiyuan	490	514	97.5	50–100	Technology test satellite
556	SRSS 1	11/3/2019	Long March 4B	Taiyuan	490	514	97.5	50–100	Sudan RS satellite
557	Xiaoxiang 1-08	11/3/2019	Long March 4B	Taiyuan	491	514	97.5	NA	French Iodine thruster Cubesat
558	(Beidou 3I)	11/4/2019	Long March 3B	Xichang	35689	35894	58.6	4200	Navigation Beidou
559	Jilin-1	11/13/2019	Kuaizhou-1A	Jiuquan	537	556	97.5	NA	Jilin RS Commercial
560	Ningxia 11	11/13/2019	Long March 6	Taiyuan	895	905	45.00	NA	RS commercial
561	Ningxia 12	11/13/2019	Long March 6	Taiyuan	893	905	45.00	NA	RS commercial
562	Ningxia 13	11/13/2019	Long March 6	Taiyuan	890	905	45.00	NA	RS commercial
563	Ningxia 14	11/13/2019	Long March 6	Taiyuan	889	904.1	45.00	NA	RS commercial
564	Ningxia 16	11/13/2019	Long March 6	Taiyuan	888	903	45.00	NA	RS commercial
565	KL-Alpha A	11/17/2019	Kuaizhou-1A	Jiuquan	1049	1068	88.9	70	CAS German Comsat -test
566	KL-Alpha B	11/17/2019	Kuaizhou-1A	Jiuquan	1051	1440	88.9	90	CAS German Comsat -test
567	Beidou 50	11/23/2019	Long March 3B	Xichang	21517	21553	55.00	1060	Navigation Beidou
568	Beidou 51	11/23/2019	Long March 3B	Xichang	21521	21549	55.00	1060	Navigation Beidou
569	Gaofen 12	11/27/2019	Long March 4C	Taiyuan	635	637	97.9	2800	Civilian RS SAR satellite
570	Jilin-1	12/7/2019	Kuaizhou-1A	Taiyuan	538	555	97.5	230	Commercial RS satellite
571	HEAD 2A	12/7/2019	Kuaizhou-1A	Taiyuan	500	500	97.4	45	Experimental IOT application
572	HEAD 2B	12/7/2019	Kuaizhou-1A	Taiyuan	500	500	97.4	45	Experimental IOT application
573	TY16	12/7/2019	Kuaizhou-1A	Taiyuan	500	500	97.4	NA	RS small satellite
574	TY17	12/7/2019	Kuaizhou-1A	Taiyuan	500	500	97.4	NA	RS small satellite

No	Satellite Name	Launch Date	Launcher	Launch site	Perigee Km	Apogee Km	Inclination Degree	Mass Kg	Comments
575	Tianqi 4A	12/7/2019	Kuaizhou-1A	Taiyuan	500	500	97.4	8	Experimental IOT application
576	Tianqi 4B	12/7/2019	Kuaizhou-1A	Taiyuan	500	500	97.4	8	Experimental IOT application
577	(Beidou 52	12/16/2019	Long March 3B	Xichang	21497	21573	55.00	1060	Navigation Beidou
578	Beidou 53	12/16/2019	Long March 3B	Xichang	21539	22115	55.00	1060	Navigation Beidou
579	CBERS 4A	12/20/2019	Long March 4B	Taiyuan	615	635	98.0	1980	Brazil ongoing RS programme
580	ETRSS 1	12/20/2019	Long March 4B	Taiyuan	615	615	98.0	65	Ethiopian RS Satellite
581	Tianqin 1/CAS 6	12/20/2019	Long March 4B	Taiyuan	615	615	98.0	35	Science Amateur radio satellite
582	Yuheng	12/20/2019	Long March 4B	Taiyuan	615	615	98.0	50 - 100	Technology test satellite
583	Shuntian	12/20/2019	Long March 4B	Taiyuan	615	615	98.0	50 - 100	Technology test satellite
584	Tianyan 01	12/20/2019	Long March 4B	Taiyuan	615	615	98.0	72	RS commercial
585	Tianyan 02	12/20/2019	Long March 4B	Taiyuan	615	615	98.0	NA	RS - Science fiction
586	Weilai 1R	12/20/2019	Long March 4B	Taiyuan	615	615	98.0	65	Technology test satellite
587	FloripaSat 1	12/20/2019	Long March 4B	Taiyuan	615	615	98.0	NA	Student test satellite Brazil
588	SJ 20	12/27/2019	Long March 5	Wenchang	35600	35818	0.1	8000	Return to flight test - 2017 failure
589	TJS 5	1/7/2020	Long March 3B	Xichang	35780	35810	0.3	5,550	Early Warning GSO
590	Jilin	1/15/2020	Long March 2D	Taiyuan	481	497	97.3	NA	Commercial RS Jilin
591	NUSAT-7	1/15/2020	Long March 2D	Taiyuan	473	491	97.3	37.5	Intl commercial RS Argentina
592	NUSAT-8	1/15/2020	Long March 2D	Taiyuan	468	484	97.3	37.5	Intl commercial RS Argentina
593	Tianqi-5	1/15/2020	Long March 2D	Taiyuan	478	495	97.3	8	DCP constellation
594	YINHE-1	1/16/2020	Kuaizhou 1A	Jiuquan	643	660	86.4	227	Orbital Comm test
595	XJS C	2/19/2020	Long March 2D	Xichang	478	488	35	NA	35-degree ELINT?
596	XJS D	2/19/2020	Long March 2D	Xichang	478	488	35	NA	35-degree ELINT?
597	XJS E	2/19/2020	Long March 2D	Xichang	478	488	35	NA	35-degree ELINT?
598	XJS F	2/19/2020	Long March 2D	Xichang	472	480	35	NA	35-degree ELINT?
599	BEIDOU 3 G2	3/9/2020	Long March 3 B	Xichang	35777	35811	2.4	4600	Beidou navigation
600	XJY 6	3/16/2020	Long March 7A	Wenchang	Failure	Failure	Failure	6,800	2nd flight ferry GSO missions
601	YAOGAN-30 R	3/24/2020	Long March 2C	Xichang	603	608	35.0	NA	Yaogan ELINT constellation
602	YAOGAN-30 S	3/24/2020	Long March 2C	Xichang	602	608	35.0	NA	Yaogan ELINT constellation

No	Satellite Name	Launch Date	Launcher	Launch site	Perigee Km	Apogee Km	Inclination Degree	Mass Kg	Comments
603	YAOGAN-30 T	3/24/2020	Long March 2C	Xichang	604	607	35.0	NA	Yaogan ELINT constellation
604	Palapa N1	4/9/2020	Long March 3B	Xichang	Failure	Failure	Failure	5550	Indonesia Satcom orbit delivery
605	XZF-SC	5/5/2020	Long March 5B	Wenchang	162	377	41.0	21,600	New crewed spacecraft
606	RCS-FC-SC	5/5/2020	Long March 5B	Wenchang	162	377	41.0	NA	Re-entry Test
607	Xingyun-2 01	5/12/2020	Kuaizhou 1A	Jiuquan	567	577	97.5	93	Orbital Comm test satellite
608	Xingyun-2 02	5/12/2020	Kuaizhou 1A	Jiuquan	563	580	97.5	93	Orbital Comm test satellite
609	XJS G	5/29/2020	Long March 11	Xichang	477	489	35	120	First CZ 11 XiChang launch ELINT
610	XJS H	5/29/2020	Long March 11	Xichang	477	489	35	120	First CZ 11 XiChang launch ELINT
611	HEAD 4	5/31/2020	Long March 2D	Jiuquan	488	506	97.3	45	AID Package for tracking ships
612	Gaofen 9-02	5/31/2020	Long March 2D	Jiuquan	494	510	97.3	NA	Civilian RS satellite Yaogan type
613	HY 1	6/10/2020	Long March 2C	Taiyuan	777	792	98.5	442	Ocean RS satellite
614	Gaofen 9-03	6/17/2020	Long March 2D	Jiuquan	493	512	97.4	2950	Civilian RS satellite Yaogan type
615	Zheda Pixing 3A	6/17/2020	Long March 2D	Jiuquan	489	509	97.4	20	University test satellite
616	HEAD 5	6/17/2020	Long March 2D	Jiuquan	487	507	97.4	45	AID Package for tracking ships
617	Beidou 3G	6/23/2020	Long March 3B	Xichang	35,778	35813	2.3	4600	Beidou constellation complete
618	Gaofen (GF)	7/3/2020	Long March 4B	Taiyuan	636	657	98.0	2,400	Civilian RS satellite
619	BY 2	7/3/2020	Long March 4B	Taiyuan	638	654	98.0	2	Student test satellite
620	Shiyan 6-02	7/4/2020	Long March 2D	Jiuquan	666	799	98.2	3,500	EW ELINT test?
621	APStar-6D	7/9/2020	Long March 3B	Xichang	35779	35809	0.1	5550	Civilian GSO Comsat
622	Jilin-1	7/10/2020	Kuaizhou-11	Jiuquan	Failure	Failure	Failure	172	Commercial RS Jilin
623	CentiSpace-1 S2	7/10/2020	Kuaizhou-11	Jiuquan	Failure	Failure	Failure	97	Navigation test satellite
624	Tianwen 1	7/23/2020	Long March 5	Wenchang	Mars	Mars orbit	Mars	4920	Mars Mission
625	ZiYuan-3 03	7/25/2020	Long March 4B	Taiyuan	502	517	97.5	2360	Operational RS series
626	Tianqi 10	7/25/2020	Long March 4B	Taiyuan	485	501	97.5	8	DCP data collection
627	Longxia Yan 1	7/25/2020	Long March 4B	Taiyuan	487	504	97.5	50	Lobster Eye science test
628	Gaofen 9-04	8/6/2020	Long March 2D	Jiuquan	493	511	97.4	3,500	Operational RS series
629	TKW	8/6/2020	Long March 2D	Jiuquan	487	510	97.4	NA	Precision mapping RS
630	Gaofen 9-05	8/23/2020	Long March 2D	Jiuquan	493	511	97.5	3500	Operational RS series

No	Satellite Name	Launch Date	Launcher	Launch site	Perigee Km	Apogee Km	Inclination Degree	Mass Kg	Comments
631	Tiantuo 5	8/23/2020	Long March 2D	Jiuquan	490	509	97.5	NA	Military test satellite
632	Shiyan	8/23/2020	Long March 2D	Jiuquan	490	510	97.5	NA	Military test small satellite
633	CSSHQ	9/4/2020	Long March 2F	Jiuquan	332	348	50.2	8500	reusable shuttle type satellite
634	Gaofen 11-02	9/7/2020	Long March 4B	Taiyuan	502	503	97.3	4,200	Civilian RS Yaogan type
635	Jilin-1	9/12/2020	Kuaizhou 1A	Jiuquan	Failure	Failure	Failure	172	RS Jilin satellite failure
636	Jilin-1	9/15/2020	Long March 11	YSLA	533	555	97.5	42	RS constellation
637	Jilin-1	9/15/2020	Long March 11	YSLA	534	554	97.5	42	RS constellation
638	Jilin-1	9/15/2020	Long March 11	YSLA	534	554	97.5	42	RS constellation
639	Jilin-1	9/15/2020	Long March 11	YSLA	534	554	97.5	42	RS constellation
640	Jilin-1	9/15/2020	Long March 11	YSLA	534	554	97.5	42	RS constellation
641	Jilin-1	9/15/2020	Long March 11	YSLA	533	555	97.5	42	RS constellation
642	Jilin-1	9/15/2020	Long March 11	YSLA	533	555	97.5	42	RS constellation
643	Jilin-1	9/15/2020	Long March 11	YSLA	533	555	97.5	42	RS constellation
644	Jilin-1	9/15/2020	Long March 11	YSLA	534	554	97.5	378	RS constellation
645	Haiyang 2C	9/21/2020	Long March 4B	Jiuquan	954	965	66	1575	Oceanography
646	HJ 2A	9/27/2020	Long March 4B	Taiyuan	646	649	98	470	RS - International
647	HJ 2B	9/27/2020	Long March 4B	Taiyuan	647	649	98	470	RS - International
648	Gaofen 13	10/11/2020	Long March 3B	Xichang	35771	35816	1.4	11,500	GSO Imaging EW?
649	YAOGAN-30 U	10/26/2020	Long March 2C	Xichang	604	607	35	2500	Yaogan ELINT
650	YAOGAN-30 V	10/26/2020	Long March 2C	Xichang	604	606	35	2500	Yaogan ELINT
651	YAOGAN-30 W	10/26/2020	Long March 2C	Xichang	600	610	35	2500	Yaogan ELINT
652	Tianqi 6	10/26/2020	Long March 2C	Xichang	590	611	35	6	DCP data collection
653	NUSAT-9	11/6/2020	Long March 6	Taiyuan	473	479	97.2	41	Argentina RS
654	NUSAT-10	11/6/2020	Long March 6	Taiyuan	469	483	97.2	41	Argentina RS
655	NUSAT-11	11/6/2020	Long March 6	Taiyuan	470	482	97.2	41	Argentina RS
656	NUSAT-12	11/6/2020	Long March 6	Taiyuan	470	485	97.2	41	Argentina RS
657	NUSAT-13	11/6/2020	Long March 6	Taiyuan	469	483	97.2	41	Argentina RS
658	NUSAT-14	11/6/2020	Long March 6	Taiyuan	469	483	97.2	41	Argentina RS

No	Satellite Name	Launch Date	Launcher	Launch site	Perigee Km	Apogee Km	Inclination Degree	Mass Kg	Comments
659	NUSAT-15	11/6/2020	Long March 6	Taiyuan	472	480	97.2	41	Argentina RS
660	NUSAT-16	11/6/2020	Long March 6	Taiyuan	471	481	97.2	41	Argentina RS
661	NUSAT-17	11/6/2020	Long March 6	Taiyuan	470	484	97.2	41	Argentina RS
662	NUSAT-18	11/6/2020	Long March 6	Taiyuan	469	483	97.2	41	Argentina RS
663	Beihangkongshi 1	11/6/2020	Long March 6	Taiyuan	468	479	97.2	16	Small satellite test
664	Tianyan 05	11/6/2020	Long March 6	Taiyuan	466	477	97.2	70	Small satellite test
665	BY 3	11/6/2020	Long March 6	Taiyuan	466	477	97.2	8	Small satellite test
666	Tiantong-1-02	11/12/2020	Long March 3B	Xichang	35774	35813	5	5400	Commercial Comsat
667	Chang'e 5	11/23/2020	Long March 5	Wenchang	NA	NA	lunar orbit	8200	Lunar sample return
668	Gaofen 14	12/6/2020	Long March 3B	Xichang	495	497	97.4	11,500	First Xichang SSO RS payload
669	GECAM A	12/9/2020	Long March 11	Xichang	594	612	29	150	X Ray Astronomy
670	GECAM B	12/9/2020	Long March 11	Xichang	593	612	29	150	X Ray Astronomy
671	XJY 7	12/22/2020	Long March 8	Wenchang	510	518	97.4	3000	First CZ 8 Launch SAR
672	Hisea-1	12/22/2020	Long March 8	Wenchang	510	518	97.4	180	C band SAR payload
673	Yuanguang	12/22/2020	Long March 8	Wenchang	510	518	97.4	20	Tribology test
674	Zhixing-1A	12/22/2020	Long March 8	Wenchang	510	518	97.4	8	Satellite for Ethiopia
675	Apocalypse 8	12/22/2020	Long March 8	Wenchang	510	518	97.4	50	DCP data collection
676	Yaogan 33R	12/28/2020	Long March 4C	Jiuquan	702	704	98.2	4200	New type 2018 launch
677	Weina 2	12/28/2020	Long March 4C	Jiuquan	702	704	98.2	NA	Small test satellite

Notes: This Table has been compiled from several open sources. It has been checked for consistency. In addition, launch data including launcher type, orbit information and other parameters such as Equatorial Crossing Time (ECT) have been used to make inferences on the mission.

The total number of satellite launchings does not include one Kaitouzhe launch which took place but is not included in many public data bases. No details are available from authentic sources for this failed launch. Taking this into account the total number of satellites that China has launched from 1970 to 2020 is 678.

The yellow highlighted launch dates indicate that satellites of that date were launched together using one launcher. Differing font colours have been used for some specific categories. The red fonts indicate either launch or mission failure. The blue font reflects the manned flights or the lunar and Mars missions that China has flown.

The abbreviation SDLC under the column Launch Site stands for Satish Dhawan Launch Centre. SHAR for Sriharikota.

The abbreviation YSLA under the launch site column stands for the Yellow Sea Launch Area that indicates launch from platforms located in the Yellow Sea.

Reference

Union of Concerned Scientists (2021) UCS Satellite Data Base, May 2021. https://ucsusa.org/res
   ources/satellite-database

Part II
Assessments

Chapter 6
Chinese Recoverable Satellites

6.1 Background

After the launch of their first two satellites in 1970 and 1971, the Chinese shifted focus towards recoverable satellites. The recoverable satellite series has the Chinese name "Fanhui Shi Weixing" or FSW in its shortened form. The two launch vehicles that China was developing at that time were the Feng Bao launcher developed by the Shanghai-based group and the CZ 2C (originally CZ 2A) launcher being developed by the Beijing-based China Academy for Launch Vehicle Technology (CALT). Almost all the early launches of these vehicles were devoted to the recoverable FSW series of satellites[1] (Harvey 2013).

There were two ostensible reasons for launching these satellites. One was of course to be able to use them for collecting reconnaissance data over other countries using camera systems. This was a trend set by both the USA and the USSR in the early years of their space programmes. The processed data from these recoverable capsules was the basis for ensuring strategic stability during the early years of the Cold War.

The other reason was of course that mastering the technology for launch, reentry, and recovery would help in creating capabilities for the human space programme.

After a string of launch failures, especially of the Feng Bao (FB) launcher, a CZ 2A launcher placed the first recoverable satellite into near earth orbit in 1975. This satellite was recovered after 3 days in orbit. Over the next two launches, the launch and recovery techniques were mastered.

After the USSR and the US, China became the third country to recover a satellite from orbit.

---

[1] pp. 105–133.

© National Institute of Advanced Studies 2022
S. Chandrashekar, *China's Space Programme*,
https://doi.org/10.1007/978-981-19-1504-8_6

6.2   The FSW Recoverable Series—The Launch Record

Table 6.1 provides details of the 25 satellites that the Chinese have launched and recovered.

Except for the Shijian 10, these satellites have been placed in low earth orbit with inclinations between 57 and 63°.

Of the 25 launches, the first launch failed. FSW 1-5 launched in October 1993 also failed. Though the launch was successful, the retro fire for reentry took place in the wrong direction. This sent the satellite into a higher orbit. The reentry of this satellite also raised concerns of possible damage on the earth just as in the case of Skylab. Fortunately, it reentered over the Atlantic Ocean.

There is some information that one of the FSW satellites had problems because of fuel venting. The second stage apparently vented fuel which covered all the windows of the reentry capsule. Though the mission went perfectly, no photographs could be taken. Though there is no definite statement on which of the FSW missions had this problem, open sources suggest that this could have been the FSW 1-1 launched in 1987.

From the launch record, the first Chinese satellites that went into Sun Synchronous Orbit (SSO) were the weather satellites of the Feng Yun 1 series. This happened in

Table 6.1   China's recoverable satellites

Satellite	Launch Date	Launcher	Mass Kg	Inclination °	Perigee Km	Apogee Km
FSW 00	11/5/1974	CZ 2C	1800	NA	NA	NA
FSW 01	11/26/1975	CZ 2C	1790	63.00	179	479
FSW 02	12/7/1976	CZ 2C	1790	59.10	171	480
FSW 03	1/26/1978	CZ 2C	1810	57.00	161	479
FSW 04	9/9/1982	CZ 2C	1780	62.90	174	393
FSW 05	8/19/1983	CZ 2C	1840	63.30	172	389
FSW 06	9/12/1984	CZ 2C	1810	67.90	174	400
FSW 07	10/21/1985	CZ 2C	1810	62.90	171	393
FSW 08	10/6/1986	CZ 2C	1770	57.00	173	385
FSW 09	8/5/1987	CZ 2C	1810	62.90	172	400
FSW 1-1	9/9/1987	CZ 2C	2070	63.00	206	310
FSW 1-2	8/5/1988	CZ 2C	2130	63.00	206	310
FSW 1-3	10/5/1990	CZ 2C	2080	57.00	208	311
FSW-1 13	8/9/1992	CZ 2D	2590	63.17	172	330
FSW-1 4	10/6/1992	CZ 2C	2060	63.00	214	312
FSW-1-5	10/8/1993	CZ 2C	2100	57.00	209	300
FSW-2 2	7/3/1994	CZ 2D	2760	63.00	174	343
FSW-2 3	10/20/1996	CZ 2D	2970	63.00	171	342
FSW 18	11/3/2003	CZ 2D	3000	63.00	194	325
FSW 32	8/29/2004	CZ 2C	3800	63.00	199	491
FSW-20	9/27/2004	CZ 2D	2600	63.00	205	297
FSW 34	8/2/2005	CZ 2C	3800	63.00	166	494
FSW 22	8/29/2005	CZ 2D	3600	63.00	203	298
Shijian-8	9/9/2006	CZ 2C	3600	63.00	178	449
Shijian 10	4/5/2016	CZ 2D	3600	43.00	220	482

1990. The first genuine remote sensing satellite carrying digital camera systems in a sun synchronous orbit was possibly the Shijian satellite carried along with the FengYun weather satellite in 1999. This was followed up with the first CBERS satellite in partnership with Brazil in the same year.

This would suggest that the FSW recoverable satellites did help to some extent in meeting reconnaissance needs of China though the system could not be considered operational. China possibly depended on other open sources of data for dealing with some of its routine reconnaissance needs.

Apart from the obvious reconnaissance functions served by these recoverable satellites, the technology base for recovering objects launched into space and for getting human beings back from orbit have many commonalities. Though a human space flight programme was visualized almost from the beginning, it went through a series of ups and downs before finally being approved in 1992. The continuity in the recoverable series of satellites despite their limited intelligence value, suggests that the Chinese have always looked at human space flight as an important component of the country's space effort.

A closer scrutiny of Table 6.1 reveals a gap between 1979 and 1982. Then there are an almost continuous series of launchings till 1996. There is a further gap of seven years followed by a string of launches from 2003 to 2006. The launchings during 2003–2006 were most probably carried out to test components for use in the human space station programme.

It is likely that these recoverable satellites served multiple purposes apart from the photo recon functions. They must have been used to test new imaging instruments as well as for experiments in materials processing and biology.

The evidence also suggests that the Chinese gradually increased the time between launch and recovery from a few days to at least 18 days over the nearly three decades of launching these recoverable satellites. The recoverable capsule was also modified to provide larger cabin space in the later launches.

The value and utility of this programme based on published information from Chinese sources suggest benefits from the biological experiments as well as from the materials processing experiments carried out on the FSW satellites. Their utility for various remote sensing applications has also been proven.

The renewal of the FSW launchings in 2003 after a gap of 7 years also suggests that these were Chinese preparations for their launch and operation of their space station. Much of the equipment used in the space station could have been tested in these launches.

It is quite likely that the early missions were largely devoted to photo reconnaissance. Later versions also carried equipment for materials processing and biological experiments.

Some of the missions were used to carry experiments on payment basis for other countries.

FSW-09 launched in 1987 carried microgravity experiments for the French company Matra.

FSW 1-2 launched in 1988 carried a remote sensing payload as well as German protein crystal-growing experiments.

FSW 1-4 launched in 1992 carried the Swedish Frejya satellite and FSW 2-3 launched in 1996 carried micro gravity materials processing experiments for Japan.

The most recent Shijian 10 satellite (2016 launch) carried a payload for the European Space Agency (ESA).

Till the Chinese launched their own digital EO satellites in 1999–2000, it is likely that these satellites performed a sporadic but useful photo recon function. In combination with other sources of data such as the SPOT and Landsat systems, they may have played an important intelligence role.

The mainstay launcher for these missions has been the CZ 2C. When the more advanced CZ 2D became available, heavier and larger satellites have been launched. It is likely that the heaviest versions launched used the three-stage variant of the CZ 2C launcher.

6.3   Trends

The overall trend in the evolution of the FSW series is shown in Fig. 6.1.

There is a large gap from 2006 to 2016. During this period the responsibility for the recoverable programme moved out of China National Space Agency (CNSA) and was transferred to the Chinese Academy of Sciences (CAS). The launch of 2016 signifies that some part of the Chinese space effort will be devoted to the basic sciences as well as for international cooperation activities. The Shijian 10 carried about 24 experiments related to basic research in Fluid Physics, Combustion, Materials Science, Radiation Biology, Gravitational Biology, and Biotechnology. It also carries a payload for the European Space Agency (eoPortal Directory).

Though the overall space programme may still be under the military and the Strategic Support Force (SSF) after the re-organization in 2015, the return of the CAS into the mainstream is indicative of a greater emphasis on basic science. The recent launch of the Quantum Satellite, the X-ray Pulsar, and the Hard X-ray satellites

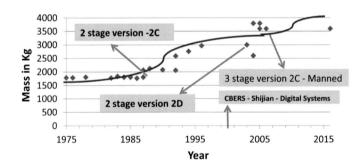

Fig. 6.1 Recoverable satellites—trends

which come under the CAS provides further evidence of this thrust towards fundamental research (Raska 2016). The FSW recoverable satellite series will provide the laboratory space for such basic science initiatives. International collaboration could also form a significant component of such efforts.

There have been no recoverable satellite launches between 2017 and 2020 suggesting that the programmes utility has come to an end with the growing maturity of China's space effort.

References

eoPortal (2010) https://directory.eoportal.org/web/eoportal/satellite-missions/s/shi-jian-10
Harvey B (2013) China in space: the great leap forward. Springer Praxis Books, New York, p 2013
Raska M (2016) China's Quantum Satellite Experiments Strategic and Military Implications, Analysis. http://www.eurasiareview.com/06092016-chinas-quantum-satellite-experiments-strategic-and-military-implications-analysis/

Chapter 7
China's Communications Satellites

7.1 Background

China has developed several launchers for placing satellites in Geostationary Orbits (GSO). These include the CZ 3 series comprising the CZ 3, CZ 3A, CZ 3B, CZ 3BE, CZ 3C as well as a CZ 2E which was an extension of the CZ 2C launcher. China has closed the CZ 3 and CZ 2E lines and only uses the others. These vehicles especially the CZ 3B, CZ 3BE as well as the CZ 3C are powerful vehicles and China has used them for other programmes such as navigation and lunar exploration missions. The 2016 test of the CZ 5 launcher will make possible the simultaneous launching of multiple heavy satellites on the same launcher. The new launch site at Wenchang also optimizes payload delivery to the GSO. Though the second test launch of the CZ 5 failed in 2017, China successfully tested it again in December 2019. In 2020 it was used for missions to the Moon and Mars as well as for testing its space station modules. The CZ 5 will now become available for placing multiple heavy satellites in GSO.

The medium lift launcher the CZ 7 that can take intermediate class communications satellite to GSO had been tested earlier in 2016 and 2017. It however failed in a launch attempt in 2020. The CZ 6 small lift launcher has also been successfully qualified. The other newly developed "green launcher" the CZ 8 made its debut in 2020.

China has a launcher fleet that can deliver any kind of communication satellite into GSO.

In the past, China has also procured launch services for placing foreign built communications satellites into GSO to overcome US imposed International Traffic in Arms Regulations (ITAR) restrictions.

It is currently engaged with several trial runs of small communication satellite constellations in Low Earth Orbit (LEO) for providing domestic and global interconnectivity for Internet applications.

© National Institute of Advanced Studies 2022
S. Chandrashekar, *China's Space Programme*,
https://doi.org/10.1007/978-981-19-1504-8_7

7.2   The Composition and Evolution of China's GSO Communications Satellites

Starting from 1984 onwards till the end of December 2020, China has launched a total of 67 Communications satellites into Geostationary Orbit (GSO). An increasing number of them are satellites built by China, launched from China for domestic use or international customers.

37 of these satellites have been built and launched by China. Fourteen have been bought from global satellite companies but launched by Chinese launchers. Seven of them have involved both the satellite and launch being bought from various international suppliers. More recently China has used its domestic satellite building and launcher capacities for in-orbit delivery of nine communications satellites to various international customers.

Figure 7.1 provides a breakup of the approaches used by China to become a globally competitive provider of communications satellite services that cater to both the domestic and international markets.

Seven satellites in the Asiasat/Apstar series have been both built and launched by foreign contractors even though they could have been launched by Chinese launchers. These actions were needed because of US export restrictions and sanctions on China under ITAR.

Europe as a whole and satellite companies in Europe have cooperated with China to make use of cheaper launch costs as well as to sell their satellites. These endeavors have also resulted in collaborative efforts in key areas of technology. The Chinese DFH 3 series of Communications satellites were built with contributions from Messerschmitt Bolkow Blohm (MBB) of Germany for the solar array, Matra Marconi for the onboard processors, Daimler-Chrysler for the antenna and Officine Galileo for the attitude control sensors.

European companies especially Thales Alenia have cooperated with China in building satellites with no US parts to bypass US imposed export restrictions and sanctions under ITAR. Apstar 6 built by Thales Alenia for the Chinese domestic

China Comsat Composition 1984-2020

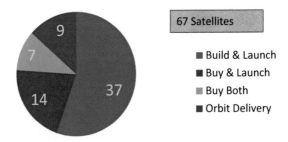

Fig. 7.1 The composition mix of China's communications satellites

market and launched by the CZ 3B from Xi Chang in 2005 was the first European satellite that did not have any US parts. After this Thales Alenia has built and launched several satellites both for China as well as for other countries like Indonesia that have been launched by Chinese launchers.

China has also leveraged and used Thales Alenia capabilities in building the complex multi-frequency multi-beam satellites for in-orbit delivery of communications satellites to countries from the developing world. This in-orbit delivery model has so far resulted in the production and delivery of nine satellites to various international customers. In 2020, the Palapa satellite built by China for in-orbit delivery to Indonesia failed to reach the GSO due to the failure of the CZ 3B rocket.

After the failure of the CZ 5 rocket in 2017, it was successfully launched from the Wenchang launch site at the end of 2019. After three other successful launches in 2020, it is now ready for taking satellites to the GSO.

Table 7.1 provides a listing of these satellites along with details.

7.3 Benchmarking Chinese Communications Satellites

Two kinds of benchmarks may be needed to assess the capabilities of today's communications satellites. One clear benchmark is the capacity of the satellite which depends on bandwidth, modulation schemes and a whole host of other technical factors such as frequency reuse, data compression methods, etc. The capacity is also dependent on the lifetime of the satellite. Composite measures that combine these can be arrived at to compare different satellites. **In simple terms this can be termed as a capacity measure**.

The other important measure is to look at how the capacity is being used to service users in different locations and the kind of services that can be provided to them. This depends on the number of beams, the geographical spread of the beams and the ability to switch between geographies and different kinds of services. **In simple terms this can be termed as a flexibility measure** (Chandrashekar 2015).

Since both measures are linked to the satellite mass, one way to compare performance is to compare the masses of the satellites. Figure 7.2 compares the masses of satellites built by China and launched by China with the masses of the satellites built by international satellite contractors and put into Geostationary Transfer Orbit (GTO) by Chinese launchers.[1]

As we can see from Fig. 7.2, there are three generations of Chinese communications satellites. The first-generation satellites were all launched by the CZ 3 launcher

---

[1] A launch vehicle places a communication satellite in an intermediate Low Earth Orbit (LEO) called the Geostationary Transfer Orbit (GTO). After this orbit is achieved, the satellite is tracked to determine its orbit. A rocket motor on the injected satellite or in some cases a stage on the coupled satellite-launcher combination is then fired to take the satellite to the Geostationary Orbit (GSO). The mass placed in GTO provides information on the final mass placed in GSO and provides a simple measure to compare satellite capabilities.

Table 7.1 List of China's communications satellites

Year	Satellite	Launcher	GTO Mass Kg	Transponders
1984	DFH 2	CZ 3	916	2
1984	DFH 2	CZ 3	916	2
1986	DFH 2	CZ 3	916	4
1988	DFH 2 A	CZ 3	962	4
1988	DFH 2 A	CZ 3	962	4
1990	DFH 2 A	CZ 3	962	24
1990	Asiasat 1 HS 376	CZ-3	1250	4
1991	DFH 2 A	CZ 3	962	
1994	DFH-3 mockup	CZ 3A		30
1994	Shijian 4	CZ 3A		Test
1994	Apstar 1 HS 376	CZ 3	1383	24
1994	DFH 3 -1	CZ 3A	2200	34
1995	Apstar 2 Hughes 601 Failure	CZ-2E	2830	34
1995	Asiasat 2 LM 7000	CZ-2E	3379	30
1996	Apstar 1A Hughes 376	CZ 3A	1383	30
1996	Zhongxing 7 Hughes 376	CZ 3A	1384	24
1997	DFH 3-2 Zhongxing 6B	CZ 3A	2200	43
1997	Apstar 2R Loral 1300	CZ-3B	3700	44
1997	Asiasat 3 Hughes 601 Failure	Proton	3480	38
1998	Zhongwei 1 A2100 A	CZ-3B	2984	38
1998	Sinosat 1 SB 3000	CZ-3B	2820	44
1999	Asiasat 3S Hughes 601	Proton	2300	
2000	Feng Huo 1 Zhongxing 22	CZ 3A	2300	
2003	Shentong 1 Zhongxing 20/21	CZ 3A	4137	44
2003	Asiasat 4 Hughes 601	Atlas 3B	4680	50
2005	Apstar 6 Alcatel / Thales Alenia	CZ 3B	2300	
2006	Feng Huo 2 Zhongxing 22A	CZ 3A	5100	24
2006	Sinosat 2/Xinnuo 2	CZ 3B	2200	24
2007	DFH 3-1 Sinosat 3 Xinnuo 3 Eutelsat 8 W D	CZ 3A	4600	38
2007	Zhongxing 6B Spacebus 4000	CZ-3B	5150	28
2007	Nigicomsat 1	CZ 3B/E	5049	28
2008	Simon Bolibar	CZ 3B/E	4500	22
2008	Zhongxing 9 Spacebus 4000	CZ-3B	3760	40
2009	Asiasat 5 SSL 1300 BUS	Proton	2300	
2010	Shentong 1-2 Zhongxing 20A	CZ 3A	5100	33
2010	Sinosat 6 Xinnuo 6 Zhongxing 6A	CZ 3B/E	5100	46
2011	Zhongxing 10 Thales Alenia Payload	CZ 3B/E	5115	30
2011	Paksat 1 R	CZ 3B/E	5600	
2011	Zhongxing 1A	CZ 3B/E	3813	46
2011	Asiasat 7 SSL 1300 BUS	Proton	5100	28
2011	Nigcomsat 1R	CZ 3B/E	5600	
2012	Shentong 2 Zhongxing 2A	CZ 3B/E	5054	56
2012	Apstar 7 Thales Alenia	CZ 3B/E	5054	47
2012	Zhonxing-12 ITAR Free Thales Alenia	CZ 3B/E	5000	45
2012	Zhonxing-11	CZ 3B	5200	32
2013	Tupac Katari	CZ 3B	4535	25
2014	Asiasat 8 SSL	Falcon	3700	28
2014	Asiasat 6 SSL	Falcon	5200	48
2015	Apstar 9 DFH- 4 Platform	CZ 3B/E	5200	46
2015	ChinaSat-2C Zhongxing-2C	CZ 3B	3800	

(continued)

Table 7.1   (continued)

Year	Satellite	Launcher	GTO Mass Kg	Transponders
2015	Laosat DFH 3B Bus	CZ 3B	5200	22
2015	Chinasat 1C Zhongxing 1C	CZ 3B/E	5200	
2016	Belintersat 1 Thales Alenia	CZ 3B	5200	38
2016	Tiantong 1	CZ 3BE	5200	
2016	Shijian 17	CZ 5	14000	Test
2017	Shijian 16 Chinasat 16	CZ 3B	5200	Ka Band
2017	Zhongxing 9A Chinasat 9A	CZ 3BE	5200	24 Ku Band
2017	Shijian 18	CZ 5	14000	Test Failure
2017	AlComsat 1 (Algeria)	CZ 3B	5200	Orbit Delivery
2018	APSTAR 6C	CZ 3B	5200	24 C & 19 Ku Ka
2019	Shentong 2 D	CZ 3B	5200	
2019	Chinasat 6C	CZ 3B	5200	
2019	Chinasat 18	CZ 3B	5200	Failure
2019	Shijian 20	CZ 5	14000	Test satellite
2020	XJY 6	CZ 7	6800	Test Failure
2020	Palapa N1	CZ 3B	5550	Failure
2020	APSTAR 6D	CZ 3B	5550	Orbit Delivery
Code		Satellite Built and Launched by China		
Code		Bought Satellite Chinese Launcher		
Code		Bought Satellite foreign launcher		
Code		Build and Launch Military satellite		
Code		In Orbit Delivery		

Notes
Satellites in red font represent failures
Blank spaces or NA indicate that the data is not available

Fig. 7.2 Make buy mass in GTO China trends

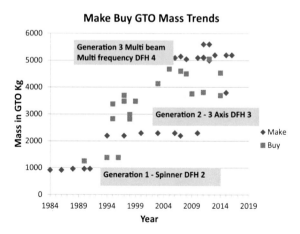

that could place about 1000 kg in Geostationary Transfer Orbit (GTO). Comparable foreign built satellites were heavier at about 1200–1300 kg. The standard, most widely prevalent satellites were built by Hughes. They were spin stabilized satellites with a de-spun antenna. The Chinese-built satellites of this period were based on this

Hughes design. Though the gap in satellite masses from the Figure is not that large even this could make a big difference in the number of transponders. The Chinese-built satellites carried a maximum of four transponders as compared to the Hughes satellites which carried 24 transponders.

When the CZ 3A launcher became available in 1994, China quickly upgraded its indigenous satellite capabilities to try and catch up with the other more advanced countries. The Chinese DFH 3 bus catered to requirements of up to 2200–2300 kg. Unlike the earlier spinner configuration, these used three-axis stabilization with solar panels deployed on either side.

Though this configuration was very similar to the configurations used by the more advanced satellite companies, as we can see from the Figure, Chinese satellites have lower masses than those of the more advanced satellite manufacturers by about a 1000 kg.

China also tried to get help in areas where it perceived itself as weak. US companies who built satellites for launch by the Chinese launchers were the early sources. After the Cox Committee Report and ITAR sanctions, China turned to European companies. China seems to be lagging in both mass and capacity during this period.

When the CZ 3B launcher became available in 1997, China could place heavier satellites weighing about 5100 kg into GTO. The satellites could be built by an outsider or jointly with a collaborator. The availability of this launcher coincided with the imposition of significant embargos and sanctions on China.

As seen in Table 7.1, starting from 1997 onwards China started using foreign launchers for placing their bought satellites in orbit. Three Asiasat satellites launched in 1997, 1999 and 2003 all used foreign launchers. In 2005 China launched Apstar 6, a satellite built by Thales Alenia that did not carry any US parts and was therefore free of the ITAR restrictions.

It is also evident from Table 7.1 that after 2005, the number of bought satellites is coming down and more domestically built satellites are being launched.

When the more powerful CZ 3B/E very heavy launcher became available in 2007 China leveraged its collaboration efforts with Europe to launch several domestic satellites. They also used these capabilities to provide in-orbit delivery of China built and China launched communications satellites to several countries.

As can be seen from Fig. 7.2 by about 2005 the Chinese had caught up with the international 5100 kg standard for satellites with their DFH 4 satellite bus. With the larger CZ 3BE launcher they can even raise this mass to 5500 kg which is slightly above western bought satellite masses.

However, mass in orbit is only one part of the competitive positioning exercise. The capacity is also important. There is therefore a need to bring in a capacity measure as well. Figure 7.3 plots the masses of Chinese-built and Chinese procured satellites with the number of transponders they carry.

The number of transponders that can be accommodated in a satellite of a given mass is dictated by many factors including the function, the frequency of operation, and the mode of operation. In general Ku and Ka Band use requires more power than

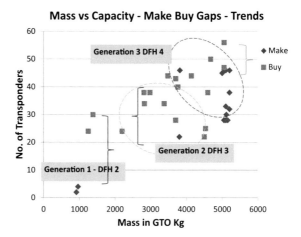

Fig. 7.3 Mass versus capacity—make buy gap China trends

C band frequencies. Typical satellites carry a mix of multi-frequency transponders with multiple beams to cater to a wide spectrum of users spread out over a given geography of interest.

From Fig. 7.3 the gap in the number of transponders for the same weight class of satellites is narrowing as China moves from the first generation through the second generation into the third generation of communications satellites.

From Figs. 7.2 and 7.3 it is obvious that through a judicious mix of built, bought and collaborative efforts China has been able to significantly lower the gap in performance between indigenously developed communications satellites and their western counterparts. However there still appears to be a gap between the Chinese-built satellites and bought satellites.

Figure 7.4 provides the cumulative totals of indigenously built and bought satellites. This provides a flavor of the strategy used by China to catch up with the rest of the world.

Fig. 7.4 Make buy choices and learning—China comsats

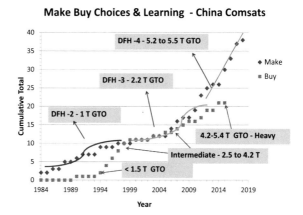

In the first period starting from 1984 to 1994 China builds rather than buys satellites. China bought its first satellite in 1990. Learning from this and realizing that a new generation of intermediate class satellites was coming into the market, they start buying satellites from internationally reputed suppliers starting from 1994.

Once the larger CZ 3A launcher becomes available, they move toward the building of the larger DFH 3 satellites from 1997. The first DFH 3 satellites catering to military needs are launched in 2000. Larger numbers are launched from 2006 to 2008. Though the larger 5100 kg GTO CZ 3B launcher is available, China's efforts at catching up are blocked by US export restrictions that cover both satellites as well as launchers carrying satellites that have US parts.

The availability of the 5100 kg GTO CZ 3B launcher and the enhancement of this capability to 5600 kg in 2007 via the CZ 3BE ensure that as far as launching capacity is concerned China can compete with the best in the world. This also enables them to close to some extent the capacity gap by building a larger DFH 4 bus.

The use of bought satellites slows down after 2004. However selective imports of satellites in the heavy category continue till 2014. This includes domestic satellites for both the Asiasat as well as the Apstar service providers. After 2012, China seems to rely more on indigenous satellites and indigenous launchers for meeting both their civilian and military needs. The pace of satellite launching increases after 2012 as China uses its acquired capabilities in satellite building, launching and delivery in orbit to service international customers. The only exception to this "made in China" rule seem to be the two Asiasat satellites bought and launched in 2014 using the Falcon Rocket.

Apstar 9—a satellite launched in 2015—is an indigenous satellite that uses the DFH-4 platform and is launched by the CZ 3BE launcher. It carries 32 C band and 14 Ku band transponders. Though still a little behind state of the art, it is adequate to cater to both the civilian and military requirements of China.

China's arrangements with European companies especially Thales Alenia have also helped them to significantly narrow the gap between their satellites and those of the more advanced countries. They have used these connections for furthering their exports of satellites and launchers to various customers around the world.

Case Study 1

China and Ion Propulsion (Chandrashekar 2015)

One major technology development that shows promise in reversing the trend toward heavier satellites has been introduced into the communications satellite market by the US company Boeing.

Over half the mass of a typical Geostationary Communications satellite is made up of propellant. Most of this propellant is consumed in moving the satellite from GTO to GSO.

Ion propulsion provides an alternative way of providing the thrust required to move a satellite from GTO to GSO. It involves the acceleration of ions typically through an electro-magnetic field and pushing it out through a suitable nozzle. These thrusters have been in use for a long time and have provided the small thrusts needed for communications satellites to maintain their orbital positions in GSO. They are normally used along with regular chemical propellant thrusters for station keeping and orbit maintenance functions. The Boeing innovation was to replace all chemical propulsion with ion propulsion saving significant weight (Clark 2015).

Though ion thrusters have high specific impulse, the thrust levels they provide are much lower than chemical propellants. Therefore, the time taken to go from GTO to GSO will take several months. Chemical propellants have lower specific impulse but provide a much higher level of thrust than ion propulsion. The journey from GTO to GSO therefore takes only hours or days instead of months.

The major gain from the use of ion propulsion is that the amount of propellant carried by the satellite that accounts for at least half of the satellite mass can be significantly reduced. Though ion propulsion also requires some mass, it is much lighter than the heavy chemical propulsion systems that spacecraft carry. This trend of replacing chemical propulsion with ion propulsion in communications satellites has become more pronounced in the last five years. Most manufacturers of communications satellites in the western world have switched over to ion propulsion.

In early March 2015 the first two all-ion propulsion satellites were launched aboard a Falcon X launcher. The two satellites were both built by Boeing. The Asia Broadcast Satellite 3A (ABS 3A) weighed 1954 kg with 24 C and 24 Ku band transponders.[2] The second satellite was the Eutelsat 115 W B satellite with a mass of 2205 kg carrying 34 Ku and 12 C band transponders.[3] Because of the significant weight saving arising from the use of ion propulsion they could both be accommodated for a dual launch on the Falcon X. This reduced the launch cost significantly making the move to an all-ion propulsion approach economically attractive. These early satellites have been followed by several others that use ion propulsion.

A similar satellite built by India, the GSAT 16, carrying 36 C and 12 Ku band transponders and launched in 2014 weighed 3182 kg. There is a significant weight saving and a direct benefit from the use of ion propulsion.

---

[2] http://www.aerospace-technology.com/projects/abs-3a-communication-satellite/.

[3] http://space.skyrocket.de/doc_sdat/eutelsat-115-west-b.htm.

China has launched several experimental satellites for testing ion propulsion systems. In 2020 it launched a Hisea 1 small satellite carrying a C Band Synthetic Aperture Radar (SAR) payload. The satellite also carries an iodine-based ion propulsion system developed by a French start-up company Thrust Me.

However so far, all communications satellites built by China have only carried chemical propulsion systems. One can expect a change in this approach as China endeavors to make the transition to ion propulsion for maintaining its competitive position in the global COMSAT industry.

7.4 Military Communications

From Table 7.1, nine, of the thirty-eight satellites launched from 2000 onwards are directly used by the military. The Tiantong 1 Mobile Communications satellite launched in 2016 is also likely to serve a military function. Like other space powers China must also be using some of the created civilian communications satellite capacities for military purposes.

7.5 In Orbit Delivery to Global Customers

Table 7.1 also reveals that China is aggressively pursuing the export of its space capabilities. To make sure that its customers do not have any export related problems because of US sanctions, China has contracted with several developing countries for in-orbit delivery of satellites. The satellite and the launch as well as in orbit operations to place it in GSO are Chinese responsibilities. China also supplies the ground stations for satellite operations and provides training so that the concerned country can operate the satellites using their own people.

In orbit delivery of satellites has taken place with Nigeria (two satellites in 2007 and 2011), Venezuela (2008), Pakistan (2011), Bolivia (2013), Laos (2015), Belarus (2016), Algeria (2017), and Indonesia (2020). Priced at about $240–$250 million they are very attractive "value for money" products for many developing countries.

7.6   Tianlian Data Relay Satellites

Five Tianlian Tracking and Data Relay satellites were launched by China in April 2008, August 2011, August 2012, November 2016, and March 2019, respectively. They were meant to support the manned Shenzhou missions. The first four satellites were launched by CZ 3C vehicles while the most recent satellite was launched by the larger CZ 3B, suggesting that it is a heavier satellite. This may the first satellite in China's second-generation Tracking and Data Relay Satellites (TDRS). The first four satellites are located at 77 degrees E, 176.8 degrees E, 16.7 E, and 76.6 degrees E[4](Harvey 2013). The Tianlian 4 launched in 2016 was probably a replacement for the first satellite. Chinese satellite masses for the first four satellites are reported to be 3800 kg for the first satellite and about 2000 kg for the next three satellites. US third-generation data relay satellites have masses of about 3200 kg (UCS 2021). They would be used for C4ISR military functions.

The most recent Chinese Tian Lian is likely to be a second-generation TDRS satellite that will progressively replace the earlier satellites. The new satellite is positioned at 79.9 E Longitude in the GSO.

7.7   A New Trend–LEO Small Satellite Constellations for Internet Needs

A small satellite revolution led by US and European companies is transforming the global communications satellite industry. Constellations of satellites in Low Earth Orbits (LEO) provide continuing connectivity catering to both domestic and international needs such as Internet services.

The use of such LEO satellites is not new. The Soviet era Strela satellite constellation that could store, and forward various kinds of data has been around for some time.[5]

The commercial world has also seen the emergence of several LEO constellations such as ORBCOM, and Globalstar that provide a variety of communications services. Most of them did not operate on direct satellite to satellite links but used suitably located ground stations to route the data via satellite to the required destination (Redding 1999).

The Iridium constellation of 66 satellites was the first orbiting satellite-based communication system that used satellite to satellite switching for providing global coverage (Maine et al. 1995). Chinese launchers were used to place 12 of them in orbit.

O3b is an operational system that uses 20 satellites operating in an 8063 km orbit above the equator to provide high speed internet connectivity to users between 45 N

---

[4] p. 220.

[5] http://www.russianspaceweb.com/strela_comsat.html.

Latitude and 45 S Latitude. It uses Ka band steerable antennae along with a network of suitably located gateway stations to provide the required connectivity. Each of the satellites weighs about 700 kg. The constellation provides a combined capacity of about 100 Gbps. They are launched four at a time by Ariane's Soyuz launcher from Kourou in French Guyana.[6]

More recently new start-up companies in the US and Europe have been launching large constellations of small satellites for providing global broad band Internet connectivity via satellite to satellite networks. Several of them have launched large numbers into various LEO orbits. Some are operational while others are likely to become operation within the next year.

The largest constellation of LEO satellites for Internet services that is currently under establishments is the Starlink network promoted by Elon Musk's SpaceX. There are currently 1441 operational satellites. They orbit the earth at an altitude of 550 km. Most of the satellites are in 53° inclination orbits. A few of them are also in higher inclination Sun Synchronous Orbit (SSO) to ensure coverage of northern latitudes. The satellites are in several orbital planes to ensure global coverage. Operational services may be offered in 2021. This constellation of 1441 satellites represents about 57% of the 2505 satellites operated by the US as of April 2021. The satellites have a launch mass of 260 kg. The satellites can also be produced in large numbers (UCS 2021).

Orbcom a US-based company has 35 operational satellites in 700 km 45-degree inclination orbits (UCS 2021).

The constellation proposed by OneWeb, a UK based company will have 648 satellites orbiting in near polar orbital planes at an altitude of about 1200 km. Such a configuration would provide excellent coverage of the polar regions of the world. Airbus Defence and Space has been selected as the contractor to roll out the satellites. The satellites weigh about 150 kg and will be launched 32 to 36 at a time on-board an Ariane Soyuz launcher[7] (UCS 2021). The satellites operate in the Ku Band for which the company has a licence. The design of the satellite is such that interference with the Ku band transmissions of GSO satellites is within acceptable limits. The satellites use phased array antenna with each satellite having a capacity of about 8 Gbps.

The satellites connect to customers through a small terminal using standard radio frequencies used for mobile services. Cell phones, laptops and tablets can be directly connected to the Internet via these terminals. These can be added suitably to existing mobile networks to directly connect to the Internet.[8]

These developments suggest that constellations of small satellites in low and medium altitude orbits can provide new kinds of telecom services for meeting both

---

[6] http://www.o3bnetworks.com/technology/.

[7] http://spaceflightnow.com/2015/07/01/oneweb-launch-deal-called-largest-commercial-rocket-buy-in-history/.

[8] The satellite Airbus company declared bankruptcy and has been bought. India's Bharti has a majority stake in the Airbus Defence & Space company.

routine as well as crisis mode operations. These could be dedicated to military use or could be a part of a civilian or dual use network.

In comparison to what is happening in the US and Europe, China has so far not been very active in the creation of small satellite constellations for providing new internet services. Over the last several years they have launched several test satellites including a Yunhai constellation in 2018. They have also supported companies for establishing such networks. However, when one compares China's efforts with what one company SpaceX is doing in the US, China is still quite some distance behind in this emerging area of applications. One should expect that this will soon change.

7.8 Conclusions

Through a judicious blend of indigenous development and satellite imports, China has managed to catch up with the rest of the world in terms of the capabilities of its communications satellites.

China has done this despite US imposed embargo regimes. It now offers in-orbit delivery of satellites to international customers.

Though there are still gaps in performance between Chinese and Western built communications satellites, China can offer satellites that are significantly cheaper than their western counterparts. It will therefore become an increasingly important player in the global market for GSO Communications satellites.

It has several communications satellites dedicated for military use. It can also use its civilian satellite fleet to complement and augment satcom functions for meeting military needs.

It has also experimented with electric propulsion systems. It may soon be able to provide satellites with such system in case they are needed by domestic and international customers.

The five data relay satellites that it has placed in orbit for data relay functions across the globe are also indicative of the Space Programme's global intentions and reach.

Though it has experimented with small satellites in Low Earth Orbit (LEO) that offer Internet connectivity the western countries especially the US are ahead in implementing operational systems. China will soon follow these trends and promote companies that offer such services to global and domestic customers.

The availability of the new CZ 5 launch vehicle will enable China to place two very high and one intermediate class satellites in Geostationary Transfer Orbit (GTO) simultaneously. The "greener" CZ 6, CZ 7, and CZ 8 launch vehicles provide a great deal of flexibility for China to provide launch services for the various segments of the GSO Communications Satellite market.

The increasingly hostile relationship between the US and China will spill over into the space domain as well. These are likely to affect Chinese efforts to capture a greater share of the global market for telecom satellites and their associated launch services.

References

Chandrashekar S (2015), Space war and security—a strategy for India, http://eprints.nias.res.in/943/1/2015-R36.pdf

Clark S (2015) Boeings first two all-electric satellites ready for launch, Spaceflight Now, March 1, 2015

Harvey B (2013) China in space: the great leap forward. Springer Praxis Books, New York

Maine K et al (1995) Overview of iridium satellite network, WESCON/'95. In: Conference record. Microelectronics communications technology producing quality products mobile and portable power emerging technologies. http://ieeexplore.ieee.org/stamp/stamp.jsp?arnumber=485428

Redding C (1999) Overview of LEO satellite systems. In: 1999 International symposium on advanced radio technologies. http://www.its.bldrdoc.gov/media/30329/red_abs.pdf

Union of Concerned Scientists (2021) UCS satellite data base. https://ucsusa.org/resources/satellite-database

Chapter 8
China's Weather Satellites

8.1 Background

After setting up ground stations to receive US weather data China decided to launch its own weather satellites. Though the programme was thought of in 1970 it did not get support till 1978.

The initial effort was directed toward placing a weather satellite in a suitable Sun Synchronous Orbit (SSO). The capacity of the CZ 2C launcher was not adequate to take a weather satellite to the required altitude. The Chinese therefore decided to improve the performance of its launcher through new developments.

The CZ 4 launcher was developed to meet the needs of larger payload delivery into polar sun synchronous orbits. Since Jiuquan was not an optimum launch site for placing satellites in Sun Synchronous Orbits (SSO) China also decided to establish a new launch complex in Taiyuan that was optimized for placing satellites in SSO.

8.2 The Weather Satellite Log

Since the time those decisions were made China has launched a total of 24 weather satellites that include SSO, Geostationary and LEO experimental weather satellites. The operational satellites are called the Feng Yun satellites meaning wind and cloud in Chinese[1] (Harvey 2013).

It launched a new Fengyun 4A satellite into GSO in 2016. Another operational FY 3D orbiting weather satellite followed in 2017. A GSO weather satellite the FY 2H was launched in 2018.

---

[1] pp. 178–184.

© National Institute of Advanced Studies 2022
S. Chandrashekar, *China's Space Programme*,
https://doi.org/10.1007/978-981-19-1504-8_8

In 2016 China launched two satellites (Yunhai 1, Tansat1) into SSO not from Taiyuan but from Jiuquan. These appear to be experimental satellites related to weather and climate. This was followed by the launch of a second Yunhai 1 satellite in 2019.

Two more developmental weather satellites were launched in 2018. One was the Gaofen 5 a domestically built experimental satellite. The other was the China France Oceanography Satellite (CFOSAT) a collaborative research project with France.

In 2019 two experimental ocean-weather small satellites the Bufeng 1A and 1B were launched from Jiuquan on the CZ 11 launcher from an offshore platform in the Yellow Sea.

Till the end of 2020 China has launched a total of 24 weather satellites out of which 17 are operational and 7 are experimental.

Table 8.1 provides the details of all the weather satellites built and launched by China.[2]

Figure 8.1 provides a timeline of operational Feng Yun Chinese weather satellites based on the details provided in Table 8.1. These include both Orbiting and Geostationary satellites.

From the **Figure** China has developed and launched two generations of orbiting satellites. The first-generation satellites had a mass of 950 kg and covered the period 1988–2002. The second-generation orbiting satellites are much larger with masses of 2300 kg. They cover the period from 2008 to the present. China also has several weather satellites in Geostationary Orbit to provide continuous coverage of weather patterns around China. The mass of these satellites is 1380 kg. In 2016 it launched its second-generation FY 4 GSO weather satellite using the larger CZ 3B launcher. This satellite has a mass of around 2600 kg in GSO.

8.3   The Polar Orbiting FY 1 and FY 3 Series

The first weather satellite that the Chinese launched was in 1988. Like the US, their first efforts were directed toward putting weather satellites in Sun Synchronous Orbit (SSO) around the earth. To do this they had to add a third stage to their CZ 2C launcher to make it the CZ 4 launcher. Since the original launch base Jiuquan was not optimized for putting satellites into SSO they also had to set up a new launch site at Taiyuan for launching these satellites.

The first two satellites were designed to be experimental missions before the satellites became operational.

The first satellite Feng Yun 1 launched on the sixth of September 1988 weighed 757 kg. It was a three-axis stabilized satellite with solar panels on either side. It carried two visible/ near infrared scanners operating in the visible and near IR bands

---

[2] All the GSO satellites used dedicated launches of the CZ 3 series. The orbiting satellites FY 1B, FY 1C and FY 1D carried four other co-passengers including two balloon and the experimental Haiyang Ocean satellite.

Table 8.1 China's weather satellites

Satellite	Launch date	Launcher	Launch site	Orbit	Comments
Orbiting satellites in SSO polar orbits					
Fengyun-1A	06-09-1988	CZ 4	Taiyuan	904 × 881 SSO	40-day life only
Fengyun-1B	03-09-1990	CZ 4	Taiyuan	900 × 885 SSO	Debris upper stage
Fengyun-1C	10-05-1999	CZ 4B	Taiyuan	868 × 849 SSO	Operational ASAT test
Fengyun-1D	15-05-2002	CZ 4B	Taiyuan	873 × 851 SSO	Operational
Fengyun-3A	27-05-2008	CZ 4C	Taiyuan	811 × 804 SSO	2nd generation
Fengyun-3B	04-11-2010	CZ 4C	Taiyuan	847 × 827 SSO	2nd generation
Fengyun-3C	23-09-2013	CZ 4C	Taiyuan	815 × 801 SSO	2nd generation
Yunhai 1	11-11-2016	CZ 2D	Jiuquan	787 × 767 SSO	Experimental?
Tansat 1	21-12-2016	CZ 2D	Jiuquan	728 × 698 SSO	Cloud & Aerosol imager
Fengyun 3D	14-11-2017	CZ 4C	Taiyuan	819 × 805 SSO	2nd generation
Gaofen 5	8-05-2018	CZ 4C	Taiyuan	709 × 706 SSO	Atmospheric Pollution
CFOSAT	29-10-2018	CZ 2C	Jiuquan	525 × 523 SSO	China French Oceansat
Yunhai 1 -2	25-09-2019	CZ 2D	Jiuquan	789 × 791 SSO	2nd satellite experiment
Orbiting satellites non-SSO low earth orbit (LEO)					
Bufeng 1A	5-06-2019	CZ 11	Yellow Sea	563 × 583	Ocean wind speed
Bufeng 1B			Yellow Sea	563 × 583	Ocean wind speed
Geostationary satellites					
Fengyun-2A	10-06-1997	CZ 3	Xichang	GSO 105° E	Only 6-month life
Fengyun-2B	25-06-2000	CZ 3	Xichang	GSO 105° E	DCP data collection
Fengyun-2C	19-10-2004	CZ 3A	Xichang	GSO 123.5° E	Multiple sensors
Fengyun-2D	08-12-2006	CZ 3A	Xichang	GSO 86.50° E	Multiple sensors
Fengyun-2E	23-12-2008	CZ 3A	Xichang	GSO 86.50° E	Multiple sensors
Fengyun-2F	13-01-2012	CZ 3A	Xichang	GSO 112.50° E	Multiple sensors
Fengyun-2G	31-12-2014	CZ 3A	Xichang	GSO 112.50° E	Multiple sensors
Fengyun 4A	10-12-2016	CZ 3B	Xichang	GSO 104.7° E	New Generation
Fengyun 2H	5-06-2018	CZ 3A	Xichang	GSO 79° E	Continuation Series 2

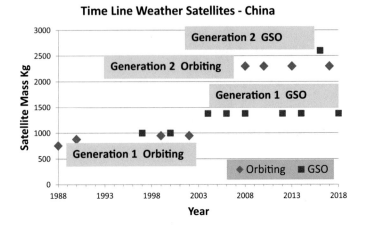

Fig. 8.1 Timeline weather satellites China

The two scanners had a combined swath of about 3000 km with a resolution of 1 km. The satellite was placed in an 881 km by 904 km 99.1° inclination Sun Synchronous Orbit (SSO).

After working for about 39 days the satellite failed. Fogging of the optical elements was possibly the reason for the failure.

The second satellite (launched in 1990) was a stretched version of the first with an increase in the mass to 880 kg. It worked for a year. It is supposed to have encountered radiation damage. Two balloon weather satellites were also launched as co-passengers.

After a lot of work to improve performance and life, the FY 1 C was launched in 1999 on the more powerful CZ 4B launcher. It carried an improved 10 channel scanner with four visible, three Near Infrared (NIR), 1 Short Wave Infrared (SWIR) and two Long Wave Infrared (LWIR) channels. FY 1 C also carried instruments for charged particle measurements and radiation measurements. It had a launch mass of about 950 kg. It was complemented by an identical FY 1 D in 2002. The orbits of the FY 1C and 1 D were slightly lower at about 850 km with an inclination of 98.80.

The FY 1 series was replaced by the FY 3 series starting from 2008. These satellites weigh about 2300 kg and are launched by the even more powerful CZ 4C launcher. They are all launched from Taiyuan and are in SSO.

The FY 3 series carry a large complement of sensors operating in the visible, infrared as well as in the microwave bands. These instruments monitor not only relevant ocean and atmospheric parameters related to weather and climate but also include sensors that monitor radiation and space weather parameters as well. The instrument package details are provided below:

- Visible and Infrared Radiometer (VIRR), 10 channels;
- Infrared Atmosphere Sounder (IRAS), 26 channels;
- Microwave Temperature Sounder (MWTS) 4 channels;

- Microwave Humidity Sounder (MWHS), 5 channels;
- Medium-Resolution Spectral Imager (MERSI), 20 channels, resolution 250 m;
- Solar Backscatter Ultraviolet Sounder (SBUS);
- Total Ozone Unit (TOU);
- Microwave Radiation Imager (MWRI);
- Atmospheric Sounding Interferometer (ASI);
- Earth Radiation Measurement (ERM);
- Space Environment Monitor (SEM);
- Solar Irradiation Monitor (SIM).

Continuity of the series has been maintained since 2008 with launches in 2010, 2013, and 2017. At any given point in time two satellites are available for the operational service. These are separated suitably to provide a morning pass and an afternoon pass. The China Meteorological Administration is the operational Agency that uses the data from these satellites (Harvey 2013).

8.4 The Geostationary FY 2 Series

In parallel with the orbiting weather satellites China also planned on placing weather satellites in Geostationary orbit once the CZ 3 launcher for placing satellites in GSO became available.

The Feng Yun 2 series of satellites were to complement the FY 1 and FY 3 orbiting weather satellites by providing continuous weather coverage over China from Geostationary Orbit (GSO).

The first satellite of the series was due to be launched in 1994 by the CZ 3 launcher. When the satellite was being loaded with propellant it exploded killing quite a few people. The Chinese took three years to redesign the whole spacecraft before launching the FY 2 A in 1997.

FY 2A was a drum-shaped satellite with a diameter 2.1 m and height of 4.5 m. The satellite mass was 1380 kg. Like the early GSO weather satellites it was a spinner. A 1.53 m long 0.9 m diameter solid rocket motor took the satellite from GTO to GSO.

It carried a five-channel visible and infrared spin scan radiometer. Visible, infrared as well as water vapor images were regularly transmitted. Vertical temperature profiles as well as wind speeds at different heights could also be obtained. It also carried a solar X-ray spectrometer and a particle detector for space weather data.

The first satellite FY 2A lasted only 6 months when control was lost. It recovered but only provided intermittent images. It was moved out of the prime slot at 105 E to another location. Service ended in 2006.

FY 2A was replaced by FY 2B in 2000. It also included a capability to transmit data received from Data Collection platforms. Both the FY 2A and FY 2B were launched by the CZ 3.

The later satellites in the series were all launched by the CZ 3A which had become available. FY 2C, FY 2D, FY 2E, FY 2F, FY 2G, and FY 2H all with masses of 1380 kg were launched in 2004, 2006, 2008, 2012, 2014, and 2018, respectively.

They provide routine 24-h weather coverage from at least four locations in GSO Over a hundred locations in China receive data from the FY series of satellites.

Apart from using the visual and infrared images, vertical temperature and humidity profiles also appear to be operational services. The monitoring of various Greenhouse gases, Ozone levels as well as the routine monitoring of space weather conditions also seem to be regular services provided by these orbiting and geostationary weather satellites.

China has agreements with eight countries for receiving the data. These include Peru, Thailand, Bangladesh, Pakistan, Mongolia, Iran, and Indonesia. It also has in place a weather information dissemination service called Fengyuncast just like Eumetcast from Europe and Geonetcast from the US.

In December 2016 China launched the first of its second-generation FY 4A weather satellite. The larger CZ 3B launcher was used to place a satellite with a mass of about 2600 kg in GSO. This will carry a complement of sensors comparable to what other countries have to offer (Harvey 2013).

All weather satellites are built by the Shanghai Academy of Space Technology (SAST).

8.5   Benchmarking Chinese Weather Satellites

Advanced space powers use a combination of geostationary and polar orbiting satellites to provide weather forecasts. The sensors and the platforms have evolved through several generations. Imaging sensors, sounders that measure vertical profiles operating in the visible, infrared and microwave regions of the spectrum along with specialized instruments for measuring substances like ozone and other pollutants provide a wealth of data. These are used for making short, medium, and long-range forecasts and for monitoring the earth environment. All these satellites come equipped with the capability to receive data from data collection platforms on the ground and reach them to a forecasting center[3] (Davis 2007).

The US operational capability at any given point in time comprises a minimum of three GOES weather satellites in GSO. Two of them provide real-time data with the third serving as an in-orbit spare. They carry imaging as well as sounding payloads operating in the visible, infrared and microwave wavelengths. They also carry instruments that provide data related to other weather and climate needs and space weather. The GOES 15 satellite weighs 3238 kg at lift-off.

The US also has a minimum of two polar orbiting satellites (there are currently 5 working satellites) in sun synchronous orbit (UCS 2021). These provide data from

---

[3] http://www.osd.noaa.gov/download/JRS012504-GD.pdf.

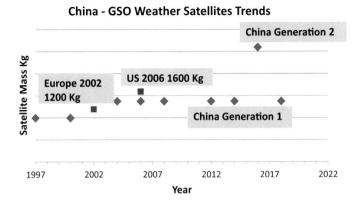

Fig. 8.2 China—GSO weather satellite trends

a standard set of instruments and allow measurements that provide a better handle on forecasts over a few days. These are in an 870 km sun synchronous orbit. Their mass is 1420 kg.[4] Data from these US satellites are provided free across the world.

Europe too has a similar combination of geostationary and orbiting satellites for providing weather satellites. The Second-Generation Meteosat (SGM) satellites have now replaced most of the first-generation GSO weather satellites. Europe now has five satellites of which one is an in-orbit spare that cover Europe, Africa, and the Indian Ocean Area. SGM satellites have a lift-off mass of around 2000 kg and a beginning of Life Mass in GSO of 1200 kg.

Europe also has two orbiting satellites in polar SSO the Metop A and the Metop B. These are in 817 km sun synchronous Orbit. Two of them are currently in orbit while the third is available for launch. The lift off mass of these satellites is 4085 kg.[5]

There is a large degree of overlap and continuity between the various instruments carried by the US and Europe. However, Europe has added additional instruments to understand climate change.

One simple benchmark to compare satellites both in GSO as well as in SSO would be look at how their masses have changed over time as they move from one generation to the next.

Figure 8.2 compares China's Geostationary Weather Satellites with their European and US counterparts.

The US is the world leader with the 2006 launch of its GOES 13 satellite—a series that went on to launch the GOES 14 and GOES 15 satellites in 2009 and 2010, respectively.

Europe with its launch of the second-generation GSO weather satellites (successors to the first-generation Meteosat Programme) with a mass of 1200 kg is behind

---

[4] http://noaasis.noaa.gov/NOAASIS/ml/genlsatl.html.

[5] http://www.eumetsat.int/website/home/Satellites/CurrentSatellites/Meteosat/index.html.

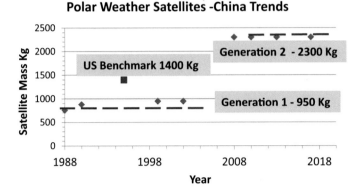

Fig. 8.3   Polar weather satellites—China trends

the US. However, it was still ahead of China till 2016. Their first of the second-generation satellites was launched in 2002 followed by launches in 2005 and 2012. Their next generation of weather satellites that is under planning may increase this mass and move it closer to the US position.[6]

China was operating only its first-generation GSO based weather satellites till 2016. Though the Chinese had the launch capacity to place large payloads in GSO with their CZ 3B launcher they have not used this capability for placing heavier weather satellites in GSO. However, in December 2016 they finally placed the first of their second-generation GSO weather satellite in orbit. The mass of 2600 kg in GSO provides them with the possibility of adding more instruments and provides greater flexibility than even the US. However, China may still be behind the US and maybe even Europe in the building of lower mass payloads for realizing a specific function.

Figure 8.3 compares China's orbiting weather satellites with their US counterparts. Unlike the GSO based weather satellites orbiting satellites generally exhibit decreasing weights for either the same or improved performance. From Fig. 8.3 US orbiting satellites with a mass of 1420 kg appear to be the leaders.

European orbiting satellites are much heavier with a mass of 4085 kg. European satellites carry instruments that cater to a variety of other research needs with many new sensors being tried out for a range of applications. These complications make comparison with other operational systems difficult.

China has moved from the first generation of satellites to the second generation. The latest satellites carry a complement of sensors similar to the US. However Chinese satellites weigh 2300 kg. These are about a 1000 kg heavier than similar US satellites.

In 2016 China launched two test satellites related to weather and climate monitoring. The Yunhai (1300 kg) and Tansat satellites were launched into SSO using the CZ 2D launcher in November and December 2016. The Tansat satellite has a mass

---

[6] http://www.eumetsat.int/website/home/Satellites/FutureSatellites/index.html.

of about 620 kg. It carries a spectrometer for the monitoring of Carbon dioxide as well as a Cloud and Aerosol Polarimetry Imager. These are sensors that are possibly being tested out for future operational missions. In keeping with global trends China will move toward smaller orbiting weather satellites with further tests of smaller satellites coupled with new sensors.

The Gaofen 5 satellite and the CFOSAT which is a collaborative ocean research effort with France also indicate new research for applications. The research and the operational component of the weather-related services that China seeks to provide are therefore closely coupled. This trend is likely to extend into the future.

Case Study 1
Weather Satellites and Catastrophes

Chinese weather satellites are also associated with several catastrophic events.

The third stage of the Long March 4 vehicle that placed the Feng Yun 1 B satellite in orbit exploded on 4th October 1990 creating a debris cloud that posed a risk to other satellites in similar orbits.

On April 2, 1994, when the first Geostationary weather satellite was being loaded with fuel there was an explosion that killed several people. It also resulted in a delay of 3 years to the Geostationary Weather Satellite Programme.

The FY 1 C orbiting weather satellite also became famous in a different way. On January 11, 2007, it became the target for an Anti-Satellite (ASAT) Test by China. The resulting debris cloud created a huge furore among various space faring countries. It has also led to many international initiatives on norms and rules of behavior on the uses of space.

References

Davis G (2007) History of the NOAA satellite program. J Appl Remote Sens 1(1):012504. http://www.osd.noaa.gov/download/JRS012504-GD.pdf

Harvey B (2013) China in space: the great leap forward. Springer Praxis Books, New York

Union of Concerned Scientists (2021) UCS satellite data base. https://ucsusa.org/resources/satellite-database

Chapter 9
China's Remote Sensing Satellites

9.1 China's Remote Sensing Satellites

As of the end of 2020 China had launched a total of 114 remote sensing satellites. Two of these satellites were procured from the UK and launched on Russian rockets. China has therefore built and launched 112 remote sensing satellites between 1970 and 2020. One of the two Gaofen satellites (the Gaofen 4 & Gaofen 13) that are in geostationary orbit is included in this list. These carry multispectral scanners in the visible and infrared bands. They are most probably Early Warning (EW) satellites for a Ballistic Missile Defence (BMD) function. The Gaofen 13 can be excluded from the LEO, SSO list of remote sensing satellites. The Gaofen 14 launched in 2020 also appears to be a military satellite whose mission is not clear and is also taken out of the remote sensing listing.

In addition to these remote sensing satellites, China has also launched 68 Yaogan military satellites. The Yaogan nomenclature is used for satellites that perform ISR[1] functions for the PLA. The series includes Electronic Intelligence (ELINT), Synthetic Aperture Radar (SAR) and Electro-Optical (EO) reconnaissance satellites that are placed in various Low Earth and Sun Synchronous Orbits. Four LKW satellites (launched in 2017, 2018) and eight XJS satellites launched between 2018 and 2020 may also perform a military recon function. The XJY 7 satellite launched in 2020 appears to be a military test SAR satellite. There are therefore a total of 191 satellites that perform a civilian remote sensing or a military recon function.

These numbers also include several small satellite constellations such as Ziyuan, Zhuhai, Tianhui, Gaofen, Haiyang Jilin, Superview, and Spark for developmental and commercial remote sensing applications. A series of CBERS remote sensing satellites in collaboration with Brazil has also been a hallmark of China's remote sensing missions. Annexure 9.1 provides a listing of these satellites sorted in several

---

[1] Intelligence, Surveillance, Reconnaissance.

© National Institute of Advanced Studies 2022
S. Chandrashekar, *China's Space Programme*,
https://doi.org/10.1007/978-981-19-1504-8_9

ways for grouping them. Several patterns that provide an insight into these satellite functions emerge from this sort. Table 9.1 provides the first set of satellites that have a military function. These include satellites with the Yaogan name as well some others with different names.

9.2 Two Types of Yaogan (YG) ELINT Constellations

Of the 68 Yaogan satellites 6 clusters of three satellites each (a total of 18) are in 63.4° 1200 km altitude orbits. These satellites are launched three at a time by the CZ 4C launcher. The three satellites fly in a triangular formation. They are launched from Jiuquan. These are broad area ELINT satellites very similar to the US Ocean Reconnaissance NOSS satellites flown in the 1970s (Chandrashekar et al. 2011). There are also 21 Yaogan satellites that have also been launched three at a time by the CZ 2C from the Xichang launch site. These are in 600 km orbits with an inclination of 35°. After launch the three satellites are moved and separated by 120° within the same orbital plane. Such an arrangement allows continuous electronic surveillance of the area under the satellite. As one satellite disappears over the horizon a second satellite replaces it (see Chap. 13 for further details).

The orbital planes of the first three triplets are separated by 120° at the equator. This ensures that as the earth rotates one triplet that covers the area of interest is replaced by a second triplet. The second triplet is then replaced by the third which in turn is replaced by the first as it comes into view over the area of interest again.

The fourth triplet that was launched occupied the same orbital plane as the third triplet thus making a total of six satellites in the same orbital plane. This would permit a spacing of 60° within the same orbital plane and improve coverage significantly.

The fifth and sixth set of triplets adds to the number of satellites in the other two planes. This architecture ensures that there are six satellites in each orbital plane making a total constellation of 18 satellites. The seventh cluster launched in October 2020 (Yaogan U, V, W) may be a replacement set for the earlier satellites. These 21 satellites constitute a theatre level continuous ELINT capability (Chandrashekar and Ramani 2018).

8 XJS satellites launched between 2015 and 2020 are also in a 35° inclination orbit. They are also all launched from Xichang. However, they are in a lower 500 km orbit and use the less powerful CZ 2D and CZ11 launchers. They are probably next generation ELINT test satellites.

These Broad Area and Theatre level ELINT constellations provide constant electronic surveillance over the maritime access routes to China's eastern coastline. They are an important component of China's A2AD military strategy (Chandrashekar and Ramani 2018). When these 39 Yaogan ELINT satellites are taken out there are 29 Yaogan satellites remaining. These are the satellites that are used for SAR and EO imaging functions.

Table 9.1 Yaogan and military reconnaissance satellites

Satellite	Launch site	Launch time UT	Launch Time CST	Apogee Km	Perigee Km	Inclination °	Launch date	ECT	Launcher	Nos
Yaogan 09 ABC	Jiuquan	04 55	12 55	1099	1083	63.41	March 5 2010	NA	CZ 4C	3
Yaogan 16 ABC	Jiuquan	04 06	12 06	1105	1085	63.39	November 25 2012	NA	CZ 4C	3
Yaogan 17 ABC	Jiuquan	19 16	03 16	1111	1076	63.41	September 1 2013	NA	CZ 4C	3
Yaogan 20 ABC	Jiuquan	05 45	13 45	1104	1086	63.40	August 9 2014	NA	CZ 4C	3
Yaogan 25 ABC	Jiuquan	19 33	03 33	1097	1089	63.41	December 10 2014	NA	CZ 4C	3
Yaogan 31 ABC	Jiuquan	04 25	12 25	1100	1090	63.40	April 10 2018	NA	CZ 4C	3
Yaogan 30 A,B,C	Xichang	04 21	12 21	609	605	35.00	September 29 2017	NA	CZ 2C	3
Yaogan 30 D,E,F	Xichang	18 10	02 10	616	607	35.00	November 24 2017	NA	CZ 2C	3
Yaogan 30 G,H,J	Xichang	19 44	03 44	610	600	35.00	December 25 2017	NA	CZ 2C	3
Yaogan 30 K,L,M	Xichang	05 39	13 39	611	602	35.00	January 25 2018	NA	CZ 2C	3
Yaogan 30 N P Q	Xichang	03 57	11 57	619	590	35.00	July 26 2019	NA	CZ 2C	3
Yaogan 30 R, S,T	Xichang	03 43	11 43	608	603	35.00	March 24 2020	NA	CZ 2C	3

(continued)

Table 9.1 (continued)

Satellite	Launch site	Launch time UT	Launch Time CST	Apogee Km	Perigee Km	Inclination °	Launch date	ECT	Launcher	Nos
Yaogan 30 U,V,W	Xichang	15 19	23 19	610	600	35.00	October 26 2020	NA	CZ 2C	3
XJS A, B	Xichang	03 30	11 30	496	488	35.00	June 27 2018	NA	CZ 2C	2
XJS C D E F	Xichang	21 07	05 07	488	478	35.00	February 19 2020	NA	CZ 2D	4
XJS G, H	Xichang	20 13	04 13	489	477	35.00	May 29 2020	NA	CZ 11	2
Yaogan 1	Taiyuan	22 48	06 48	626	624	97.80	April 26 2006	06 00	CZ 4C	1
Yaogan 3	Taiyuan	22 48	06 48	624	613	97.80	November 11 2007	06 00	CZ 4C	1
Yaogan 10	Taiyuan	22 48	06 48	632	624	97.82	August 9 2010	06 00	CZ 4C	1
Yaogan 29	Taiyuan	21 24	05 24	619	615	97.84	November 26 2015	04 30	CZ 4C	1
Yaogan 33	Taiyuan	22 49	06 49	NA	NA	NA	May 22 2019	06 00	CZ 4C	1
Yaogan 13	Taiyuan	18 50	02 50	511	505	97.11	November 29 2011	01 56	CZ 2C	1
Yaogan 23	Taiyuan	18 53	02 53	513	492	97.33	November 14 2014	02 00	CZ 2C	1
Yaogan 18	Taiyuan	02 50	10 50	511	492	97.55	October 29 2013	09 56	CZ 2C	1
Yaogan 6	Taiyuan	02 55	10 55	521	486	97.63	April 22 2009	10 01	CZ 2C	1
Yaogan 7	Jiuquan	08 42	16 42	659	623	97.84	December 9 2009	15 00	CZ 2D	1
Yaogan 2	Jiuquan	07 12	15 12	655	630	97.85	May 25 2007	13 30	CZ 2D	1

(continued)

Table 9.1 (continued)

Satellite	Launch site	Launch time UT	Launch Time CST	Apogee Km	Perigee Km	Inclination °	Launch date	ECT	Launcher	Nos
Yaogan 24	Jiuquan	07 12	15 12	653	630	97.91	November 20 2014	13 30	CZ 2D	1
Yaogan 4	Jiuquan	04 42	12 42	652	634	97.92	December 1 2008	11 00	CZ 2D	1
Yaogan 11	Jiuquan	02 42	10 42	657	624	98.00	September 22 2010	09 00	CZ 2D	1
Yaogan 30	Jiuquan	02 43	10 43	655	626	98.07	May 15 2016	09 01	CZ 2D	1
LKW 3	Jiuquan	07 10	15 10	509	491	97.30	January 13 2018	13 30	CZ 2D	1
LKW 4	Jiuquan	07 10	15 10	510	497	97.30	March 17 2018	13 30	CZ 2D	1
LKW 1	Jiuquan	04 11	12 11	510	496	97.46	December 4 2017	10 30	CZ 2D	1
LKW 2	Jiuquan	04 14	12 14	510	496	97.50	December 23 2017	10 33	CZ 2D	1
Yaogan 14	Taiyuan	07 06	15 06	474	470	97.24	May 10 2012	14 14	CZ 4B	1
Yaogan-28	Taiyuan	07 06	15 06	482	460	97.24	November 8 2015	14 14	CZ 4B	1
Yaogan 5	Taiyuan	03 22	11 22	492	481	97.40	December 15 2008	10 30	CZ 4B	1
Yaogan 12	Taiyuan	03 21	11 21	491	484	97.41	November 9 2011	10 29	CZ 4B	1
Yaogan 21	Taiyuan	03 22	11 22	494	480	97.42	September 8 2014	10 30	CZ 4B	1

(continued)

Table 9.1 (continued)

Satellite	Launch site	Launch time UT	Launch Time CST	Apogee Km	Perigee Km	Inclination °	Launch date	ECT	Launcher	Nos
Yaogan 26	Taiyuan	03 22	11 22	491	484	97.44	December 27 2014	10 30	CZ 4B	1
Yaogan 33 R	Jiuquan	15 44	23 44	704	702	98.2	December 28 2020	22 44	CZ 4C	1
Yaogan 32 A,B	Jiuquan	02 43	10 43	704	703	98.30	October 10 2018	09 00	CZ 2C	2
XJY 7	Wenchang	04 37	12 37	518	510	97.4	December 22 2020	11 37	CZ 8	1
Yaogan 15	Taiyuan	07 31	15 31	1206	1202	100.13	May 29 2012	14 30	CZ 4C	1
Yaogan 22	Taiyuan	06 31	14 31	1209	1196	100.32	October 20 2014	13 30	CZ 4C	1
Yaogan 27	Taiyuan	02 31	10 31	1206	1194	100.46	August 27 2015	09 29	CZ 4C	1
Yaogan 19	Taiyuan	03 31	11 31	1207	1201	100.48	November 20 2013	10 29	CZ 4C	1
Yaogan 8	Taiyuan	02 31	10 31	1204	1193	100.50	December 15 2009	09 29	CZ 4C	1

9.3 Broad Area Coverage Yaogan (YG) Electro-Optical (EO) Satellites

Yaogan 8, 15, 19, 22, and 27 are all in a near circular 1200 km Sun Synchronous Orbit. They are all launched from Taiyuan using the most powerful CZ 4C launcher. From the orbital characteristics and the Equatorial Crossing Times (ECT) of these satellites they carry broad area coverage 5–10 m resolution optical sensors. They are a vital part of the Yaogan constellation of satellites that China has deployed as a part of its Anti-Access and Area Denial Strategy to deter US Aircraft Carriers from coming close to the Chinese mainland in case of a crisis (Chandrashekar et al. 2011).

9.4 Yaogan (YG) SAR Satellites

SAR Cluster 1

Yaogan 1, 3, 10 with an ECT of 06 00 h and Yaogan 29 with an ECT of 04 30 h launched by the more powerful CZ 4C launcher from Taiyuan are all SAR satellites. They are all in a 620 km SSO. The Yaogan 33 launched from Taiyuan in May 2019 has launch parameters very similar to Yaogan 1, 3, 10. The launch was a failure. It would have replaced the Yaogan 10.

SAR Cluster 2

Yaogan 13 and Yaogan 23 with an ECT of 02 00 h launched from Taiyuan by the slightly less powerful CZ 2C launcher are also SAR satellites. Yaogan 6 and 18 with an ECT of 10 00 h are very similar to the Yaogan 13 and 23 satellites. They are also SAR satellites. Yaogan 23 is a replacement for the Yaogan 13 while the Yaogan 18 is a substitute for Yaogan 6. Unlike the earlier SAR satellites these satellites are in a lower 500 km SSO.

9.5 High-Resolution Yaogan (YG) EO Satellites

Group 1

The next pattern that can be seen consists of a group of six Yaogan satellites. All of them are in 640 km Sun Synchronous Orbits. They are all launched from Jiuquan on the same CZ 2D launcher. Yaogan 2 and 24 with an ECT of 13 30 are replacement satellites for each other. Yaogan 11 and 30 with an ECT of 09 00 h are also successors to each other. They are complemented by the Yaogan 4 (ECT 11 00 h) and Yaogan 7 (ECT 15 00 h).

From Table 9.1 **it is evident** that four LKW satellites are very similar to the high-resolution EO satellites launched by the CZ 2D launcher from Jiuquan. They are also high-resolution Electra-optical (EO) satellites used for military purposes.

Group 2

The next group of satellites that show a common pattern consist of Yaogan 5, 12, 14, 21, 26, and 28. All of them are in 470–500 km Sun Synchronous Orbits. They are all launched by the CZ 4B launcher from Taiyuan.

Based on their ECTs and orbit characteristics these are also high-resolution EO satellites. Yaogan 5, 12, 21, and 26 with an ECT of 10 30 h are successive replacement satellites. Yaogan 28 with an ECT of 14 14 h is a replacement for the Yaogan 14 satellite.

9.6 The Yaogan (YG) 32 A, 32 B and Yaogan 33 Satellites—A New Series?

On October 10, 2018, China launched two satellites on a CZ 2C launcher from Jiuquan. The satellites were injected into an SSO orbit with at an altitude of 700 km. They appear to be the first of a new series of military satellites. Another Yaogan 33 R satellite was also launched by a CZ 4C launcher from Jiuquan in December 2020 with similar orbital characteristics. They resemble the Shijian 11 series of military satellites and could be early warning satellites for BMD. The XJY 7 satellite launched in December 2020 is reported to be a classified SAR satellite. It may be the test satellite for the next generation SAR satellites.

9.7 Development Oriented Civilian Remote Sensing Satellites

After the 81 Yaogan and other military recon satellites are taken out there are still 110 satellites remaining. China has several well-established national systems of remote sensing data supply for a wide range of applications. Many of them cater to development-oriented needs. Some may also provide complementary or supplementary data for meeting military or security requirements.

Table 9.2 provides details of these "public good" satellites that China operates on a routine basis. There are 54 satellites that can be identified as providing a public or development service.

Table 9.2 Public good Chinese remote sensing satellites

Satellite	Launch site	Launch time UT	Launch Time CST	Apogee Km	Perigee Km	Inclination °	Launch date	ECT	Launcher	Nos
CBERS 1	Taiyuan	03 15	11 15	750	731	98.50	October 14 1999	10 30	CZ 4B	1
CBERS 2	Taiyuan	03 16	11 16	750	731	98.50	October 21 2003	10 30	CZ 4B	1
CBERS 2 B	Taiyuan	03 26	11 26	773	775	98.60	September 19 2007	10 40	CZ 4B	1
CBERS 3	Taiyuan	03 26	11 26	NA	NA	Failure	December 9 2013	NA	CZ 4B	1
CBERS 4	Taiyuan	03 26	11 26	751	742	98.50	December 7 2014	10 45	CZ 4B	1
CBERS 4 A	Taiyuan	03 22	11 22	635	615	98.00	December 20 2019	10 20	CZ 4B	1
Huanjing 1 A	Taiyuan	03 25	11 25	660	626	98.00	September 6 2008	10 30	CZ 2C	1
Huanjing 1 B	Taiyuan	03 25	11 25	672	627	98.00	September 6 2008	10 30	CZ 2C	1
Huanjing 1 C	Taiyuan	22 53	06 53	503	487	97.40	November 18 2012	05 59	CZ 2C	1
Huanjing 2A	Taiyuan	03 23	11 23	649	647	98	September 27 2020	10 23	CZ 4B	1
Huanjing 2B	Taiyuan	03 23	11 23	649	647	98	September 27 2020	10 23	CZ 4B	1
Haiyang 1 A	Taiyuan	01 50	09 50	795	792	98.80	May 15 2002	08 55	CZ 4B	1

(continued)

Table 9.2 (continued)

Satellite	Launch site	Launch time UT	Launch Time CST	Apogee Km	Perigee Km	Inclination °	Launch date	ECT	Launcher	Nos
Haiyang 1 B	Taiyuan	03 27	11 27	814	793	98.60	April 11 2004	10 30	CZ 2C	1
Haiyang 1 C	Taiyuan	03 15	11 15	793	777	98.60	September 7 2018	10 20	CZ 2C	1
Haiyang 2 A	Taiyuan	22 57	06 57	963	963	99.40	August 15 2011	06 00	CZ 4B	1
Haiyang 2 B	Taiyuan	22 57	06 57	975	972	99.40	October 24 2018	06 00	CZ 4B	1
Haiyang 2 C	Jiuquan	05 40	13 40	965	954	66	September 21 2020	NA	CZ 4B	1
Tansuo 1	Xi Chang	15 59	23 59	615	600	97.60	April 18 2004	12 00	CZ 2C	1
Tansuo 2	Xi Chang	10 45	18 45	711	694	98.10	November 18 2004	07 00	CZ 2C	1
Tansuo 4	Jiuquan	00 15	08 15	804	783	98.50	November 20 2011	06 31	CZ 2D	1
Tansuo 3	Jiuquan	00 15	08 15	804	786	98.50	November 5 2008	06 31	CZ 2D	1
Tianhui 1	Jiuquan	07 10	15 10	504	488	97.30	August 24 2010	13 29	CZ 2D	1
Tianhui 2	Jiuquan	07 10	15 10	506	490	97.40	May 6 2012	13 29	CZ 2D	1
Tianhui 3	Jiaquan	07 10	15 10	511	495	97.40	October 26 2015	13 29	CZ 2D	1
Tianhui 2-01A	Taiyuan	22 52	06 52	524	522	97.5	April 29 2019	05 52	CZ 4B	1
Tianhui 2-01B	Taiyuan	22 52:	06 52	524	522	97.5	April 29 2019	05 52	CZ 4B	1
Ziyuan-2	Taiyuan	03 25	11 25	499	483	97.40	September 1 2000	10 30	CZ 4B	1

(continued)

Table 9.2 (continued)

Satellite	Launch site	Launch time UT	Launch Time CST	Apogee Km	Perigee Km	Inclination °	Launch date	ECT	Launcher	Nos
Ziyuan-2B	Taiyuan	03 17	11 17	483	470	97.40	October 27 2002	10 25	CZ 4B	1
Ziyuan 2-3	Taiyuan	03 10	11 10	483	473	97.30	November 6 2004	10 20	CZ 4B	1
Ziyuan-2C	Taiyuan	03 26	11 26	774	773	98.50	December 22 2011	10 30	CZ 4B	1
Ziyuan-3A	Taiyuan	03 17	11 17	506	498	97.50	January 9 2012	10 25	CZ 4B	1
Ziyuan III-02	Taiyuan	03 17	11 17	511	507	97.50	May 30 2016	10 24	CZ 4B	1
Ziyuan 1-2D	Taiyuan	03 26	11 26	781	781	98.60	September 12 2019	10 30	CZ 4B	1
Ziyuan-3 03	Taiyuan	03 13	11 13	517	502	97.5	July 25 2020	10 13	CZ 4B	1
Gaofen 1	Jiuquan	04 13	12 13	653	630	98.10	April 26 2013	10 30	CZ 2D	1
Gaofen 9	Jiuquan	04 42	12 42	664	617	98.00	September 14 2015	10 59	CZ 2D	1
Gaofen 6	Jiuquan	04 13	12 13	655	642	98.10	June 2 2018	10 30	CZ 2D	1
Gaofen 9-02	Jiuquan	08 53	16 53	510	494	97.3	May 31 2020	15 10	CZ 2D	1
Gaofen 9-03	Jiuquan	07 19	15 19	512	493	97.4	June 17 2020	13 35	CZ 2D	1
Gaofen 9-04	Jiuquan	04 01	12 01	511	493	97.4	August 6 2020	10 17	CZ 2D	1
Gaofen 9-05	Jiuquan	02 27	10 27	511	493	97.5	August 23 2020	08 45	CZ 2D	1
Gaofen 8	Taiyuan	06 22	14 22	481	469	97.30	June 26 2015	13 30	CZ 4B	1
Gaofen 11-02	Taiyuan	05 57	13 57	503	502	97.3	September 7 2020	13 00	CZ 4B	1

(continued)

Table 9.2 (continued)

Satellite	Launch site	Launch time UT	Launch Time CST	Apogee Km	Perigee Km	Inclination °	Launch date	ECT	Launcher	Nos
Gaofen 7	Taiyuan	03 22	11 22	514	490	97.50	November 3 2019	10 22	CZ 4B	1
Gaofen 2	Taiyuan	03 15	11 15	631	608	98.00	August 19 2014	10 20	CZ 4B	1
Gaofen DM	Taiyuan	03 10	11 10	657	636	98.0	July 3 2020	10 10	CZ 4B	1
Gaofen 11	Taiyuan	03 00	11 00	701	255	97.40	July 31 2018	~10 00	CZ 4B	1
Gaofen 3	Taiyuan	22 55	06 55	759	758	98.40	August 9 2016	06 00	CZ 4C	1
Gaofen 10	Taiyuan	18 55	04 55	NA	NA	Failure	August 31 2016	04 00	CZ 4C	1
GAOFEN 10R	Taiyuan	18 51	02 51	635	630	97.80	October 4 2019	01 50	CZ 4C	1
Gaofen 12	Taiyuan	23 52	08 52	637	635	97.90	November 27 2019	07 52	CZ 4C	1
Gaofen1-02,03,04	Taiyuan	03 22	11 22	642	638	98.00	March 30 2018	10 30	CZ 4C	3

9.8   China Brazil Earth Resources Satellite (CBERS) Series of Remote Sensing Satellites

In Table 9.2 there are six CBERS Satellites that have orbital and launch characteristics that are very similar. Published information from official Chinese sources indicate that these are civilian satellites for meeting development and commercial needs. Sometimes the Ziyuan nomenclature is also used to categorize a few of these satellites.

The first all-digital remote sensing satellite launched by China was the CBERS satellite. The satellite was the result of a collaborative effort between China and Brazil. The first two CBERS satellites were in orbits of about 740 km. The third CBERS 2B satellite operated at a slightly higher altitude of 770 km. The CBERS 3 satellite failed while the CBERS 4 satellite launched in 2014 reverted to a 740 km orbit.[2] The orbit of the latest CBERS 4A satellite is much lower at an altitude of 620 km. The Ziyuan 2C is also very similar to these satellites.

All the satellites carried CCD cameras as well as multispectral scanners with some infrared channels. The satellites are box shaped with the solar panel deployed on only one side. This configuration may have been dictated by the need to prevent direct sunlight from getting into the Infrared detector and cooling area on the satellite. The spent upper stage that placed the first CBERS satellite in orbit exploded creating debris. After completion of its mission CBERS 1 is also reported to have exploded creating some debris.

With launches in 1999, 2003, 2007, 2011, 2013, 2014, and 2019 these satellites provide continuity of coverage. The series has been extremely successful and is likely to continue. According to various published reports and papers this collaboration has resulted in significant benefits for both parties[3] (Harvey 2013). The collaboration involves financial contributions from both Brazil and China.

9.9   The Huanjing (HJ) Series of Optical and SAR Satellites for APSCO

The Huanjing series of satellites are China's contribution to an international programme on disaster mitigation. Huanjing means environment. This programme was approved in 2003. In 2007 the China National Space Agency (CNSA) became a member of the International Charter "Space and Major Disasters".

---

[2] The CBERS 3 satellite did not reach orbit. The CBERS 4 was launched within a year of this mishap to provide continuity of coverage. All of them were initially placed in a slightly lower orbit and then maneuvered into the right orbit.

[3] pp. 189–197.

The first phase of this programme involved the launch of a constellation of three satellites. HJ 1A carries a CCD camera and an infrared camera. HJ 1B carries a CCD camera and a hyperspectral imaging sensor.[4] HJ 1C carries an S-band SAR payload.

These HJ satellites are China's contribution to the Asia Pacific Space Cooperation Organization (APSCO).[5]

The HJ 1A and HJ 1B satellites are much smaller than the CBERS satellites weighing about 475 kg. The HJ 1A and HJ 1B were launched together on a CZ 2C from Taiyuan in June 2008. HJ 1C was launched along with FN 1A, FN1B, and a XY-1 satellite in November 2012.[6] The orbits of the HJ 1A and 1B are very similar to the orbits of the Yaogan series of high-resolution EO satellites in a 640 km SSO.

The HJ 1C satellite is also a part of the series of satellites that China is offering for use by the APSCO countries. As mentioned earlier the lower orbiting Yaogan series of SAR satellites share similar orbital characteristics as the HJ 1C.

The Equatorial Crossing Time (ECT) of the HJ 1A and HJ 1B is 10 30 h. The ECT of the HJ 1C is 06 00 h. HJ 1C SAR satellite has orbital characteristics very similar to the Yaogan series of SAR satellites. Despite problems with the SAR antenna in the HJ 1C all the satellites of the HJ series seem to be functioning as of the end of 2016.[7]

In 2020 China launched another pair of Huanjing satellites the HJ 2A and the HJ 2B. They are in slightly lower orbits than the HJ 1A and HJ 1B. The orbital and launch parameters indicate that these are electro-optical imaging satellites. They could be the second-generation successor satellites to the HJ 1A and HJ 1B.

9.10   Haiyang (HY) Ocean Series

The first of the Haiyang series of ocean satellites was launched by China in May 2002 by a CZ 4B launcher from Taiyuan. It was launched along with the FY 1 D weather satellite as a co-passenger into an 860 km SSO orbit. Since this orbit was not suitable for ocean monitoring the onboard propulsion unit of the Haiyang 1 satellite was used to lower it into 795 km SSO[8] (Harvey 2003).

The satellite carried a 10-band ocean color scanner and a four-band CCD scanner and weighed 367 kg. The Ocean color scanner has 8 visible and near IR bands as well as two bands in the far infrared. The CCD scanner is optimized for Coastal Zone imaging. The satellite functioned for nearly two years[9] (Harvey 2003). The satellite was built by the China Aerospace & Industry Corporation (CASC).

---

[4] http://claudelafleur.qc.ca/Spacecrafts-2008.html#HJ-1A.

[5] The Member Countries of APSCO include Bangladesh, China, Iran, Mongolia, Pakistan, Peru, Thailand, and Turkey.

[6] http://claudelafleur.qc.ca/Spacecrafts-2012.html#HJ-1C.

[7] https://directory.eoportal.org/web/eoportal/satellite-missions/h/hj-1.

[8] pp. 140–141.

[9] https://directory.eoportal.org/web/eoportal/satellite-missions/h/hy-1a.

This was followed by the slightly heavier Haiyang IB satellite (420 kg) launched by the CZ 2C launcher into a near 800 km SSO in 2007. The satellite functioned till 2012.[10]

The Haiyang 2A satellite launched in August 2011 (the project was cleared in 2007) by a CZ 4B launcher into a slightly higher 960 km SSO is the first of the second-generation ocean satellites. The satellite carries both optical as well as microwave payloads. Comparisons with similar sensors on-board other satellites established the basic accuracies of the sensor measurements. The mass of the satellite is about 1500 kg.[11] Because of the infrared sensing capabilities the satellite will most probably have only a single solar panel like the CBERS series of satellites.

The Haiyang 2B satellite launched in October 2018 is a replacement for the Haiyang 2A satellite. The Haiyang 2C satellite launched in September 2020 may supplement and complement the coverage of the Haiyan 2B satellite. Unlike the other Haiyang satellites the Haiyang 2C is not in Sun Synchronous Orbit (SSO) but in a 66° inclination orbit. This choice may have arisen from the need to improve coverage over China's coastal seas and ocean areas.

9.11 The Tansuo (TAN) Series

Tansuo means Experimental. The Tansuo series of satellites were small land mapping satellites. Tansuo 1 was a small (204 kg) terrain mapping satellite developed by the University of Technology Harbin along with the Photomechanical Institute of the CAS. The European Astrium company was a co-developer. The Tansuo 1 carried a 10 m stereo resolution 120 km swath sensor. The satellite was launched in a due north direction from Xi Chang (the first time such a launch was carried out from Xi Chang) in April 2004. Launched into a 600 km SSO it had a 12 noon Equatorial Crossing Time (ECT).

Tansuo 1 was quickly followed by Tansuo 2 launched in November 2004. This was also launched northward from the Xi Chang launch center. This was a slightly heavier 300 kg satellite placed in a lower orbit with an average altitude of about 500 km.[12] With a 07 00 h ECT it seems to complement the Tansuo 1 in terms of coverage. Apart from the same sensor complement as Tansuo 1 it seems to have carried some additional experimental payloads. This could have been a transponder for the collection of Data from remote platforms.

The Tansuo 3 and the Tansuo 4 launched in 2008 and 2011 represent the operational part of the Tansuo series. They were both launched along with co-passenger

---

[10] https://directory.eoportal.org/web/eoportal/satellite-missions/h/hy-1b.

[11] https://directory.eoportal.org/web/eoportal/satellite-missions/h/hy-2a.

[12] Some sources suggest that the orbit is much higher. It is possible that there is some confusion with the later Tansuo 3 and 4 satellites.

Chuangxin satellites. The main satellite Tansuo carried the complement of optical sensors while the two Chuangxin satellites hosted a payload for collecting data from remote platforms. The two satellites together provided the data required for mapping. These satellites were launched from Jiuquan in the southward direction.

9.12   Tianhui (TH) Series

The Tansuo series was followed by the Tianhui series. These seem to be improved second-generation operational satellites for mapping purposes. Three of these satellites have been launched in 2010, 2012, and 2015, respectively. They have all been launched from Jiuquan with the same equatorial crossing time of 13 30 h into similar 490 × 510 Sun Synchronous Orbits. They are all also launched by CZ 2D launcher. The dedicated launch also suggests a much heavier payload than the Tansuo series. The PAN camera has a resolution of 5 m while the CCD camera has a resolution of 10 m. The CCD camera can be swiveled through an angle of 25°.[13]

In April 2019 two more Tianhui satellites were launched from Taiyuan on a CZ 4B launcher. They were placed in a slightly higher 520 km SSO. The time of launch suggests that the satellite will cover China and its surroundings in the early morning hours. These satellites may also have some military function.

The lower orbits of these satellites with reported stereo resolution of 5 m suggest that they are military satellites. The satellite construction seems to have moved from the prototype lab development into a proper industrial production capability at CAST and SAST respectively[14] (Harvey 2003).

9.13   The Ziyuan (ZY) Series

Eight satellites have been launched under the Ziyuan name. All of them have Equatorial Crossing Times (ECT) around 10 30 h. The early launches of the Ziyuan series such as the Ziyuan 2, 2B, and 2-3 satellites appear to be the precursor test satellites for the high-resolution EO satellites of the Yaogan series in a 500 km SSO. The Ziyuan 3A, Ziyuan III-02, and the Ziyuan 3-03, launched in 2012, 2016, and 2020 are also very similar to the Yaogan EO satellites (YG 13, YG 23, YG 18, YG 6). They may be used both for civilian and military purposes based on their similarities with the Yaogan 500 km SSO satellites.

The Ziyuan 2C launched in December 2011 and the Ziyuan 1-2D launched in 2019 are in a very different 770 km orbit very similar to the orbits of the CBERS

---

[13] https://www.nasaspaceflight.com/2015/10/long-march-2d-lofts-tianhui-1-satellite/.

[14] http://claudelafleur.qc.ca/Spacecrafts-2004.html#Shiyan-1.

series of satellites. The Ziyuan name can therefore be used for either civilian or military applications. Six of the eight satellites in the series may also serve military functions.

9.14 The Gaofen (GF) Series of Satellites

There are several satellites with the name Gaofen. They are in several different orbits that include Sun Synchronous and Geostationary Orbits.

According to Chinese sources the Gaofen (GF) satellites will consist of a constellation of 7–9 satellites. The constellation will have both orbiting as well as geostationary components. The constellation is meant to provide 24-h data. The satellite data will be combined with balloon and aircraft data to cater to a variety of needs including agriculture, disasters as well as security and scientific applications. The Gaofen series will carry CCD imagers, SAR as well as multispectral cameras. The programme conceived in 2006 received the go-ahead in 2010. Seven satellites were to be launched between 2013 and 2016.[15] These satellites may be the civilian counterparts (at least partially) to the military Yaogan series of satellites.

The Gaofen 1 is the first of a series of High-Definition Earth Observation Satellites (HDEOS) later renamed CHEOS for China High-Definition Earth Observation System. According to information available in the public domain GF 1 will carry two kinds of sensors. The first sensor is a two-camera payload with a 2 m PAN and an 8 m Multispectral capability with a combined swath of 69 km. It also carries a wide field of view camera system with four cameras that can provide a 16 m resolution over a swath of 830 km.

GF 2 will have a high-resolution camera system that will provide a 0.8 m PAN image and a 3.2 m multispectral image. The swath would be 45 km. The GF 3 would carry a C Band SAR payload with a resolution of 1 m. The GF 4 will have a 50 m resolution staring camera that will be placed in GSO. The camera will cover an area of 7000 km by 7000 km with each scene covering an area of 400 km by 400 km.

The GF 5 will carry a complement of 6 sensors. These include a visible SWIR hyperspectral camera, a spectral imager, a Greenhouse Gas Detection system, an Infrared atmospheric sounder with high resolution, a differential absorption spectrometer for trace gases as well as a multi-angle polarization sensor. The GF 6 will be a replacement satellite for the GF 1. The GF 7 will carry a hyperspectral stereo imaging payload.[16]

---

[15] These sensors are similar to those performing a military ISR function.

[16] https://directory.eoportal.org/web/eoportal/satellite-missions/g/gaofen-1.

Though this was the purported plan the realized launch sequence reveals several variations

23 Gaofen satellites have been launched between 2013 and 2020. There has been a significant increase in the pace of launchings of Gaofen satellites in the last 3 years (2018–2020). Technical details of the satellite and sensors carried by these satellites are available in the public domain[17] (eoportal).

Unlike the other GF satellites which are in SSO, the GF 4 and the GF 13 were launched from Xi Chang by the heavy CZ 3B launcher in December 2015 into a Geostationary Orbit (GSO). They carry a stop and stare multi-spectral Electro-optical scanner system. The system operates in the visible, shortwave, medium-wave, and longwave infrared regions of the Electro-Magnetic spectrum. These satellites are possible precursors to the deployment of a Missile Early Warning Satellite in GSO. The GF 4 and the GF 13 are therefore categorized as military satellites and taken out of the remote sensing grouping of satellites.

The Gaofen 14 was also launched into a 496 km SSO from Xichang in December 2020 using a CZ 3B launcher. This was the first launch from Xichang into a Sun Synchronous Orbit (SSO). No details of the mission are available in the public domain. It is most probably a military satellite and can also be removed from the remote sensing satellite list. Table 9.2 provides a listing of the remaining 20 Gaofen Remote Sensing satellites. They have been sorted by launch site, altitudes as well as dates in the Table presentation.

Seven of the 20 Gaofen satellites have been launched from the Jiuquan launch site using the CZ 2D launcher. These are the Gaofen 1, Gaofen 9, Gaofen 6 and the Gaofen 3-02, 3-03 and 3-03 satellites. The Gaofen 1, 9, and 6 satellites are all in 650 × 625 km, 98.10°, Sun Synchronous Orbits with Equatorial Crossing Times (ECT) between 10 30 and 11 00. Their orbital parameters are very similar to the Yaogan series of satellites (Yaogan 7, 2, 24, 4, 11, 30) which are also launched from Jiuquan by the CZ 2D.

The remaining four satellites of this series (Gaofen 9-02, 9-03, 9-04, 9-05) are also launched by the CZ 2D from the Jiuquan launch site into lower 510 × 493 km Sun Synchronous Orbits. Their orbits are very similar to the orbits of the LKW series of military satellites that are launched by the CZ 2D from the Jiuquan site. These satellites can therefore be categorized as dual use Electro-optical satellites that perform a complementary or supplementary military remote sensing function.

Six satellites the Gaofen 2, the Gaofen 8, the Gaofen 11, the Gaofen 7, Gaofen, and Gaofen 11–02 are all launched from Taiyuan by the CZ 4B rocket. The Gaofen 8 has orbital parameters that are very similar to the Yaogan 14, 18, 5, 12, 21, and 26 which are all in a similar 490 × 470 km orbit. They are all launched by the same CZ 4B launcher from the same Taiyuan launch site. The Gaofen 7 and the Gaofen

---

[17] https://directory.eoportal.org/web/eoportal/satellite-missions/g/gaofen-1.

11-02 also have orbital parameters that are very similar to the Yaogan 13, 23, 18 and Yaogan 6 satellites. These satellites also perform a dual use remote sensing function.

The other three satellites of the Gaofen series that are launched by the CZ 4B rocket from Taiyuan are the Gaofen 2, Gaofen D, and the Gaofen 11. Their launch details and orbital parameters do not match any of the Yaogan or other military remote sensing satellites. These satellites can therefore be grouped as civilian satellites.

The Gaofen 3, 10, 10R, and the Gaofen 12 are all launched by the powerful CZ 4C from Taiyuan. The Equatorial Crossing Times (ECTs) of the GF 3, GF 10, and the GF 10R are 06 00, 04 00, and 01 50 respectively. These ECTs indicate that they are SAR satellites. The orbital parameters of the GF 12 are very similar to those of the GF 10R. It is also a SAR satellite. The GF 10 failed to reach orbit due to a launch failure. The GF 10R is a replacement satellite.

The Gaofen 3 is at an altitude of 759 km. This is different from the altitudes of the other GF satellites in this group. This would make the GF 3 a civilian satellite. The operating altitudes of the GF 10R and the GF 12 are 630 km. Their launch and orbital parameters make them very similar to the first-generation Yaogan satellites that include YG 1, YG 3, YG 10, YG 29, and YG 33. Three out of the four satellites in this set are dual use satellites which also have a military function.

In March 2018, three satellites the GF1-02, GF-03 and the GF-04 were launched together by a CZ 4C launcher from Taiyuan into a 640 km SSO with Equatorial Crossing Times around 10 30 AM. This is the first launching of such a triplet under the Gaofen name. The launch and orbital parameters do not match any of the other military remote sensing satellites. They are probably civilian satellites for various applications. This analysis suggests that 13 out of the 20 Gaofen remote sensing satellites serve a dual use function.

9.15 Commercial Small Satellite Remote Sensing Constellations—An Emerging Trend

In keeping with emerging global trends China has launched several commercial small satellite remote sensing constellations. Table 9.3 provides a listing of the smaller remote sensing satellites by China till the end of 2020. These have been sorted in terms of names and launch dates.

The Jilin (JN) Constellation

The largest commercial small satellite remote sensing constellation operated by China as of end 2020 is the Jilin constellation with 27 satellites.

Table 9.3 Commercial small satellite remote sensing constellations end 2020

Satellite	Launch site	Launch time UT	Launch time CST	Apogee Km	Perigee Km	Inclination °	Launch date	ECT	Launcher	Nos
Jilin 1A	Jiuquan	04 13	12 13	656	656	98.00	October 7 2015	10 30	CZ 2D	1
Lingqiao A	Jiuquan	04 13	12 13	656	656	98.00	October 7 2015	10 30	CZ 2D	1
Lingqiao B	Jiuquan	04 13	12 13	656	656	98.00	October 7 2015	10 30	CZ 2D	1
Jilin Linye-1-03	Jiuquan	04 11	12 11	558	553	97.54	January 9 2017	10 30	Ku 1A	1
Jilin 1 03 04,05,06	Taiyuan	04 50	12 50	542	530	97.50	November 21 2017	12 00	CZ 6	4
Jilin 07, 08	Jiuquan	04 12	12 12	547	523	98.50	January 19 2018	10 30	CZ 11	2
Jilin-1-09	Jiuquan	13 40	21 40	549	527	97.5	January 21 2019	20 40	CZ 11	1
Jilin-1-10	Jiuquan	13 40	21 40	552	523	97.5	January 21 2019	20 40	CZ 11	1
Jilin 1 03A	YLSA	04 06	12 06	586	569	45.00	June 5 2019	NA	CZ 11	1
Jilin-1 02A	Jiuquan	03 40	11 40	556	537	97.50	November 13 2019	10 40	Ku 1A	1
Jilin-1 02B	Taiyuan	02 55	10 55	555	538	97.50	December 12 2019	09 55	Ku 1A	1
Jilin	Taiyuan	02 53	10 53	497	481	97.3	January 15 2020	09 50	CZ 2D	1
Jilin-1	Jiuquan	04 17	12 17	NA	NA	NA	July 10 2020	11 17	Ku 1A	1
Jilin-1	Jiuquan	05 02	13 02	NA	NA	NA	September 12 2020	12 02	Ku 1A	1
Jilin 1-9	YSLA	01 23	09 23	555	533	97.5	September 15 2020	08 23	CZ 11	9
Spark-01	Jiuquan	19 22	03 22	726	698	98.20	December 21 2016	01 39	CZ 2D	1

(continued)

Table 9.3 (continued)

Satellite	Launch site	Launch time UT	Launch time CST	Apogee Km	Perigee Km	Inclination °	Launch date	ECT	Launcher	Nos
Spark-02	Jiuquan	19 22	03 22	726	698	98.20	December 21 2016	01 39	CZ 2D	1
Yijian	Jiuquan	19 22	03 22	728	697	98.20	December 21 2016	01 39	CZ 2D	1
SuperView 1-01	Taiyuan	03 23	11 23	524	214	97.60	December 28 2016	10 30	CZ 2D	1
SuperView 1-02	Taiyuan	03 23	11 23	524	214	97.60	December 28 2016	10 30	CZ 2D	1
SuperView1-03,04	Taiyuan	03 24	11 24	535	528	97.60	January 9 2018	10 30	CZ 2D	2
ZuhaiOVS 2A to E	Jiuquan	04 42	12 42	524	508	97.40	April 26 2018		CZ 11	5
ZHUHAI-1 03A	Jiuquan	06 42	14 42	530	510	97.40	September 19 2019	13 40	CZ 11	1
ZHUHAI-1 03B	Jiuquan	06 42	14 42	527	513	97.40	September 19 2019	13 40	CZ 11	1
ZHUHAI-1 03C	Jiuquan	06 42	14 42	526	514	97.40	September 19 2019	13 40	CZ 11	1
ZHUHAI-1 03D	Jiuquan	06 42	14 42	526	513	97.40	September 19 2019	13 40	CZ 11	1
ZHUHAI-1 03E	Jiuquan	06 42	14 42	519	498	97.40	September 19 2019	13 40	CZ 11	1
Ningxia 1 1	Taiyuan	06 35	14 35	905	895	45.00	November 13 2019	13 35	CZ 6	1

(continued)

Table 9.3 (continued)

Satellite	Launch site	Launch time UT	Launch time CST	Apogee Km	Perigee Km	Inclination °	Launch date	ECT	Launcher	Nos
Ningxia 1 2	Taiyuan	06 35	14 35	905	893	45.00	November 13 2019	13 35	CZ 6	1
Ningxia 1 3	Taiyuan	06 35	14 35	905	890	45.00	November 13 2019	13 35	CZ 6	1
Ningxia 1 4	Taiyuan	06 35	14 35	904	889	45.00	November 13 2019	13 35	CZ 6	1
Ningxia 1 6	Taiyuan	06 35	14 35	903	888	45.00	November 13 2019	13 35	CZ 6	1
Xiaoxiang 1-07	Jiuquan	23 41	07 31	619	599	97.80	August 30 2019	06 30	Ku 1A	1
Xiaoxiang 1-08	Taiyuan	03 22	11 22	514	491	97.5	November 3 2019	10 22	CZ 4B	1
Tianyi 16	Taiyuan	08 52	16 52	500	500	97.40	December 12 2019	15 50	Ku 1A	1
Tianyi 17	Taiyuan	08 52	16 52	500	500	97.40	December 12 2019	15 50	Ku 1A	1
Tiankun 1	Jiuquan	23 53	07 53	404	374	96.90	March 2 2017		Ka 2	1
Tianyan 01	Taiyuan	03 22	11 22	635	615	98.00	December 20 2019	10 20	CZ 4B	1
Hisea-1	Wenchang	04 37	12 37	518	510	97.4	December 22 2020	11 37	CZ 8	1

Jilin is a remote sensing data mission initiated by the Jilin Province. The purpose of this mission is to provide near real and eventually real-time data to commercial users. The first part of this constellation comprising four satellites that were launched together on a CZ 2D launcher from Jiuquan in July 2015.

The Jilin 1 is a 420 kg satellite that provides 0.72 m PAN imagery along with 4 m resolution spectral imagery. The Lingqiao A and B satellites provide 1 m resolution video over a strip of 4.3 × 2.4 km. They weigh 95 kg each. The fourth satellite LQSAT is a micro satellite weighing 54 kg. It carries an Amateur Radio Payload and is not included in this assessment of remote sensing satellites.

Another Jilin satellite (1-03) was launched by the Kuaizhou launcher in 2017 followed by the launches of the Jilin 04, 05, and 06 using the newly developed CZ 6 launcher. Two more Jilin satellites (Jilin 07, Jilin 08) were launched using the CZ 11 launcher in 2018.

Five Jilin satellites were launched in 2019 followed by further 12 in 2020. A CZ 11 rocket placed 9 Jilin satellites from a sea-based platform in the Yellow Sea to raise the constellation size to 23. The calendar year 2020 also saw the successive failures of two Kuaizhou launchers carrying Jilin satellites thus reducing the constellation size to 21 operational satellites.

The Jilin satellite is developed by the Chang Guang Satellite Technology Company under the Changchun Institute of Optics, Fine Mechanics & Physics of the CAS.

The current constellation is the first step of a larger plan. Between 2020 and 2030 the constellation is expected to grow to 60 satellites with a 30-min update period. The plan beyond 2030 is to have a constellation of 138 satellites that will provide all-day and all-weather coverage with a revisit period of 10 min.[18] All the satellites in the current set carry electro-optical imaging sensors.

Spark Small Satellite Constellation

A three-satellite Spark constellation was created by the Shanghai Small Satellite Center using a CZ 2D launcher from Jiuquan in 2016. The Small Satellite Center may be connected to the Shanghai Academy of Space Technology. The satellites are in a 710 km SSO with an ECT of 01 40 h. The ECT would mean that these satellites cover the Chinese mainland during a night pass. The satellites may provide optical imagery during the daytime ascending pass and not during the night-time descending pass or they could be tests for some new sensors for future small satellites.

Superview (SV) Small Satellite Constellation

Another small satellite constellation of four was created by the Siwei Star Company a subsidiary of the China Aerospace Science & Technology Corporation (CAST). They were launched two at a time using a CZ 2D launcher from Taiyuan. The first launch took place in 2016. The satellites were injected into a lower orbit than planned.

---

[18] https://www.nasaspaceflight.com/2015/10/china-launches-jilin-1-mission-long-march-2d/.

Onboard thrusters were used to correct the orbit. The second launch took place in 2018. The four satellites are currently in a 520 SSO with an ECT of around 10 30 h.

The Zhuhai (ZU) Small Satellite Constellation

A Zuhai constellation of five satellites was placed into a 500 km SSO by a CZ 11 launcher in April 2018. A second group of another 5 satellites were placed in a similar orbit by another CZ 11 launcher in September 2019. There are therefore 10 Zhuhai satellites providing data.

The satellites are owned by the Zhuhai Orbita Aerospace Science & Technology Ltd based in the southern Chinese city of Shenzen. These commercial launches were preceded by two small satellite launches in 2017 (CAS 4A and CAS 4B). There is a plan to expand the constellation to 38 satellites.

The Ningxia Constellation

Five Ningxia satellites were launched in November 2019 using a CZ 6 rocket. These satellites are not in Sun Synchronous Orbits (SSOs) but in a 45° inclination 900 km orbit. It is possible that these are test satellites for future ELINT missions.

Hisea 1

The Hisea 1 C band SAR satellite was launched in December 2020 by the Chinese company Spacety. It is now offering SAR imagery for commercial uses. Spacety is also involved in several other small satellite remote sensing and communications missions.

Other Groupings

2019 also saw the orbiting of several other small satellites for remote sensing. These include two Xiaoxiang satellites, two Tianyi satellites as well as a Tiankun and Tianyan satellite.

Global Trends

The world is witnessing a small satellite revolution that encompasses remote sensing. US companies are leading this race. The San Francisco based Planet Labs company currently has 175 small Dove satellites in orbit that can provide commercial services on a daily basis for many civilian applications. The Finnish company Iceye has 7 small SAR satellites providing regular data. Another US start-up Capella has three SAR satellites with more being planned (UCS 2021).

China is keeping up with this emerging trend and is well-placed to use this opportunity for catering to both domestic and international market needs.

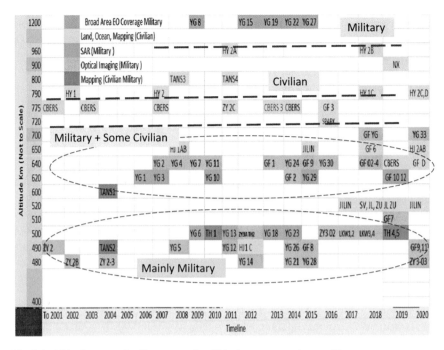

Fig. 9.1   Altitude versus timeline—grouping China's remote sensing satellites

9.16   A Remote Sensing Satellite Timeline

A timeline of these satellites provides additional insights. Using the data in Tables 9.1, 9.2, and 9.3 a timeline versus altitude chart for all imaging and mapping satellites can be generated. These include the names Yaogan (YG), Huanjing (HJ), Haiyang (HY), Ziyuan (ZY), Gaofen (GF), Tansuo (TANS), Tianhui (TH), Jilin (JN), Superview (SV) Zhuhai (ZU), and Ningxia (NX).

Figure 9.1 provides such a timeline. Nomenclatures and imaging functions can be related to each other. Similarities can also be identified. Using official pronouncements and other published sources a consolidated assessment of Chinese capabilities in imaging applications can be made.

9.17   The Dominance of the Military Function

A quick perusal of the **Tables** as well as the **Figure** reveals certain features of the Chinese Earth Observation programme. There are several broad groupings based on the altitude of operations.

The first set of satellites all operates in the range of 480–520 km Sun Synchronous Orbits. They carry EO sensors for surveillance and mapping as well as SAR payloads for all weather and day night surveillance. Some of them may also supply data for commercial purposes. Out of the 40 satellites in this set, 34 satellites serve a military function. These include satellites under the generic names Yaogan, LKW, XJY, Tianhui, Ziyuan, and Gaofen.

The HJ 1C SAR satellite, which is also in this group, is a part of China's contribution to the Asia Pacific Space Cooperation Organization (APSCO). Five satellites appear to be tested for future deployment.

There are 35 small satellites that include commercial constellations operating at altitudes between 520 and 570 km. These include the Jilin, Superview, and the Zhuhai series of satellites that provide data for civilian applications. Their resolutions may be good enough however to cater to some military applications.

The third set of 29 satellites function in Sun Synchronous Orbits between 600 and 650 km. They include both SAR and hi-resolution EO satellites. These include 10 Yaogan, 1 Tansuo, 8 Gaofen, four HJ, 3 Jilin, and two other satellites. The latest CBERS commercial satellite launched in 2019 also falls within this group. Nineteen of these satellites appear to be military satellites.

Thirteen satellites are in orbits of 700–770 km. These include 3 Yaogan, 4 CBERS, one Tansuo, two Gaofen, and three Spark satellites. 6 satellites may have a military function.

The fifth group of 15 satellites operate in Sun Synchronous Orbits between 770 and 980 km altitude. They include six Haiyang (HY) ocean satellites, two Tansuo mapping satellites, two Ziyuan and 5 Ningxia commercial satellites. Four of them may be military test satellites.

The last group consists of 5 Yaogan satellites in a 1200 km SSO. These are Broad Area Coverage Electro-optical satellites. They are military satellites.

A total of 144 satellites that may carry imaging sensors were launched into Sun Synchronous Orbit (SSO). Taking out five failures there were a total of 139 satellites. 62 out of 139 or nearly 45% of all remote sensing satellites perform ISR functions. It is possible that some of the data from the civilian satellites can also be used for some military applications and vice versa.

9.18 Civilian Uses of Remote Sensing

Figure 9.2 provides an overview of the civilian organization set up for using data from various earth observation satellites.[19] Though the space programmme under the civilian China National Space Administration (CNSA) is responsible for building and launching the satellites the ocean related data and applications are handled by the

---

[19] https://directory.eoportal.org/web/eoportal/satellite-missions/g/gaofen-1.

Fig. 9.2 The organization of remote sensing activities in China

State Ocean Administration (SOA). Weather satellite data reception and distribution appears to be the responsibility of the China Meteorological Administration (CMA). CNSA is only responsible for data reception and distribution of satellite data devoted to land applications.

Published reports along with the launch of the first commercial Jilin constellation of high-resolution satellites seem to suggest that China has achieved near complete autonomy related to the acquisition and distribution of high-resolution imaging data from satellites. Figure 9.3 shows this transition from imported to indigenous sources.[20]

China has also used the international collaboration route to bring state-of-art capabilities into the country. The Jilin 1 domestic constellation of remote sensing satellites will directly compete with data supplied by the Twenty-First Century Aerospace Company. This company has bought all the capacity available on the three-satellite remote sensing constellation owned by the Surrey Small Satellite Technology Ltd. Company located in England. The resulting competition is likely to lower prices further and increase the demand for civilian uses of satellite imagery (**For more details see the chapter on International Footprint of China's Space Programme**).

---

[20] http://spacenews.com/china-launches-high-resolution-commercial-imaging-satellite/#sthash.aJPQz6yx.dpu.

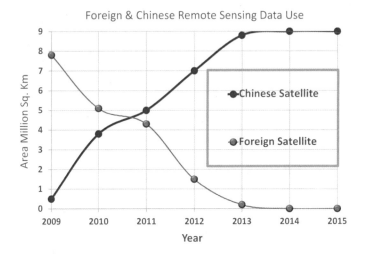

Fig. 9.3 Achieving strategic autonomy in remote sensing data

9.19   Conclusions

A large part of the remote sensing satellite efforts of China has been devoted toward the use of these satellites for military ISR functions. Though this is so, China also has in place enough domestic capacities for meeting civilian use requirements as well. It has used international collaboration judiciously for building technological and organizational capacities as well as to cater to the needs of domestic civilian users. Its HJ1 series of satellites is also directed toward cooperating with countries in the Asia Pacific Region.

Most of the remote sensing satellites appear to be completely indigenously built. Functionally they are reasonably close to meeting the needs of both reconnaissance and civilian applications.

China is now extending its military satellite capabilities into the civilian domain. It is also allowing companies to enter the market for high-resolution images. By creating competition in the supply of high-resolution images it is trying to promote innovation in the applications and use of satellite data for meeting societal needs. Over time, it has also replaced imported high-resolution satellite data with indigenous supply.

Annexure 9.1: China's Remote Sensing Satellites

Satellite	Launch site	Launch Time UT	Launch Time CST	Apogee Km	Perigee Km	Inclination degree	Launch date	ECT	Launcher	Nos
Yaogan 09 ABC	Jiuquan	04 55	12 55	1099	1083	63.41	March 5 2010	NA	CZ 4C	3
Yaogan 16 ABC	Jiuquan	04 06	12 06	1105	1085	63.39	November 25 2012	NA	CZ 4C	3
Yaogan 17 ABC	Jiuquan	19 16	03 16	1111	1076	63.41	September 1 2013	NA	CZ 4C	3
Yaogan 20 ABC	Jiuquan	05 45	13 45	1104	1086	63.40	August 9 2014	NA	CZ 4C	3
Yaogan 25 ABC	Jiuquan	19 33	03 33	1097	1089	63.41	December 10 2014	NA	CZ 4C	3
Yaogan 31 ABC	Jiuquan	04 25	12 25	1100	1090	63.40	April 10 2018	NA	CZ 4C	3
Yaogan 30 A,B,C	Xichang	04 21	12 21	609	605	35.00	Sept 29 2017	NA	CZ 2C	3
Yaogan 30 D,E,F	Xichang	18 10	02 10	616	607	35.00	November 24 2017	NA	CZ 2C	3
Yaogan 30 G,H,J	Xichang	19 44	03 44	610	600	35.00	December 25 2017	NA	CZ 2C	3
Yaogan 30 K,L,M	Xichang	05 39	13 39	611	602	35.00	January 25 2018	NA	CZ 2C	3
Yaogan 30 N P Q	Xichang	03 57	11 57	619	590	35.00	July 26 2019	NA	CZ 2C	3
Yaogan 30 R, S, T	Xichang	03 43	11 43	608	603	35.00	March 24 2020	NA	CZ 2C	3
Yaogan 30 U,V,W	Xichang	15 19	23 19	610	600	35.00	October 26 2020	NA	CZ 2C	3
XJS A, B	Xichang	03 30	11 30	496	488	35.00	June 27 2018	NA	CZ 2C	2
XJS C D E F	Xichang	21 07	05 07	488	478	35.00	February 19 2020	NA	CZ 2D	4
XJS G, H	Xichang	20 13	04 13	489	477	35.00	May 29 2020	NA	CZ 11	2
Yaogan 1	Taiyuan	22 48	06 48	626	624	97.80	April 26 2006	06 00	CZ 4C	1
Yaogan 3	Taiyuan	22 48	06 48	624	613	97.80	November 11 2007	06 00	CZ 4C	1
Yaogan 10	Taiyuan	22 48	06 48	632	624	97.82	August 9 2010	06 00	CZ 4C	1
Yaogan 29	Taiyuan	21 24	05 24	619	615	97.84	November 26 2015	04 30	CZ 4C	1
Yaogan 33	Taiyuan	22 49	06 49	NA	NA	NA	May 22 2019	06 00	CZ 4C	1
Yaogan 13	Taiyuan	18 50	02 50	511	505	97.11	November 29 2011	01 56	CZ 2C	1
Yaogan 23	Taiyuan	18 53	02 53	513	492	97.33	November 14 2014	02 00	CZ 2C	1
Yaogan 18	Taiyuan	02 50	10 50	511	492	97.55	October 29 2013	09 56	CZ 2C	1

Satellite	Launch site	Launch Time UT	Launch Time CST	Apogee Km	Perigee Km	Inclination degree	Launch date	ECT	Launcher	Nos
Yaogan 6	Taiyuan	02 55	10 55	521	486	97.63	April 22 2009	10 01	CZ 2C	1
Yaogan 7	Jiuquan	08 42	16 42	659	623	97.84	December 9 2009	15 00	CZ 2D	1
Yaogan 2	Jiuquan	07 12	15 12	655	630	97.85	May 25 2007	13 30	CZ 2D	1
Yaogan 24	Jiuquan	07 12	15 12	653	630	97.91	November 20 2014	13 30	CZ 2D	1
Yaogan 4	Jiuquan	04 42	12 42	652	634	97.92	December 1 2008	11 00	CZ 2D	1
Yaogan 11	Jiuquan	02 42	10 42	657	624	98.00	Sep 22 2010	09 00	CZ 2D	1
Yaogan 30	Jiuquan	02 43	10 43	655	626	98.07	May 15 2016	09 01	CZ 2D	1
LKW 3	Jiuquan	07 10	15 10	509	491	97.30	January 13 2018	13 30	CZ 2D	1
LKW 4	Jiuquan	07 10	15 10	510	497	97.30	March 17 2018	13 30	CZ 2D	1
LKW 1	Jiuquan	04 11	12 11	510	496	97.46	December 4 2017	10 30	CZ 2D	1
LKW 2	Jiuquan	04 14	12 14	510	496	97.50	December 23 2017	10 33	CZ 2D	1
Yaogan 14	Taiyuan	07 06	15 06	474	470	97.24	May 10 2012	14 14	CZ 4B	1
Yaogan-28	Taiyuan	07 06	15 06	482	460	97.24	November 8 2015	14 14	CZ 4B	1
Yaogan 5	Taiyuan	03 22	11 22	492	481	97.40	December 15 2008	10 30	CZ 4B	1
Yaogan 12	Taiyuan	03 21	11 21	491	484	97.41	November 9 2011	10 29	CZ 4B	1
Yaogan 21	Taiyuan	03 22	11 22	494	480	97.42	September 8 2014	10 30	CZ 4B	1
Yaogan 26	Taiyuan	03 22	11 22	491	484	97.44	December 27 2014	10 30	CZ 4B	1
Yaogan 33 R	Jiuquan	15 44	23 44	704	702	98.2	December 28 2020	22 44	CZ 4C	1
Yaogan 32 A,B	Jiuquan	02 43	10 43	704	703	98.30	October 10 2018	09 00	CZ 2C	2
XJY 7	Wenchang	04 37	12 37	518	510	97.4	December 22 2020	11 37	CZ 8	1
Yaogan 15	Taiyuan	07 31	15 31	1206	1202	100.13	May 29 2012	14 30	CZ 4C	1
Yaogan 22	Taiyuan	06 31	14 31	1209	1196	100.32	October 20 2014	13 30	CZ 4C	1
Yaogan 27	Taiyuan	02 31	10 31	1206	1194	100.46	August 27 2015	09 29	CZ 4C	1
Yaogan 19	Taiyuan	03 31	11 31	1207	1201	100.48	November 20 2013	10 29	CZ 4C	1
Yaogan 8	Taiyuan	02 31	10 31	1204	1193	100.50	December 15 2009	09 29	CZ 4C	1
CBERS 1 Ziyuan 1	Taiyuan	03 15	11 15	750	731	98.50	October 14 1999	10 30	CZ 4B	1
CBERS 2	Taiyuan	03 16	11 16	750	731	98.50	October 21 2003	10 30	CZ 4B	1

Satellite	Launch site	Launch Time UT	Launch Time CST	Apogee Km	Perigee Km	Inclination degree	Launch date	ECT	Launcher	Nos
CBERS 2 B	Taiyuan	03 26	11 26	773	775	98.60	Sep 19 2007	10 40	CZ 4B	1
CBERS 3	Taiyuan	03 26	11 26	NA	NA	Failure	December 9 2013	NA	CZ 4B	1
CBERS 4	Taiyuan	03 26	11 26	751	742	98.50	December 7 2014	10 45	CZ 4B	1
CBERS 4 A	Taiyuan	03 22	11 22	635	615	98.00	December 20 2019	10 20	CZ 4B	1
Huanjing 1 A	Taiyuan	03 25	11 25	660	626	98.00	September 6 2008	10 30	CZ 2C	1
Huanjing 1 B	Taiyuan	03 25	11 25	672	627	98.00	September 6 2008	10 30	CZ 2C	1
Huanjing 1 C	Taiyuan	22 53	06 53	503	487	97.40	November 18 2012	05 59	CZ 2C	1
Huanjing 2A	Taiyuan	03 23	11 23	649	647	98	Sept 27 2020	10 23	CZ 4B	1
Huanjing 2B	Taiyuan	03 23	11 23	649	647	98	Sept 27 2020	10 23	CZ 4B	1
Haiyang 1 A	Taiyuan	01 50	09 50	795	792	98.80	May 15 2002	08 55	CZ 4B	1
Haiyang 1 B	Taiyuan	03 27	11 27	814	793	98.60	April 11 2004	10 30	CZ 2C	1
Haiyang 1 C	Taiyuan	03 15	11 15	793	777	98.60	September 7 2018	10 20	CZ 2C	1
Haiyang 2 A	Taiyuan	22 57	06 57	963	963	99.40	August 15 2011	06 00	CZ 4B	1
Haiyang 2 B	Taiyuan	22 57	06 57	975	972	99.40	October 24 2018	06 00	CZ 4B	1
Haiyang 2 C	Jiuquan	05 40	13 40	965	954	66	Sept 21 2020	NA	CZ 4B	1
Tansuo 1	Xi Chang	15 59	23 59	615	600	97.60	April 18 2004	12 00	CZ 2C	1
Tansuo 2	Xi Chang	10 45	18 45	711	694	98.10	November 18 2004	07 00	CZ 2C	1
Tansuo 4	Jiuquan	00 15	08 15	804	783	98.50	November 20 2011	06 31	CZ 2D	1
Tansuo 3	Jiuquan	00 15	08 15	804	786	98.50	November 5 2008	06 31	CZ 2D	1
Tianhui 1	Jiuquan	07 10	15 10	504	488	97.30	August 24 2010	13 29	CZ 2D	1
Tianhui 2	Jiuquan	07 10	15 10	506	490	97.40	May 6 2012	13 29	CZ 2D	1
Tianhui 3	Jiuquan	07 10	15 10	511	495	97.40	October 26 2015	13 29	CZ 2D	1
Tianhui 2-01A	Taiyuan	22 52	06 52	524	522	97.5	April 29 2019	05 52	CZ 4B	1
Tianhui 2-01B	Taiyuan	22 52:	06 52	524	522	97.5	April 29 2019	05 52	CZ 4B	1
ZiYuan-2	Taiyuan	03 25	11 25	499	483	97.40	September 1 2000	10 30	CZ 4B	1
Ziyuan-2B	Taiyuan	03 17	11 17	483	470	97.40	October 27 2002	10 25	CZ 4B	1
Ziyuan 2-3	Taiyuan	03 10	11 10	483	473	97.30	November 6 2004	10 20	CZ 4B	1

Satellite	Launch site	Launch Time UT	Launch Time CST	Apogee Km	Perigee Km	Inclination degree	Launch date	ECT	Launcher	Nos
Ziyuan-2C	Taiyuan	03 26	11 26	774	773	98.50	December 22 2011	10 30	CZ 4B	1
Ziyuan-3A	Taiyuan	03 17	11 17	506	498	97.50	January 9 2012	10 25	CZ 4B	1
Ziyuan III-02	Taiyuan	03 17	11 17	511	507	97.50	May 30 2016	10 24	CZ 4B	1
Ziyuan 1-2D	Taiyuan	03 26	11 26	781	781	98.60	Sept 12 2019	10 30	CZ 4B	1
Ziyuan-3 03	Taiyuan	03 13	11 13	517	502	97.5	July 25 2020	10 13	CZ 4B	1
Gefen 1	Jiuquan	04 13	12 13	653	630	98.10	April 26 2013	10 30	CZ 2D	1
Gaofen 9	Jiuquan	04 42	12 42	664	617	98.00	Sept 14 2015	10 59	CZ 2D	1
Gaofen 6	Jiuquan	04 13	12 13	655	642	98.10	June 2 2018	10 30	CZ 2D	1
Gaofen 9-02	Jiuquan	08 53	16 53	510	494	97.3	May 31 2020	15 10	CZ 2D	1
Gaofen 9-03	Jiuquan	07 19	15 19	512	493	97.4	June 17 2020	13 35	CZ 2D	1
Gaofen 9-04	Jiuquan	04 01	12 01	511	493	97.4	August 6 2020	10 17	CZ 2D	1
Gaofen 9-05	Jiuquan	02 27	10 27	511	493	97.5	August 23 2020	08 45	CZ 2D	1
Gaofen 8	Taiyuan	06 22	14 22	481	469	97.30	June 26 2015	13 30	CZ 4B	1
Gaofen 11-02	Taiyuan	05 57	13 57	503	502	97.3	Sep 7 2020	13 00	CZ 4B	1
Gaofen 7	Taiyuan	03 22	11 22	514	490	97.50	November 3 2019	10 22	CZ 4B	1
Gaofen 2	Taiyuan	03 15	11 15	631	608	98.00	August 19 2014	10 20	CZ 4B	1
Gaofen DM	Taiyuan	03 10	11 10	657	636	98.0	July 3 2020	10 10	CZ 4B	1
Gaofen 11	Taiyuan	03 00	11 00	701	255	97.40	July 31 2018	~10 00	CZ 4B	1
Gaofen 3	Taiyuan	22 55	06 55	759	758	98.40	August 9 2016	06 00	CZ 4C	1
Gaofen 10	Taiyuan	18 55	04 55			Failure	August 31 2016	04 00	CZ 4C	1
GAOFEN 10R	Taiyuan	18 51	02 51	635	630	97.80	October 4 2019	01 50	CZ 4C	1
Gaofen 12	Taiyuan	23 52	08 52	637	635	97.90	November 27 2019	07 52	CZ 4C	1
GF 1- 02,03,04	Taiyuan	03 22	11 22	642	638	98.00	March 30 2018	10 30	CZ 4C	3
Jilin 1A	Jiuquan	04 13	12 13	656	656	98.00	October 7 2015	10 30	CZ 2D	1
Jilin Linye-1 - 03	Jiuquan	04 11	12 11	558	553	97.54	January 9 2017	10 30	Ku 1A	1
Jilin 1 03 04,05,06	Taiyuan	04 50	12 50	542	530	97.50	November 21 2017	12 00	CZ 6	4
Jilin 07, 08	Jiuquan	04 12	12 12	547	523	98.50	January 19 2018	10 30	CZ 11	2

Satellite	Launch site	Launch Time UT	Launch Time CST	Apogee Km	Perigee Km	Inclination degree	Launch date	ECT	Launcher	Nos
Jilin-1 - 09	Jiuquan	13 40	21 40	549	527	97.5	January 21 2019	20 40	CZ 11	1
Jilin-1 - 10	Jiuquan	13 40	21 40	552	523	97.5	January 21 2019	20 40	CZ 11	1
Jilin 1 03A	YSLA	04 06	12 06	586	569	45.00	June 5 2019	NA	CZ 11	1
Jilin-1 02A	Jiuquan	03 40	11 40	556	537	97.50	November 13 2019	10 40	Ku 1A	1
Jilin-1 02B	Taiyuan	02 55	10 55	555	538	97.50	December 12 2019	09 55	Ku 1A	1
Jilin	Taiyuan	02 53	10 53	497	481	97.3	January 15 2020	09 50	CZ 2D	1
Jilin-1	Jiuquan	04 17	12 17	NA	NA	NA	July 10 2020	11 17	Ku 1A	1
Jilin-1	Jiuquan	05 02	13 02	NA	NA	NA	Sept 12 2020	12 02	Ku 1A	1
Jilin 1 -9	YSLA	01 23	09 23	555	533	97.5	Sept 15 2020	08 23	CZ 11	9
Spark-01	Jiuquan	19 22	03 22	726	698	98.20	December 21 2016	01 39	CZ 2D	1
Spark-02	Jiuquan	19 22	03 22	726	698	98.20	December 21 2016	01 39	CZ 2D	1
Yijian	Jiuquan	19 22	03 22	728	697	98.20	December 21 2016	01 39	CZ 2D	1
SuperView - 01	Taiyuan	03 23	11 23	524	214	97.60	December 28 2016	10 30	CZ 2D	1
SuperView - 02	Taiyuan	03 23	11 23	524	214	97.60	December 28 2016	10 30	CZ 2D	1
SuperView-03,04	Taiyuan	03 24	11 24	535	528	97.60	January 9 2018	10 30	CZ 2D	2
ZuhaiOVS 2A to E	Jiuquan	04 42	12 42	524	508	97.40	April 26 2018		CZ 11	5
ZHUHAI-1 03A	Jiuquan	06 42	14 42	530	510	97.40	Sept 19 2019	13 40	CZ 11	1
ZHUHAI-1 03B	Jiuquan	06 42	14 42	527	513	97.40	Sept 19 2019	13 40	CZ 11	1
ZHUHAI-1 03C	Jiuquan	06 42	14 42	526	514	97.40	Sept 19 2019	13 40	CZ 11	1
ZHUHAI-1 03D	Jiuquan	06 42	14 42	526	513	97.40	Sept 19 2019	13 40	CZ 11	1
ZHUHAI-1 03E	Jiuquan	06 42	14 42	519	498	97.40	Sept 19 2019	13 40	CZ 11	1
Ningxia 1 1	Taiyuan	06 35	14 35	905	895	45.00	November 13 2019	13 35	CZ 6	1
Ningxia 1 2	Taiyuan	06 35	14 35	905	893	45.00	November 13 2019	13 35	CZ 6	1
Ningxia 1 3	Taiyuan	06 35	14 35	905	890	45.00	November 13 2019	13 35	CZ 6	1
Ningxia 1 4	Taiyuan	06 35	14 35	904	889	45.00	November 13 2019	13 35	CZ 6	1
Ningxia 1 6	Taiyuan	06 35	14 35	903	888	45.00	November 13 2019	13 35	CZ 6	1
Lingqiao A	Jiuquan	04 13	12 13	656	656	98.00	October 7 2015	10 30	CZ 2D	1

Satellite	Launch site	Launch Time UT	Launch Time CST	Apogee Km	Perigee Km	Inclination degree	Launch date	ECT	Launcher	Nos
Lingqiao B	Jiuquan	04 13	12 13	656	656	98.00	October 7 2015	10 30	CZ 2D	1
Xiaoxiang 1-07	Jiuquan	23 41	07 31	619	599	97.80	August 30 2019	06 30	Ku 1A	1
Xiaoxiang 1-08	Taiyuan	03 22	11 22	514	491	97.5	November 3 2019	10 22	CZ 4B	1
Tianyi 16	Taiyuan	08 52	16 52	500	500	97.40	December 12 2019	15 50	Ku 1A	1
Tianyi 17	Taiyuan	08 52	16 52	500	500	97.40	December 12 2019	15 50	Ku 1A	1
Tiankun 1	Jiuquan	23 53	07 53	404	374	96.90	March 2 2017		Ka 2	1
Tianyan 01	Taiyuan	03 22	11 22	635	615	98.00	December 20 2019	10 20	CZ 4B	1
Hisea-1	Wenchang	04 37	12 37	518	510	97.4	December 22 2020	11 37	CZ 8	1

Notes:

The red fonts indicate launch, satellite failure

Inclination is in degrees

ECT is the Equatorial Time of Crossing which is the local time when a satellite in Sun Synchronous Orbit (SSO) crosses the Equator.

YSLA stands for Yellow Sea Launch Area.

References

Chandrashekar S, Ramani N (2018) China's space power & military strategy—the role of the Yaogan satellites, ISSSP Report No. 02–2018. Bangalore: International Strategic and Security Studies Programme, National Institute of Advanced Studies (NIAS, July 2018). http://isssp.in/wp-con tent/uploads/2018/07/ISSSP-Report-July-2018.pdf

Chandrashekar S et al (2011) China's anti-ship ballistic missile game changer in the Pacific Ocean, international strategic & security studies programme (ISSSP). National Institute of Advanced Studies. http://isssp.in/wp-content/uploads/2013/01/2011-november-r-5-chinas-anti-ship-ballistic-missile-report2.pdf

Eoportal https://directory.eoportal.org/web/eoportal/satellite-missions/g/gaofen-1

Harvey B (2003) China's space programme from conception to manned space flight. Springer

Harvey B (2013) China in space: the great leap forward. Springer Praxis Books, New York

Union of Concerned Scientists (2021) UCS satellite data base. https://ucsusa.org/resources/satell ite-database

Chapter 10
China's Navigation Satellite Constellation

10.1 The Beidou Navigation System is Operational

As of the end of 2020 China had launched 61 navigation satellites. These include 15 satellites in GSO, 12 satellites in an Inclined Geosynchronous Orbit (IGSO) as well as 32 satellites in MEO. China also launched two experimental Centispace satellites using its Kuiazhou launcher into SSO in 2018 and 2020. The second launch in 2020 failed. A record 16 MEO 55° inclination satellites were placed in orbit in 2018 as China accelerated its Beidou navigation system toward completion. With the launches in 2019 and 2020 the Beidou navigation system is now operational (See Annexure 10.1 for details).

Two satellites in the Beidou system experienced failures. These were the Beidou 1D (GSO) satellite launched in February 2007 and the Compass 2 G2 (GSO) satellite launched in April 2009. One of the experimental Centispace navigation satellite did not reach orbit because of a launch failure in 2020. All the navigation satellites of the Beidou system have been launched using the CZ 3A, CZ 3BE as well as the CZ 3C launchers. Most of the MEO satellites have been launched two at a time using the CZ 3BE launcher. A total of 43 launchers have been used to place the Beidou constellation in orbit.

10.2 Origins and Evolution of China's Navigation Satellite Programme

The idea of setting up a navigation system goes back to a 1983 proposal to develop a regional navigation satellite service using two satellites in GSO. The concept was tested out in 1989 using two in-orbit DFH 2 communications satellites. The formal programme started in 1993. The first two experimental satellites were placed in GSO in 2000 followed by two other satellites in 2003 and 2007, respectively.

© National Institute of Advanced Studies 2022
S. Chandrashekar, *China's Space Programme*,
https://doi.org/10.1007/978-981-19-1504-8_10

The concept to use satellites in GSO for the navigation service proposed by the Chinese is different from the concept adopted by the US, Russia as well as Europe.[1] These countries used a constellation of satellites in different inclination Medium Earth Orbits (MEO) that guaranteed that any user on the Ground could see at least four of them simultaneously. Using the precise ephemeris and time information transmitted by these satellites the user could fix position accurately. Two kinds of transmissions are used. One provides a relatively coarse fix and the other provides a more accurate position location. The use of atomic clocks on these satellites and precision tracking of the satellite constellation from the ground make for very accurate position fixing. Since the satellites were in different planes the innate accuracy of the system is high. In this system the receiver is passive and emits no signals making it immune to interference.

The early Chinese system did away with this concept completely. The satellites did not carry any atomic clocks. A ground control center periodically sent interrogation signals to users via the two GSO satellites. On receiving such a signal from any one of the GSO satellites the user terminal sent a reply signal to both satellites which then send them on to the control station on the ground. Using the Time Difference of Arrival (TDOA), the user terminal position is fixed in two coordinates of x and y. Using this information and suitable terrain maps the complete position information is determined and sent back to the user via the GSO satellites. The position determination accuracy achieved using this approach is about 100 m. With Ground-based augmentation from a set of stations this accuracy can reach up to 20 m.[2,3] In this approach since many to and fro transmissions are involved the system vulnerability to interference is high.

After the first phase which was experimental the Chinese went into a second phase. In this phase the first task was to make the GSO architecture operational. After this the idea was to extend the capabilities of the system for offering improved services. Following the launch of the fourth Beidou GSO satellite in 2007 China launched a MEO satellite into a 21,150 km 55° inclination orbit using a dedicated CZ 3A launcher. This was a precursor test satellite for an operational system. It indicated very clearly that China was moving toward an operational system that might comprise both GSO and MEO satellites. These launches were followed by five Beidou satellite launches in 2010 followed by three more in 2011. Five of these satellites were in a 55° inclination Geosynchronous orbit while three of them were in GSO.

In 2012 China launched two more GSO satellites and four MEO satellites. The MEO satellites were launched two at a time using a CZ 3BE launcher with the third stage engine being fired at least twice so that the satellites could be separated in the same orbital plane. 2015 saw the launch of four Beidou satellites. One was placed

---

[1] Lack of capabilities in key areas such as space based atomic clocks could also be a reason although Chinese writings do suggest capabilities in atomic clocks. The GSO and IGSO combination approach was also proposed by Japan. India's IRNSS is also based on the same approach. With inputs from GPS satellites accuracies can be improved significantly.

[2] https://directory.eoportal.org/web/eoportal/satellite-missions/c-missions/cnss.

[3] http://spaceflight101.com/spacecraft/beidou-3/.

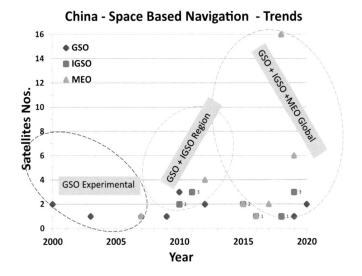

Fig. 10.1 China space-based navigation—trends

in an inclined Geosynchronous Orbit, another in GSO while the remaining two were placed in 55° inclined MEO orbit of 21,150 km. In 2016, 2017 and 2018 three, two and 18 Beidou satellites were launched into MEO, IGSO, and GSO, respectively. The constellation was completed with the launch of 10 satellites in 2019 and two in 2020. Figure 10.1 plots the timeline of the various navigation satellites launched by China.

As seen from Fig. 10.1 the experimental phase of China's navigation system started in 2000 with the launch of the two satellites needed for a minimal navigation function. To improve accuracy a third satellite was launched in 2003 followed by a fourth in 2007. In 2007 the Chinese also launched an experimental MEO navigation satellite into a 21,500 km, 55° inclination orbit. Since the 2007 GSO satellite developed problems, the Chinese launched a replacement satellite into GSO in 2009 which also seemed to have run into operational difficulties.[4] The launch of five Inclined Geosynchronous satellites between 2010 and 2011 suggests a move toward using a combination of GSO and IGSO satellites to provide navigation services over the Asia Pacific region. This was possibly the first phase in which a limited service was being provided.

The launch of 4 MEO 55° inclination satellites in 2012 followed by two more in 2015 and one in 2016 indicates that the Chinese were moving toward a GPS/Glonass/Galileo type orbiting constellation of navigation satellites in inclined MEO orbits. This tempo picked up steam with the launch of 18 MEO satellites in 2017 and 2018. Further launches into MEO, GSO, and IGSO completed the Beidou

---

[4] It is possible though unlikely that the Chinese were experimenting with different locations in the GSO to optimize the system. For details see http://www.zarya.info/Diaries/China/Beidou/Bei dou.php.

navigation constellation. The system caters to both legacy and new users The reception of MEO signals will provide accuracies very similar to those provided by the GPS system.

The original plan was to have a constellation of 35 satellites with five of them in GSO, three in Inclined Geosynchronous Orbit (IGSO) and 27 of them in 21,500 km 55° inclination orbits. There are currently 6 satellites in GSO, 7 satellites in IGSO and 27 satellites in MEO making a total of 40 operational satellites. A few of these may be orbital spares for any failures. The space component of the Beidou system is operational.

From the launch record the first four satellites are all GSO satellites launched by the CZ 3A in a dedicated mode. Since the CZ 3 A can place about 2300 kg in GTO, these first-generation satellites all had masses of about 2300 kg. The GSO satellite launched in 2009 also belongs to this generation.

The second generation of five GSO satellites has been launched from 2010 onwards. They are all launched by the bigger CZ 3C vehicle which can place 3800 kg in GTO.

The first generation of five IGSO satellites has used the CZ 3A launcher. Their masses in GTO will also be about 2300 kg.

On March 30, 2015, a CZ 3C launcher with the addition of another special restartable upper stage (YZ 1) placed another Beidou satellite in 55° inclination IGSO. The restartable upper stage can fire multiple times and take the satellite directly into orbit without using the satellite propulsion system. If such a stage is used the satellite mass can come down since no fuel is required to take the satellite from transfer to geosynchronous orbit. Reports suggest that this more compact Beidou satellite will have a mass of about 1000 kg. It will be different from the first five IGSO satellites because it will not carry the nearly 1000 kg of fuel that is required to take the satellite from GTO to IGSO. However, in both cases the effective payload in the final IGSO orbit may be about 1000 kg.

The second-generation IGSO satellites however appear to be different. The first of them was launched on September 29, 2015 by the larger CZ 3 B launcher (5100 kg GTO). Reports suggest that it carries a hydrogen maser based atomic clock and involves transmission signals that are compatible with GPS. The mass of the satellite at launch is reported to be about 4500 kg.[5]

After experiments with its first MEO Beidou satellite in 2007, China launched 4 Beidou satellites into 55° inclination 21,150 km MEO orbits in 2012. These were launched two at a time by China's most powerful CZ 3BE launcher (5500 kg in GTO). On July 25, 2015, another pair of MEO satellites were launched by CZ 3B with an YZ 1 upper restartable stage that could directly inject the satellites into the

---

[5] http://www.spaceflightinsider.com/missions/commercial/china-launches-long-march-3b-rocket-with-beidou-3-navigation-satellite/.

Table 10.1 Navigation satellite mass

Constellation component	GTO mass (Kg)	GSO mass (Kg)	IGSO mass (Kg)	MEO mass (Kg)
GSO generation 1	2300	1200		
GSO generation 2	3800	2000	2000	
IGSO generation 1	2300		1200	
IGSO generation 2	4500	2200		
MEO generation 1	2400			1200

required MEO. This would translate into satellite masses of about 1200 kg in their final MEO orbits.

From the analysis of the launch data and other sources of information certain inferences can be made about the masses of the various Beidou satellites. Table 10.1 provides a summary of these masses.

All generation 1 satellites have an in-orbit GSO mass of around 1200 kg. All second-generation GSO and IGSO satellites have an in-orbit mass of around 2000 kg.

While the orbiting part of the constellation mimics the US GPS System, the GSO and the IGSO component provides continuity with the earlier navigation services that may also include some civilian uses. One would expect that over time the GSO and IGSO components to be phased out and replaced by a MEO constellation of satellites like the GPS.

Publicly available reports suggest that all the first-generation Beidou satellites were based on the DFH-3 satellite bus. Second generation GSO and IGSO Beidou satellites may be based on the larger DFH 4 bus. The use of a restartable upper stage YZ 1 on the CZ 3C and CZ 3B launchers makes direct delivery to the required orbit possible. This will mean that the second-generation satellites may be based on components of the DFH 3 bus but will be a more compact satellite without the large propellant tanks that are required for moving satellites from transfer to final orbits. The new restartable YZ 1 upper stage is also compatible with the much bigger CZ 5 launcher that is under development. This launcher is now available for the Beidou satellite launches.

In 2014 the Beidou Navigation Satellite System officially became a part of the Global Radio navigation System. China thus became the third country after the US and Russia to offer a global satellite-based navigation service to customers. As a part of this move to make sure that the Beidou system is fully compatible with GPS and Galileo, the civil signals will shift to the same frequency bands and use the same modulation schemes as GPS and Galileo.

10.3   Assessment of System Capabilities (Chandrashekar 2015)

The preliminary evaluation of system performance is consistent with other comparable systems in the world (Montenbruck 2013). The parameters for assessing performance of satellite-based navigation systems would be linked to position accuracy, velocity accuracy, and timing accuracy. The position accuracy along with the mass of the satellite can be used as the basis for evaluating capability[6] (Beutler 2003).

In view of the importance of navigation for meeting security needs many space powers are keen to establish their own system. Currently the US GPS and the Russian Glonass system offer global coverage.[7] The European and Chinese systems are also moving toward global coverage with the additional launchings of many MEO satellites.

The use of satellites for navigation purposes has been around for quite some time. The US GPS system set a new standard for accurate position fixing using a constellation of 24 satellites in Medium Earth Orbit. It has now become a global standard. Here too the distinctions between civilian and military use are becoming increasingly blurred. Both the GPS and Glonass systems work on the principle that at least four satellites whose positions in space are known accurately are always available to a receiver on the ground. Using the signals along with the known positions of the satellite the position of the receiver can be estimated.

The GPS satellites are in a near circular orbit at an altitude of about 20,000 km with an inclination of 55°. Three generations of satellites comprise the current constellation. The masses of the satellites at lift-off range from 1630 to 2200 kg. The current generation has a lift-off mass of 1630 kg. The lifetime of these satellites is 10 years (UCS 2021).

The Russian Glonass system consists of 29 satellites in the current constellation (UCS 2021). They are in 19,100 km orbit with an inclination of 64.8°. Starting in 1982, the first complete constellation of 24 satellites was established in 1995 about a year behind the US GPS system. After the breakup of the Soviet Union the Russian operational capability got diluted. The programme got a revamp in 2001. Complete coverage was restored in 2011.

The Glonass programme has also seen three generations of navigation satellites. The first-generation satellites had lift-off masses of about 1260 kg and were launched three at a time on the Proton launcher. Direct injection into the required orbit is a feature of the Proton Soyuz launches. The propellant mass on the satellite would therefore be less resulting in lower satellite mass in comparison to the launches from Western countries. The second-generation satellites had a lift-off mass of 1415 kg. The mass of the third-generation satellites is 930 kg.[8] The Glonass system is available

---

[6] pp 3–19.

[7] https://www.princeton.edu/~alaink/Orf467F07/GNSS.pdf.

[8] For details see http://www.russianspaceweb.com/uragan.html.

for civilian use. Most commercial navigation receivers are compatible with both GPS and Glonass. The higher inclination orbits provide better coverage at higher latitudes.

Europe is planning to set up its own Galileo Network of orbiting satellites to provide navigation services to both Europe and the rest of the world. It is in the process of establishing a constellation of 30 satellites with 24 operational and six backup in-orbit spares. The satellites are distributed equally in three orbital planes. They orbit the earth at an altitude of 23,222 km with an inclination of 56°. 25 of these satellites are already in orbit. The space segment is now operational. The satellites have a lift-off mass of 733 kg. They are launched two at a time by the Soyuz Rocket from Kourou in French Guiana. With modifications being carried out to the Ariane 5 launcher they could be launched four at a time on future flights of the Ariane launcher.[9]

Japan is a country where high mountains and high buildings make access to good quality reception from GPS satellites difficult. This resulted in a different architecture for their satellite constellation. Rather than competing directly with the GPS they chose an approach that would complement and enhance GPS capabilities within their region. The system should also serve as a backup in case there were problems with the availability of the GPS system.

The Japanese system involves a constellation of seven satellites. Four of them are in inclined geosynchronous orbits. The orbit inclination is chosen in such a way that at least one satellite will always be visible to users in Japan. These along with three satellites in GSO will provide the four signals for locating the position on the ground. These satellites will transmit position and time and error correction in the same bands and frequencies as the GPS system[10] (Takahashi 2004).

Japan has so far launched 4 Quasi-Zenith Satellite System (QZSS). The first satellite was launched in 2010 followed by 3 launches in 2017. All of them were launched by the H2A launcher. The lift-off mass of the satellites is 4000 kg. Three of them are in geosynchronous orbit with an inclination of 40° while the fourth is in a Geostationary Orbit (UCS 2021). Reports suggest that with this satellite in place and with data that comes from precisely located receivers on the ground whose separation distances are accurately known error corrections received by users via the geosynchronous satellite has improved location accuracies from the meter to the centimeter level. Japan proposes to augment its capacity with the launch of 3 more inclined geosynchronous satellites as well the three satellites in GSO in the next few years to complete the system.[11]

India's efforts at improved navigation uses two approaches. In the first approach the existing civilian capabilities using GPS are augmented to improve the accuracy of position fixing and navigation. The Indian system called Gagan uses a precisely positioned network of ground stations to eliminate some of the systemic errors in the

---

[9] http://www.esa.int/Our_Activities/Navigation/The_future_-_Galileo/What_is_Galileo.

[10] pp 259–264.

[11] http://www.theregister.co.uk/2014/04/24/japan_satellite_qzss_enhanced_gps/.

Table 10.2 Global satellite navigation systems—a comparison

Country	USA	Russia	Europe	China	Japan	India
System	GPS	Glonass	Galileo	Beidou	QZSC	IRNSS
First launch	1975	1982	2005	2000	2010	2013
Service area	Global	Global	Global	Global	Region	Region
Operations	1995	1996	2015	2015	2010	2016
Satellites	31	24	27	35	7	7
GSO				5	3	3
IGSO				3	4	4
MEO	31	24	27	27	NA	NA
No of planes	6	3	3	3	NA	NA
Inclination°	55°	64.8°	56°	55°	40°	30°

reception of GPS signals. These error corrections are transmitted continuously to all users within the region via a geostationary satellite who then correct the GPS based positions. This system is being deployed in India to facilitate precision landing of aircraft[12] (Kibe 2003).

India is also in the process of setting up an Indian Regional Navigation Satellite Service (IRNSS). When completed, the system will have four satellites in a 30° inclination geosynchronous orbit and three other satellites in Geostationary Orbit. Five satellites are currently operational (UCS 2021). Together they will provide an India oriented navigation service that caters to both civilian and military uses. The system being set up may also be compatible with additional enhancements based upon improvements from a precision-based ground network set up for the Gagan service.[13] The Indian system is very similar to the Japanese and the initial Chinese systems that use a combination of Geostationary and inclined geosynchronous satellites.

Table 10.2 provides an overview of the current and future satellite navigation systems that are likely to be fielded by various countries.[14]

The US GPS system is the oldest and the most advanced global system. It is also the system that has the largest user base in the world which includes very large numbers of commercial users. The Russian as well as the European and Chinese systems when completed will provide comparable accuracies to the GPS system. China may phase out its GSO / IGSO system over time as its MEO satellites become operational. The Japanese regional system when complete is not a stand-alone system but depends on the GPS for its applications. The Indian regional system if used in a stand-alone mode may not provide the same level of accuracies as the other global systems. This may force them to set up a system like the GPS.

---

[12] pp 1405–1411.

[13] http://www.isro.gov.in/sites/default/files/pdf/pslv-brochures/PSLVC22.pdf.

[14] Compiled from various sources including https://directory.eoportal.org/web/eoportal/satellite-missions/c-missions/cnss.

10.4 The Standards Battle—Who Will Win?

Though military and strategic needs have influenced the development of these systems, their extension into the civilian sphere will have a major impact on the sustainability of each system. Given the importance of a large civilian base of users for sustaining any system, what will the outcomes of increased competition between these systems at the global level?

The competitive dynamics between these alternative systems are governed by what is commonly termed as "network economics"[15] (Shapiro 1999). In such situations the standard that has the largest user base has a significant advantage and will attract a larger share of the users requiring such services. This would suggest that the GPS standard will continue to dominate the global navigation services market. Reversing this advantage will require a major effort and may also involve a significant subsidy to promote civilian applications.

Given this scenario, China's Beidou/Compass Navigation System is the most likely challenger to the current dominant position of the US GPS System. China has made all the right moves by registering its standard and becoming a part of the Global Radio Navigation System. It has also made its service compatible with both the GPS and Galileo Standards. By doing so it has signaled its intentions of becoming a global provider of satellite-based navigation services. Once its system becomes fully operational one could expect some aggressive moves from it to promote its standard and wooing customers away from the ubiquitous GPS standard.

Whether these moves will payoff remains a moot question. China is possibly the main challenger to the continued US domination of the global satellite navigation services market.

Case Study 1: International Collaboration and Technology Acquisition in Navigation

Like in other areas China before embarking on its own navigation satellite programme also explored possibilities of international collaboration to obtain key technologies.

In 2003 it signed an agreement with the European Union (EU) for the joint development of the EU Galileo Satellite-based Navigation Programme. The Chinese decided to provide 200 million Euros ($228 million at that time) to help the EU jointly develop the Galileo System (Lague 2013). The US raised several objections to this agreement regarding the transfer of sensitive technologies. It has continued to voice these concerns since that time.

The EU which had serious internal problems in raising the money required for the Programme saw China's financial offering as a heaven-sent opportunity to push ahead with the project. China set up a special company called

---

[15] The best-known example of such competitive dynamics and its outcomes is the operating system battle of standards in the PC market.

China Galileo Industry to coordinate the projects under the collaboration. Shareholders in this company included China Aerospace Science and Industry Corporation (CASC) a major satellite and missile producer, as well as the China Academy of Space Technology (CAST). Major European companies involved in the projects included EADS as well as Thales. The National Remote Sensing Center provided project oversight over 12 contracts worth 33 million Euros.

The Galileo project experienced major delays arising from the need to get a viable consensus on funding and management of the Project. Whether it was because of these delays or whether it was the original Chinese plan, China went ahead and started deploying its own indigenously developed system.

While it is not very clear whether any major technology transfer took place, there is evidence to suggest that China did gain considerably in getting technology for making atomic clocks.

Apparently, their effort to get Rubidium clocks from EADS was not successful. However, they did buy 20 Rubidium atomic clocks between 2003 and 2007 from the Swiss company Temex Time now known as Spectra time. Reports also suggest that China had initiated a Rubidium atomic clock development programme at the No. 203 Institute a research institute under the China Aerospace Corporation starting from 2004 onwards. These reports also state that ten of these clocks have been built and are flying on Beidou satellites.

Galileo became a 100% taxpayer financed project in 2010. This made sure that collaboration and partnerships between EU and China on navigation satellite systems effectively came to an end.

Apart from the technology transfer issue since both the Beidou and Galileo systems use the same frequencies for transmission there are also interference issues between the two parties. It is also clear that Beidou and Galileo will both compete in the civil applications market. China with its lower costs and state subsidized pricing may have an advantage.

Once again it appears that China has a clear long-term strategy for addressing many key areas of use that involve the development of hi-tech products.

10.5   Pushing Innovation

On November 9, 2016, a new generation CZ 11 launcher placed a 240 kg X-ray Pulsar Navigation Satellite in a 500 km SSO. This launch was from a canister carried on a mobile transporter erector vehicle. The purpose behind this satellite is to use X-ray emissions from pulsars for on orbit satellite navigation. This is the first satellite of its kind to test out this novel system which could replace an atomic clock-based system. The launching of this satellite typifies a new Chinese approach to becoming a "state-of-the-art" player in space systems.[16]

In 2018 and 2020, China also launched two experimental navigation satellites into an SSO orbit using the Kuaizhou launcher. These small satellites are reported to have been developed by the Innovation Academy of Microsatellites of the CAS. These satellites are to test low orbit navigation enhancement being developed by the Beijing Future Navigation Technology Company.

Annexure 10.1: List of Navigation Satellites Launched by China till End 2020

Satellite	Launch date	Launcher	Launch site	Orbit
Beidou 1A	10/30/2000	CZ 3A	Xichang	GSO
Beidou 1B	12/20/2000	CZ 3A	Xichang	GSO
Beidou 1C	5/24/2003	CZ 3A	Xichang	GSO
Beidou 1D	2/2/2007	CZ 3A	Xichang	GSO
Compass-M1	4/13/2007	CZ 3A	Xichang	MEO
Compass-2 G2	4/14/2009	CZ 3A	Xichang	GSO
Compass-2 IG1	7/31/2010	CZ 3A	Xichang	IGSO
Compass-2 IG2	12/17/2010	CZ 3A	Xichang	IGSO
Compass-2 IG3	4/9/2011	CZ 3A	Xichang	IGSO
Compass-2 IG4	7/26/2011	CZ 3A	Xichang	IGSO
Compass-2 IG5	12/1/2011	CZ 3A	Xichang	IGSO
Beidou 3 M1	7/25/2015	CZ 3B	Xichang	MEO
Beidou 3 M2	7/25/2015	CZ 3B	Xichang	MEO
Beidou 12 S	9/29/2015	CZ 3B	Xichang	IGSO
Compass-2 G1	1/16/2010	CZ 3C	Xichang	GSO
Compass-2 G3	6/2/2010	CZ 3C	Xichang	GSO
Compass-2 G4	10/31/2010	CZ 3C	Xichang	GSO
Compass-2 G5	2/24/2012	CZ 3C	Xichang	GSO
Beidou G6	10/25/2012	CZ 3C	Xichang	GSO
Beidou 3 I1	3/30/2015	CZ 3C	Xichang	IGSO
Beidou M3	4/29/2012	CZ 3BE	Xichang	MEO

(continued)

---

[16] https://chinaspacereport.com/launch-vehicles/cz11/.

(continued)

Satellite	Launch date	Launcher	Launch site	Orbit
Beidou M4	4/29/2012	CZ 3BE	Xichang	MEO
Beidou M2	9/18/2012	CZ 3BE	Xichang	MEO
Beidou M5	9/18/2012	CZ 3BE	Xichang	MEO
Beidou-3 M3	2/1/2016	CZ 3C	Xichang	MEO
Beidou-2 I3-S	3/29/2016	CZ 3A	Xichang	IGSO
Beidou-2 G7	6/12/2016	CZ 3C	Xichang	GSO
BeiDou-3 M1	11/5/2017	CZ 3BE	Xichang	MEO
BeiDou-3 M2	11/5/2017	CZ 3BE	Xichang	MEO
BEIDOU 3M7	11/1/2018	CZ 3BE	Xichang	MEO
BEIDOU 3M8	11/1/2018	CZ 3BE	Xichang	MEO
BEIDOU 3M3	12/2/2018	CZ 3BE	Xichang	MEO
BEIDOU 3M4	12/2/2018	CZ 3BE	Xichang	MEO
BD-3 M9	29/3/2018	CZ 3BE	Xichang	MEO
BD-3 M10	29/3/2018	CZ 3BE	Xichang	MEO
BeiDou-2-I7	9/7/2018	CZ 3A	Xichang	IGSO
BD-3 M5	29/7/2018	CZ 3BE	Xichang	MEO
BD-3 M6	29/7/2018	CZ 3BE	Xichang	MEO
BEIDOU 3 M11	24/8/2018	CZ 3BE	Xichang	MEO
BEIDOU 3 M12	24/8/2018	CZ 3BE	Xichang	MEO
BD-3 M13	19/9/2018	CZ 3BE	Xichang	MEO
BD3 M14	19/9/2018	CZ 3BE	Xichang	MEO
CentiSpace1S1	29/9/2018	Kuaizhou-1A	Jiuquan	SSO
BD-3 M15	15/10/2018	CZ 3BE	Xichang	MEO
BD-3 M16	15/10/2018	CZ 3BE	Xichang	MEO
BEIDOU 3G1	01/11/18	CZ 3BE	Xichang	GSO
BD-3 M17	18/11/2018	CZ 3BE	Xichang	MEO
BD-3 M18	18/11/2018	CZ 3BE	Xichang	MEO
Beidou 3I	20/04/2019	CZ 3BE	Xichang	IGSO
Beidou 2G8	22/5/2019	CZ 3C	Xichang	GSO
Beidou 46	24/6/2019	CZ 3BE	Xichang	IGSO
BD-3 M23 (47)	22/9/2019	CZ 3BE	Xichang	MEO
BD-3 M24 (48)	22/9/2019	CZ 3BE	Xichang	MEO
BD3- 3I	4/11/2019	CZ 3BE	Xichang	IGSO
BD-3 M21 (50)	23/11/2019	CZ 3BE	Xichang	MEO
BD-3 M22 (51)	23/11/2019	CZ 3BE	Xichang	MEO

(continued)

(continued)

Satellite	Launch date	Launcher	Launch site	Orbit
BD-3 M19 (52)	16/12/2019	CZ 3BE	Xichang	MEO
BD-3 M20 (53)	16/12/2019	CZ 3BE	Xichang	MEO
BEIDOU 3 G2	9/3/2020	CZ 3BE	Xichang	GSO
BD 3G	23/6/2020	CZ 3BE	Xichang	GSO
CentiSpace-1S2	10/7/2020	Kuaizhou-1A	Xichang	Failure

References

Beutler Gerhard (2003) Satellite navigation system for earth and space sciences. International Space Sciences Institute, Spatium No. 10, June 2003, http://www.issibern.ch/PDF-Files/Spatium_10.pdf

Chandrashekar S, Space war and security—a strategy for India, http://eprints.nias.res.in/943/1/2015-R36.pdf

Kibe SV (2003) Indian plan for satellite-based navigation for civil aviation. Curr Sci 84(11)

Lague David (2013) Special report—in satellite tech race china hitched a ride from Europe at http://www.reuters.com/article/breakout-beidou-idUSL4N0JJ0J320131222

Montenbruck O et al (2013) Initial assessment of the COMPASS/BeiDou-2 regional navigation system, http://saegnss1.curtin.edu.au/Publications/2013/Montenbruck2013Initial.pdf

Shapiro C, Hal VR (1999) Information rules—a strategic guide to the network economy. Harvard Business School Press, Boston, Massachusetts

Takahashi Hideto (2012) Regional navigation satellite system the JRANS concept. J Global Posit Syst 3, 1–2, http://www.sage.unsw.edu.au/wang/jgps/v3n12/v3n12p32.pdf

Union of Concerned Scientists (2021) UCS satellite data base, May 2021, https://ucsusa.org/resources/satellite-database

Chapter 11
The Human Space Flight Programme

11.1 Background

As discussed earlier a human space programme had been an integral part of China's Space ambitions almost from the very beginning. However, it was only in 1992 that there was a final congruence of opinion that gave a go-ahead to the human space flight effort. The Human Space Flight Programme codenamed 921 Project was kept under wraps for quite some time. The original roadmap involved a very ambitious approach culminating in the realization of a Soviet MIR type of space station by 2010. The programme went through several iterations that lowered the scope as well as the schedule of the 921 Project.

The Human Space Flight Programme as it is currently configured involves three clear phases.

The first stage involves placing humans in LEO using Shenzhou spacecraft.

The second phase involves the development and testing of the technologies and techniques needed for building a space station. These include EVA, orbital rendezvous and docking capabilities. It also involves two human tended single module space laboratories launched as technology demonstrators for the future space station.

The final phase involves the establishment of a 60-tonne space station in LEO.[1] This station would be able to support the stay of 3 astronauts for up to 6 months in orbit. Shenzhou space craft would be used to ferry crews to and fro from the space station. Unmanned cargo vessels would be used for replenishing supplies to the space station. According to the plan this should happen by 2020. More recent indications are that this would happen by 2022 or even a little later.

China has made significant progress and completed phases 1 and 2 of the programme. The realization of phase 3 involved the development of a new launcher the CZ 5 as well the construction of the final space station in orbit. The failure of the second flight test of the CZ 5 launcher in July 2017 delayed in the establishment

---

[1] More recent reports suggest that the final space station will be about 60 tonnes.

© National Institute of Advanced Studies 2022
S. Chandrashekar, *China's Space Programme*,
https://doi.org/10.1007/978-981-19-1504-8_11

of the Space Station. Its return to service in 2019 was followed by the successful launching and testing of an improved crew module though the test of an inflatable heat shield during this same mission was not successful, China is now in position to go-ahead with its establishment of a permanent station in Low Earth Orbit (LEO). The operational date is likely to be 2022–2023.

The Tiangong 1 Space Lab which was a part of the second phase of the programme also made an uncontrolled but safe reentry in April 2018 over the South Pacific (Phillips and Bonnie 2016). The event did create concerns of the likely damage on earth from large pieces of space debris.

China launched a second Tiangong Space laboratory very similar to the first one but with some modifications on September 15, 2016. This was followed by the launch of a Shenzhou spacecraft with a crew of two astronauts on October 16, 2016 (Barbosa 2016). The spacecraft docked with the Tiangong 2, and the astronauts spent 30 days in the space station. This is double the time that the astronauts on-board the Shenzhou 10 spacecraft spent at the Tiangong 1 station. A small 40 kg Banxing 2 satellite, equipped with a full frame 25 Megapixel visible camera was launched from the Tiangong 2 space station. The purpose of this satellite is to provide images of the space station as well as for SSA functions. The satellite was developed by CAST for the CAS.

The Tianzhou unmanned ferry to China's future space station was successfully launched by a CZ 7 rocket in April 2017. It successfully docked several times with the unmanned Tiangong 2 space laboratory and launched a small satellite before making a planned destructive reentry. Though the successful rendezvous and docking clears most of the hurdles for the realization of the space station, a repeat launch of the CZ 7 in March 2020 failed. This may have some impact on the schedule for the space station. The return to flight of the CZ 5 however does ensure that the space station will become a reality in the next two years.

11.2   The Shenzhou Spacecraft

Though the Shenzhou spacecraft look like the Russian Soyuz spacecraft, it is a completely indigenous Chinese design. China did get some help from Russia on key technologies required for the human space flight effort. However, all the capabilities required for a sustained human space flight effort appear to have been created indigenously (please see Sect. 11.6 of this book).

Like the Soyuz, the Shenzhou consists of three modules. The lowermost module is the service or propulsion module. It houses the reentry rockets as well as the maneuvering engines. There are two solar panels mounted on either side that provide power. It also houses the radiators to dissipate the heat generated.

The descent module with space for three astronauts is mounted on top of the service module and constitutes the middle section of the Shenzhou spacecraft. It is shaped like a beehive. It has a hatch at the top, two portholes as well as a sighting window. The main and reserve parachutes needed for the descent phase are housed within this module. The temperature within the cabin will be in the range of 17–25 degrees Centigrade with a humidity level that is between 30 and 70%.

The orbital module comes on top of the descent cabin. It is sufficiently large to cater to the basic needs of the astronauts. Sleeping bags mounted on the walls of the orbital module are available for use. The astronauts can use a sealed plastic tent in the module for taking a shower. The orbital module also has place for performing experiments. Separate solar panels mounted on either side provide an independent source of power. It also has independent propulsion capabilities. When the human part of the mission is over, the orbital module is separated from the descent module and remains in orbit for a longer time. This enhances the capabilities of these missions not only to conduct a variety of experiments in the space environment but also enables the testing of new technologies.

Table 11.1 below provides the basic features of the Shenzhou spacecraft and the

Table 11.1 Comparison of Shenzhou and Soyuz spacecraft

Parameters	Shenzhou Spacecraft	Soyuz TM spacecraft
Complete spacecraft		
Mass (tonnes)	7.8 tonnes	7.21 tonnes
Length (meters)	9.15 m	6.98 m
Diameter (meters)	2.8 m	2.6 m
Propulsion module		
Total mass (tonnes)	3 Tonnes	2.95 Tonnes
Propellant mass (tonnes)	1.1 Tonnes	0.9 Tonnes
Length (meters)	2.94 m	2.3 m
Diameter (meters)	2.8 m	2.2 m
Base (meters)	2.8 m	2.72 m
Solar panels (nos. & area)	Two panels −24 sq. m area	Two panels
Descent module		
Mass (tonnes)	3.2 Tonnes	3 Tonnes
Length (meters)	2.5 m	1.9 m
Diameter (meters)	2.5 m	2.17 m
Orbital module		
Mass (tonnes)	2 Tonnes	1.3 Tonnes
Length (meters)	2.8 m	2.2 m
Diameter (meters)	2.8 m	2.25 m
Solar panels (aos. & area)	Two panels −12 sq. m area	None

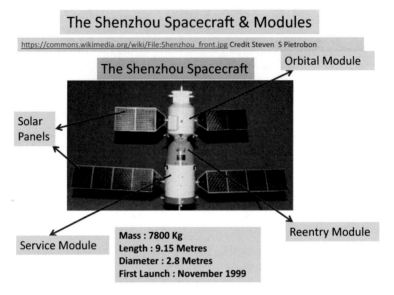

Fig. 11.1 The Shenzhou spacecraft with modules

three modules that comprise it. Corresponding data for the Soyuz spacecraft is also provided for comparison purposes[2] (Harvey 2013).

Figure 11.1 provides an overview of how the final assembled Shenzhou spacecraft functions in orbit.

11.3   The Tiangong Space Laboratory

The second phase of China's human space effort involves the docking of the Shenzhou spacecraft with the Tiangong space station. This would be followed by astronauts moving into the station staying there for some time before coming back to earth. During their stay they perform several activities that include the conduct of experiments.

The first Tiangong Space Station was put into orbit on September 29, 2011. It remained functional till March 2016. In September 2016 a second Tiangong Space Module was placed in LEO followed by a docking of the Shenzhou spacecraft carrying two astronauts. A second Banxing satellite was also launched from the space station. Like the Shenzhou spacecraft, the Tiangong consists of three sections—an aft service module, a transition section, and the final habitable space module.[3]

---

[2] pp. 267–271.

[3] http://spaceflight101.com/spacecraft/tiangong-1/.

The Tiangong service module is cylindrical in shape with a length of about 3.3 m and a diameter of 2.5 m. There are two solar arrays mounted on the exterior. They generate an average power of 2500 Watts with a peak power of 6000 Watts. Silver Zinc batteries are also located in this section to provide power during eclipse conditions.

There are two main engines that operate on Mono Methyl Hydrazine (MMH) and Nitrogen Tetroxide (N2O4). Vernier engines mounted along the periphery are used for carrying out fine maneuvers. Additional thrusters are available for roll, pitch, and yaw control. There are four propellant tanks each of which can carry 1000 kg of propellant. There are also smaller high pressure gas tanks for pressurizing the propellants.

Between the service module and the orbital module there is a small transition section that flares outward in diameter to connect the 2.5 m diameter service module with the 3.35 m diameter habitable module. This space holds the oxygen and nitrogen gas tanks as well as the water tanks that provide air and water to the astronauts in the habitable module.

The Habitable module provides working and living space for visiting crews. The habitable module is 5 m long with a diameter of 3.5 m. It contains basic scientific equipment and provides space for mounting mission specific equipment.

External radiators mounted on the aft end of the module dissipate the nearly 2000 Watts heat generated inside the space station. Two sleep cubicles are available for use by visiting astronauts. If there is a third astronaut, he or she would have to sleep in the Shenzhou spacecraft.

The module uses a two-color paint scheme representing the ground and the sky to help astronauts orient themselves when working. Tiangong may not have toilet or kitchen facilities. However, facilities available in the Shenzhou could be used by the astronauts.

High-resolution cameras provide live coverage to Mission control. A docking system in the front of the module enables the linking up of the Tiangong with the Shenzhou spacecraft.

The basic Tiangong comes equipped with a set of scientific instruments that can be operated without crews. A suite of earth observation instruments including a hyperspectral imager are mounted on the outside.

There is also a material science facility for studies related to crystal growth in space. Imagery and video links provide information on the progress of crystal growth to scientists on the ground. Tiangong 1 also carries instruments for detecting and analyzing energetic particles from the sun as well as instruments for measuring atmospheric properties and ionospheric disturbances.

Figure 11.2 provides an overview of the Tiangong Shenzhou composite operations.

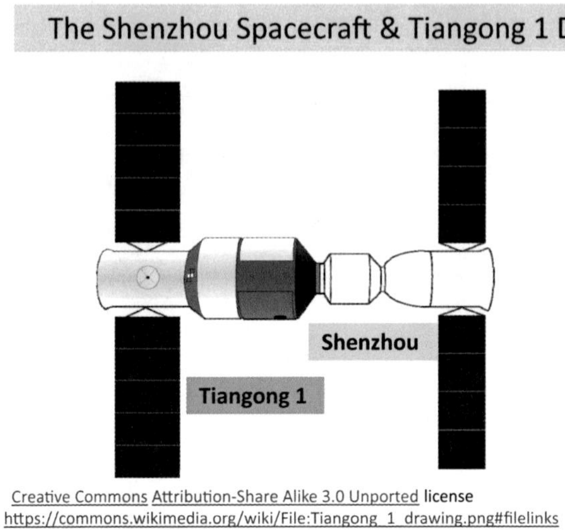

The Shenzhou Spacecraft & Tiangong 1 Docking

Shenzhou

Tiangong 1

Fig. 11.2 The space station tiangong 1 docking

11.4   The Space Station

Phase 3 of the current Chinese Human Flight programme involves the establishment of a permanent station in near earth orbit. Most of the elements needed for constructing the station have been tested out.

These include several Shenzhou missions that tested all the human activities for the mission as well as the launch of the Tianzhou cargo ferry and its docking and undocking with the Tiangong 2 Space Laboratory.

All other support activities including space walks, experimental set ups, etc. have been tested out. The successful return to flight of the CZ 5 launcher will now accelerate the establishment of the station.

11.5   The CZ 2F, CZ 5, and CZ 7 Launchers[4]

The launcher used for the programme was based on the proven pedigree of the CZ 2 series of launchers. The basic CZ 2C launcher had been modified with the addition of liquid strap on boosters for launching communications satellites into GTO. Figure 11.3 shows a typical lift-off of a CZ 2F launcher with a Shenzhou spacecraft.

The CZ 2F launcher used for the programme was a modified form of this launcher. The major addition was the addition of an escape tower based on a Russian Soyuz

---

[4] For more details see http://spaceflight101.com/spacerockets/long-march-2f/.

CZ 2F Liftoff with the Shenzhou Spacecraft

Source: Creative Commons Attribution 3.0 Unported license. Credit: China News Service

The Shenzhou 12 launch by the CZ 2F Rocket on June 17, 2021

Shenzhou 12 carried 3 Taikonauts

Docked with the Tianhe core module of the Tiangong Space Station.

The Reentry Module of the Shenzhou spacecraft returned to earth on September 17, 2021, after 92 days in orbit

https://commons.wikimedia.org/wiki/File:Shenzhou_12_launch_(cropped).jpg

Fig. 11.3   CZ 2F launch of shenzhou spacecraft

design to take care of any problems in the initial takeoff and ascent portion of the flight. Engineering changes were also made to the original configurations to enhance the reliability of the launcher for human flight. A new fairing to accommodate the Shenzhou spacecraft and the escape tower had also to be developed and tested.

Most of these changes and the testing were completed before the Shenzhou spacecraft had been readied for its first flight. A new launch pad for the CZ 2F was also constructed at Jiuquan for the Shenzhou and Tiangong launches.

The launcher to be used for the construction of the space station is the new CZ 5 launcher. The launcher for ferrying supplies to the Space Station will be the CZ 7. For technical details of these launchers please refer to Chap. 14.

11.6   Tracking Telemetry and Command (TT&C) Capabilities

The human space flight effort also involved a significant augmentation of China's TT&C facilities. China initially depended on a large network of domestically located TT&C stations for various satellite, missile, and launcher missions. These were augmented suitably over the years depending on the needs. For TT&C operations needed outside of China, it largely depended upon shipborne TT&C stations that could be placed at suitable locations on the high seas. These shipborne TT&C terminals were called Yuan Wang.

For catering to the human space flight requirements China had to significantly augment its shipborne TT&C stations. It also had to establish or use TT&C facilities located in other countries.

Two Yuan Wang TT&C ships (Yuan Wang 1 & 2) catered to Chinese needs in the 1980s. For the initial phase of the human programme two more Yuan Wang (Yuan Wang 3 & 4) were added in 1995 and 1999. Two additional ships entered service in 2007 and 2008 for the phase 2 part of the human programme that involved docking with the Tiangong space station.

To resolve some of the operational issues and to reduce the cost and complexity of operations China also decided to set up or use TT&C facilities located outside of China. After operating a tracking station in the South Tarawa Atoll from 1997 to 2003 China established its first overseas station at Swakopmund in Namibia. This location is crucial for monitoring the Shenzhou reentry operations. This was followed by an agreement with Sweden for using TT&C stations located in Sweden and Norway. A station in Karachi (monitoring reentry) followed in 2003. In the same year the Italian Space Agency also offered the use of their Malindi site in Kenya for TT&C operations. Later another site Dongara in Australia was added to the network[5] (Harvey 2013).

In addition to these ground-based and sea-based tracking facilities, China also decided to augment its capabilities to monitor and communicate with the Shenzhou and Tiangong spacecraft by placing Tracking & Data Relay Satellites (TDRS) in Geostationary Orbit (GSO).

Three Tianlian Tracking and Data Relay satellites were launched by China in April 2008, August 2011, and August 2012 respectively. They were meant to support the human Shenzhou missions. The three satellites are located at 77 degrees E, 176.8 degrees E and 16.7 degrees E. In November 2016 and March 2019 China launched additional Tianlian Tracking and Data Relay Satellites to replace the earlier generation of satellites.

11.7   The Organization of the Human Space Flight Programme

The decision to launch a human into space needed a major expansion in capabilities across several areas[6] (Aliberti 2015). The major areas that needed to be addressed were

- The development and testing of the Shenzhou spacecraft;
- The development and testing of the Tiangong space station;
- The development of the CZ 2F launcher;
- New Facilities at the Launching site;

---

[5] pp. 60–65.
[6] pp. 83–84.

- Facilities at the Landing site;
- Expansion of the TT&C networks including the establishment of shipborne and overseas stations;
- Developing and launching the Tianlian Tracking and Data Relay Satellites;
- The selection and Training of astronauts;
- Development of the Space Science and Application Payloads;
- Mission Planning;
- Project and Programme Management.

It is evident from the above requirements that a complex organization structure is needed to make sure that the various components mesh to realize the final mission. A new organization structure was put in place so that the project could be realized quickly. Figure 11.4 provides the organization structure put in place to manage the human space flight programme.

The human programmme has a two-layered management structure that has been designed to make sure that all relevant organizations and entities needed for the formulation and implementation of the programme are involved in providing guidance for the programme.

The human programme first came under the immediate purview of the General Armaments Division (GAD) of the Central Military Commission (CMC) which represented the PLA at the highest level of decisión-making in China. Along with the Politburo and the State Council the CMC is involved in all major decisions taken by China.

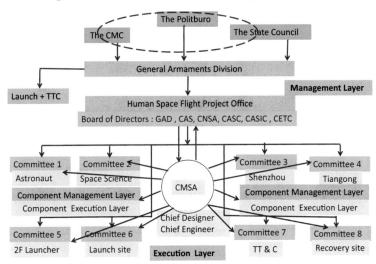

Fig. 11.4 Organization structure—human space flight programme

For the human space programmme a Human Space Flight Office was created under the GAD. Its major component is a Board of Directors that is responsible for overall management of the human programme. Its main function is to provide guidance and direction to the programmme. It may also be the entity that formulates and provides strategic direction to the programmme as it evolves over time.

The re-organization of the PLA in 2015 saw the abolition of the GAD. It is likely that the Board of Directors and the Human Space Flight Office now directly come under the CMC.

This Board comprises high-level administrative officials representing the different stake holders involved in the programme. A Deputy Director from the GAD/ CMC, high-level officials from the Chinese Academy of Sciences (CAS), top representatives of the China National Space Agency (CNSA), representatives from the China Aerospace Science & Technology Corporation (CASC), China Aerospace Science & Industry Corporation (CASIC), and the China Electronics Group Corporation (CETC). This Board through the Human Space Flight Office reports to the State Council and the Politburo. There are 8 Committees directly under the Board that are each responsible for a major component of the human space programmme. Each of these components has a lead agency that is responsible for the implementation of the project.

Each Committee also has two layers associated with the implementation of the project. One layer is the management layer that takes all direction and strategy decisions. The other layer is a technology and project implementation layer that has a Chief Designer as well as a Chief Engineer heading it.

Both layers will also have representatives from other entities involved in the project. A China Manned Space Agency ensures coordination within and between the various component projects. It most probably has an overall Chief Designer and a Chief Engineer to provide the key connections with the major components of the programme. It directly links up with the Board and the Human Space Flight Office to provide the necessary inputs for overall guidance and decisión-making. The CMSA and the 8 Committees are the major entities responsible for the implementation of the Project while the Board provides strategic direction. Table 11.2 provides details of the 8 major components of the human space programme and the lead organizations involved.

Annexure 1 provides details of the various Shenzhou missions.

11.8 Benchmarking China's Space Station

While China's human spaceflight has evoked admiration and excitement around the world it may be useful to compare China's achievements with those of the other two major space powers the USA and Russia. While there are several parameters that may be needed for making this comparison a simple yardstick to examine the capabilities would be the mass of the station that is placed in orbit.

Table 11.2 Activities and lead centers for the human spaceflight programme

Programme component	Activity	Lead Agency
Astronaut Project	Selection and Training of Astronauts; medical monitoring and support; spacesuit development and testing; spacecraft life support and environment control	PLA 507th Institute
Space Applications Project	Onboard scientific application packages and payloads	Chinese Academy of Sciences (CAS)
Shenzhou Project	Development of the Shenzhou Spacecraft	CAST and SAST (Service Module)
Tiangong Laboratory	Development of the Tiangong Spacelab	CAST and SAST (Service Module)
CZ 2F Launcher	Development of the CZ 2F launch vehicle	China Academy of Launch Vehicle Technology (CALT)
Launch site Project	Construction of manned launch site	Jiuquan, Wenchang Launch Centers
TT&C Project	TT&C Network and Operations	Xi'an Satellite Control Center
Recovery Site Project	Recovery Operations	General Armaments Division (GAD)

Figure 11.5 provides a plot of the mass of the various station types versus the year in which they were placed in orbit.

From the Figure, when China placed the Tiangong 1 in orbit in 2011 (mass 8.5 tones) it appears to be only about 50% of the mass of 18 tonnes of the first 1971

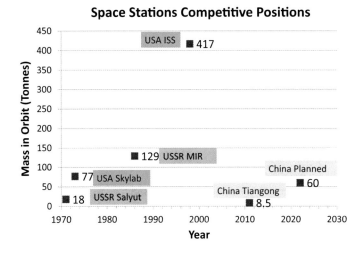

Fig. 11.5 Space station trends

Soviet Salyut 1 Space station. It is also significantly lower than the 77 tonne US Skylab station that was orbited in 1973.

Even as per current Chinese plans when they place in orbit their new planned station of 60 tonnes in orbit by 2022 it will still be below the USSR MIR station of vintage 1986 and Skylab of vintage 1973. The US International Space Station with an orbital mass of 417 tonnes is clearly far ahead.

Despite this, China's Human Space Fight effort and the way they have gone about achieving this shows a great deal of ingenuity. From a purely political point of view, it does signal China's entry into the rank of most powerful nations. There is no reason to doubt that they will build upon what they have already achieved and improve their capabilities and capacities to sustain human presence in space.

11.9   The Future

With a significant improvement in the capacities of China's new launchers to place heavy payloads into LEO China's immediate plan appears to be to establish and operate a 60-tonne space station. The dates projected for this have been progressively shifted. Currently it appears that this will be accomplished by about 2022–2023.

China has also initiated a major programme for lunar exploration. It is likely that at some stage the two programmmes could be combined for a possible human voyage to the moon. Beyond that China has also toyed with scientific missions to Mars.[7]

In 2020 it successfully carried out a mission to Mars. It is likely that after the moon this could be the target for a series of exploratory and scientific missions followed by a possible human mission to Mars.

A major feature of the Chinese human space flight programme is the remarkable alignment between the political, military, and scientific communities within China that this programme was needed for China to establish its credentials on the world stage.

Ever since the return of Quian to China, through different political eras starting from Mao, and Deng, to the current crop of Chinese leaders there is remarkable consistency in persisting with a human space flight programme. Though realistic constraints of capabilities and capacities for a human spaceflight effort have had an impact on the immediate execution of plans, there is a uniform thread of continuing support for a human in space programme. This seems to cut across political, scientific, and military lines as well as across time.

This alignment of forces over such a long period in a world of shifting needs and priorities appears to be quite unique to China. It is a possible reflection of the value system of China's elite that China must do everything necessary to claim its rightful position in the comity of nations.

---

[7] In 2011 China launched a Yinghuo 1 probe to Mars. The 110 kg satellite was launched along with the Russian Fobos Grunt mission to Phobos. Due to problems with the upper stage the mission ended in a failure.

China will also leverage its human space flight capabilities to showcase its credentials as a responsible space power on the global stage. Direct bilateral approaches, approaches through multilateral fora like the Asia Pacific Space Cooperation Organization, (APSCO) as well as other international organizations like the UN will be used to project responsible peaceful behavior.

Case Study 1

International Cooperation and the Chinese Space Station

China appears to assign an important role to the UN in its efforts to project itself as a firm promoter of international cooperation in the peaceful uses of outer space. At the 59th Plenary of the UN Committee on the Peaceful Uses of Outer Space (COPUOS) a Human Space Technology Initiative was announced by the UN Office for Outer Space Affairs (UNOOSA). According to this, UNOOSA will work together with the CMSA for promoting international cooperation in the utilization of China's Space Station (United Nations Office for Outer Space Affairs 2016).

Annexure 11.1: Timeline of Shenzhou and Tiangong Missions

Mission	Date	Launch site	Mass (Kg)	Purpose
Shenzhou 1	19–11-1999	Jiuquan	7600	Unmanned test of CZ 2F and Shenzhou spacecraft—only launch no manoeuvre
Shenzhou 2	09–01-2001	Jiuquan	7600	Unmanned test of CZ 2F and Shenzhou spacecraft. Manoeuvre of the orbital lab also demonstrated after separation. Descent module never exhibited
Shenzhou 3	25–03-2002	Jiuquan	8600	Unmanned test of CZ 2F and Shenzhou spacecraft. Orbital laboratory release. Medium resolution camera, ELINT tests reported on orbital module
Shenzhou 4	29–12-2002	Jiuquan	7600	Last unmanned test of the Shenzhou system—very similar to the Shenzhou 2nd and 3rd missions
Shenzhou 5	15–10-2003	Jiuquan	7970	Crewed flight with China's first astronaut Yang Liwei
Shenzhou 6	12–10-2005	Jiuquan	8100	Two week-long-flight with two astronauts

(continued)

(continued)

Mission	Date	Launch site	Mass (Kg)	Purpose
Shenzhou 7	25–09-2008	Jiuquan	7890	Three astronauts. Zhai Zhigang and Liu Boming performed China's first space walk
Banxing 1	27–09-2008	Jiuquan	40	The Shenzhou 7 in its 31st orbit released the Banfei Xiaoweixing-1 (BX 1) satellite carried by the spaceship. It took pictures and video of the spaceship. On October 30, 2009, the BX-1 satellite re-entered the atmosphere
Tiangong 1	29–09-2011	Jiuquan	8500	Tiangong 1 space laboratory launch by CZ 2FT. Unmanned Shenzhou 8 spacecraft automatically docked with Tiangong 1 twice during its six-week mission
Shenzhou 8	31–10-2011	Jiuquan	8082	Tiangong 1 space laboratory launch by CZ 2FT. Unmanned Shenzhou 8 spacecraft automatically docked with Tiangong 1 twice during its six-week mission
Shenzhou 9	16–06-2012	Jiuquan	8000	Shenzhou 9 with three crew members docked with the Tiangong 1 space station for an 11-day stay. Crew included first Chinese female astronaut, Liu Yang
Shenzhou 10	11–06-2013	Jiuquan	7700	Shenzhou 10 with crew of three docked with Tiangong 1 from June 13 to June 26
Tiangong 2	15–09-2016	Jiuquan	8600	Replacement for the Tiangong 1 which failed in March 2016
Banxing 2	15–09-2016	Jiuquan	47	Launched from the Tiangong space station
Shenzhou 11	16–10-2016	Jiuquan	8082	Two astronauts dock with Tiangong 2 for a month long stay
Tianzhou 1	20–04-2017	Wenchang	12,910	The Tianzhou docked and undocked with the Tiangong 2 several times. A small Silk Road satellite was also deployed

References

Aliberti M (2015) When China goes to the moon, studies in space policy Volume 11, European Space Policy Institute (ESPI), Springer

Barbosa CR (2016) Chinese duo launched to begin Shenzhou 11 mission, NASA SPACE FLIGHT.COM October 16, https://www.nasaspaceflight.com/2016/10/chinese-duo-launch-she nzhou-11/

Harvey B (2013) (2013), China in space: the great leap forward. Springer Praxis Books, New York

Phillips T, Bonnie M (2016) China's Tiangong space station out of control and will crash to Earth, the Guardian, https://www.theguardian.com/science/2016/sep/21/chinas-tiangong-1-space-station-out-of-control-crash-to-earth

United Nations Office for Outer Space Affairs (UNOOSA) (2016) https://www.unoosa.org/oosa/en/informationfor/media/2016-unis-os-468.html

Chapter 12
Breakthrough Research in Space Science and Applications

12.1 Background

As mentioned in the first part of this book the return of Quian to China served as a fillip for boosting space sciences in the country. Quian's affiliations to both the Chinese Academy of Sciences (CAS) as well as to the 5th Academy that dealt with missiles made him a key link of the first satellite project that was initiated within the CAS. The onset of the Cultural Revolution however resulted in a shift of all space activities away from the CAS into COSTND and its successor COSTIND.

Early Chinese satellites were technology demonstrators or satellites needed for meeting military and societal needs. The move toward economic and development goals were strengthened as the Chinese space programme transitioned from the Mao to the Deng era. Though this was so, CAS scientific payloads may have flown as piggyback payloads or as packages in some of the recoverable satellite missions.[1]

This trend away from the sciences was reversed to some extent when the 863-programme received high-level support from the political system. The renewal of the human space flight programme provided new opportunities to the space sciences community for experiments as well as for the development of instrument payloads needed for various scientific missions. The CAS has thus once again become a major player in the national space ecosystem.[2]

---

[1] There is little authentic public information on some of the early Chinese missions with the generic Shijian name. While some of them may have carried instruments for radiation monitoring, the orbital parameters of the satellites and other supporting data suggest that most of them may have been test satellites for Technology development, ELINT or ISR purposes.

[2] The CAS is identified as the lead organization for the development of experimental packages and payloads for both the Shenzhou spacecraft's orbital laboratory as well as for the Tiangong space station.

© National Institute of Advanced Studies 2022
S. Chandrashekar, *China's Space Programme*,
https://doi.org/10.1007/978-981-19-1504-8_12

12.2 A Push for Space Science

The CAS however does not appear to be a monolithic organization with a single-minded focus. Within the CAS there appear to be different groups that seem to pursue different interests. They also seem to be well-connected to the political and decision-making system within the Chinese space programme. When the launch of the first Shenzhou spacecraft ran into technical problems after all other flight systems were ready, there was apparently a move for using the available CZ 2F launcher for a mission to the moon. This reached the very highest political levels within the Chinese establishment before it was turned down[3] (Kulacki 2009).

After the human space flight programme had been cleared in 1992, Chinese space scientists were instrumental in getting a major lunar exploration programme cleared by the Government. Like the human space effort, a "return to the moon" confers great prestige on the nations undertaking such activities. The space science establishment seems to have been quite successful in harnessing these political aspirations with the requirements of an in-depth scientific programme for exploring the moon.[4] There has also been an effort to explore Mars. China launched a Yinghuo mission to Mars on the Russian Phobos mission in 2011. However, the failure of the launcher to put the satellite on the right orbit to get to Mars seems to have resulted in a lower priority for a Mars initiative.

These developments have resulted in greater attention being paid to basic research problems in the space sciences and related fields. This trend has become particularly pronounced after 2000. Two initiatives that involved a major space component have been approved by the Chinese Government. The first major programme was the Doublestar Programme, a collaborative effort with the European Space Agency (ESA). The details of the origins and how this collaboration translated into a major cooperative effort are provided in Case Study 1. The other major programme that has captured the attention of the international community has been China's Lunar Exploration Programme (CLEP).

Case Study 1
CNSA ESA Collaboration—The Double Star Cluster Programme

One of the important research areas identified by Chinese scientists has been to get a deeper understanding of how the sun influences the earth environment via the solar wind and the earth's magnetosphere. The European Space Agency (ESA) initiated a programme called Cluster to investigate and understand the

---

[3] pp. 26–27.

[4] A return to the moon and the exploration of the solar system seems to have a lot of appeal for emerging space powers. India too has carried out programmes for exploring the Moon and Mars. Apart from the science objectives harnessing the resources of the moon such as Helium 3 appear to be one reason for justifying the mission.

influence of solar activity on the earth's environment. A constellation of four satellites were placed in suitable orbits around the earth in such a way that critical measurements could be carried out simultaneously in the regions between the earth and the sun that could be influenced by the sun's activity. Chinese scientists participated as investigators in this ESA initiative (Liu et al. 2005).

In 1997 during a visit of an ESA delegation to China, Chinese scientists presented a proposal to ESA that involved two Chinese satellites—one in a highly eccentric polar orbit and the other in a highly eccentric equatorial orbit. The orbits were chosen in such a way that measurements that complement the ESA Cluster measurements could be made in the critical regions of coupling between the solar wind and the earth's magnetosphere. After detailed discussions ESA agreed to this programme and provided the backup instrument payloads that had been developed for the Cluster satellites. The Chinese also added their own instruments to the scientific package to build their own capability and for calibration. The programme was approved in 2000.

The equatorial satellite Double Star 1 was launched from Xi Chang by a CZ 2C launcher in 2003. It was followed by the Double Star 2 the next year. This was also launched using a CZ 2C rocket from Taiyuan into an eccentric polar orbit.

The two Chinese satellites worked well with the ESA Cluster constellation. Double Star 1 reentered the earth's atmosphere in 2007. Double Star 2 worked well till 2009 after which it encountered certain technical problems resulting in loss of control.

The cooperative effort between ESA and the CNSA resulted in over 1000 articles in peer reviewed journals.[5] The International Academy of Astronautics also conferred the "Laurels for Team Achievement Award" to the Double Star Cluster Team in 2010.[6]

The coupling between the Sun and the Earth and its influence on the earth's environment continues to be a major area of interest that has once again been highlighted in a recent perspective plan outlined by the CAS in 2013.

12.3 The Lunar Exploration Programme

Though there were plans for a lunar exploration effort even in the 1970s, it was only in the early 1990s that the idea of a robotic exploration of the lunar surface gained momentum in China. The launch of the Japanese Hiten lunar probe in 1990 followed by NASA's launch of the Clementine Lunar exploration mission in 1994 may have

---

[5] http://sci.esa.int/cluster/45044-cluster-double-star-1000-publications/.

[6] http://sci.esa.int/cluster/47789-laurels-for-cluster-double-star-teams/.

acted as triggers for Chinese scientists. Prominent scientists felt that unless China did something they would fall far behind the other advanced countries.

Case Study 2

Chronology of the Lunar Exploration Programme

In 1992 a Group of prominent scientists proposed sending a metal emblem to the Moon's surface in 1997 to commemorate the handing over of Hong Kong to China.

In 1993 the newly established China National Space Agency (CNSA) conducted a feasibility study on the use of a CZ 3A launcher to reach the moon.

Both proposals were turned down on the grounds that they lacked scientific merit.

In 1995 the CAS initiated a report on the need and feasibility of a Lunar Exploration Programme. They proposed the use of the DFH 3 platform for performing the function of a lunar orbiter.

In 1997 three members of the CAS once again submitted a report with recommendations for a national Lunar Exploration Programme.

Another initiative by scientists of the CAS resulted in another study by COSTIND in 1998. The report addressed overall design and the key technologies for a Lunar Exploratory Robot. It also set out a set of scientific objectives for a future lunar probe.

The State Council and the State Planning Commission formally approve the scientific objectives for the Lunar Exploration Programme in 2000. This was followed by a series of Conferences by the various organizations involved in the space effort.

The Lunar Exploration Programme was incorporated in the 2000 Chinese White Paper on space. The White Paper identified "Deep Space Exploration centering on the Moon" as an area of research to be pursued.

The Programme received official approval in 2003. Prime Minister Wen Jiabao formally approved the programme in 2004[7,8] (Aliberti 2015; Besha 2010).

China's Lunar Exploration Programme (CLEP) has three main components that progressively improve its ability to explore the lunar surface. These components are as follows:

- To successfully launch and orbit a satellite to map the lunar surface;
- To land an unmanned rover on the surface of the moon;
- To carry out a lunar sample return mission.

---

[7] pp. 97–98.

[8] pp. 214–221.

With the successful launch of the Chang'e 1 and Chang'e 2 missions in 2007 and 2010 the first phase of the programme has been completed.

The Chang'e 1 carried five payloads for accomplishing the various scientific objectives set out for the first phase of the CLEP. A multispectral cum stereo imaging camera would provide imagery for the mapping mission. A laser altimeter would provide additional topographic information that would complement the image data. It would also provide information on the thickness of the soil and features on and below the surface of the moon. An X-ray cum Gamma-ray spectrometer along with a microwave radiometer would make measurements to evaluate the influence of the sun and the solar wind on the moon and its surface.[9] The mission's major technology objectives were to establish capabilities to build satellites that could orbit the moon and return data to the ground using a Deep Space Network.

The mission was a resounding success. After 16 months of operation the satellite was destroyed through a planned crash landing on the lunar surface. This accomplishment boosted China's image as a major space power. Newspapers as well as major political leaders like Prime Minister Wen Jinbao lauded the accomplishments of Chinese space scientists and engineers. A complete high-resolution, three-dimensional map of the lunar surface was published by China in December 2009 making China only the third country in the world to do so.

In 2009 Chinese scientists decided that before going on to the more ambitious rover and sample return missions they needed more experience with certain key technologies to reduce the risk for future missions. They decided that they needed another Chang'e 2 mission using the backups they had built for the first mission. The availability of the more powerful CZ 3C launcher would facilitate a direct injection into earth-moon orbit without having to go the transfer orbit route. They also needed high-definition imagery of the planned landing site for the Chang'e 3 landing mission. Experiments with lower orbits, braking technology, X-band Communications, etc. were also needed for the refinement of the Chang'e 3 mission. On the first of October Chang'e 2 was launched. It reached the Moon four days later. It went into a 12-h lunar orbit. Twice during the period when it was in orbit around the Moon, the spacecraft carried out maneuvers to go into a lower 100 km by 15 km elliptic orbit. This enabled the spacecraft to get closer to the surface and get higher resolution imagery of the landing area needed for the next mission.

In June 2011, Chang'e 2 left lunar orbit for the L2 earth sun Lagrangian point. After a 77-day trajectory it went into a parking orbit around L2. After scientific observations at L2 and the validation of capabilities to conduct TT&C operations for 235 days Chang'e ventured further into deep space. The Chinese are using the

---

[9] One of the objectives used for obtaining clearance for the lunar programme is the prospect of obtaining Helium 3 a potential critical resource for a fusion reactor. Since the moon does not have a magnetosphere Helium 3 contained in the solar wind may be trapped in the lunar surface. The instruments chosen for the first two missions could shed some light on this.

Table 12.1 Capability enhancement Chang'e 1 to Chang'e 2

Mission component	Chang'e 1	Chang'e 2
Launcher	CZ 3A	CZ 3C
Spacecraft mass	2350 kg	2480 kg
Trajectory	Earth orbit to moon trajectory	Direct Earth to Moon trajectory
Time to reach Moon	12 days	5 days
Moon Orbit (nominal)	200 km × 200 km	100 km × 100 km, 100 km × 15 km
Resolution of Camera	120 m	7 m & 1.3 m
TT&C	S Band TT&C	S Band & X Band TT&C
Mission	Moon only	Moon, L2 and Deep Space

Chang'e 2 as a test bed for future missions to Mars and deep space. The enhancement of capabilities from the Chang'e 1 to the Chang'e 2 missions is presented in Table 12.1[10] (Aliberti 2015).

The launch of the Chang'e 3 spacecraft on a CZ 3B launcher on December 2, 2013, marked the beginning of Phase 2 of the CLEP. The major scientific objectives for the mission were

- Survey the topography and geology of the Moon;
- Understand and analyze the distribution and composition of various minerals and elements on the Moon;
- Survey and monitor the space environment between the Earth, the Moon, and the Sun;
- Carry out optical astronomy experiments from the lunar surface.

The Chang'e 3 spacecraft consists of a service module and a lander module. After going into a 100 km × 100 km lunar orbit the lander separated from the service module and descended into a 100 km × 15 km 45° inclination elliptical orbit. At the 15 km perigee point thrusters were fired to reduce the altitude to 4 km. The engines were then shut down for a free fall landing on the lunar surface. Immediately after this the lander deployed a six wheeled lunar rover named Yutu (Jade Rabbit) for exploring the lunar surface.

Yutu has a total mass of 120 kg with a payload mass of 20 kg. It has a design life of 90 lunar days and could survey a total area of 3 sq. Km. It could travel a maximum distance of 10 km. It can navigate the lunar surface on its own. It can identify the most promising areas for exploration. It has a ground penetrating radar

---

[10] pp. 99–100.

for sub-surface observations and carries an optical telescope for astronomy from the lunar surface. It has several cameras for imaging the lunar surface. It also has a robotic arm to collect samples from the surface for analysis. Image and other data can be sent to the Earth on a real-time basis. Yutu continued to operate till August 3, 2016, setting a record of 31 months of operation on the lunar surface. More than 100 papers have been published from this mission.[11]

On October 23 China launched another Chang'e 5 T mission to the moon. This mission was intended to test and validate the reentry of the capsule carrying lunar material for the planned lunar sample return mission. After going into orbit around the Moon the service module placed the reentry module on a trajectory to the Earth. The capsule was successfully recovered on October 31, 2014, with reentry speeds that were significantly greater than reentry speeds from earth orbit. Meanwhile the service module in orbit around the moon carried out a series of maneuvers to simulate the rendezvous with a lunar ascent module and the exchange of samples between them.[12]

Case Study 3

An Overview of the Chang'e 5 Sample Return Lunar Mission

The Chang'e 5 spacecraft will consist of four modules—a service module, a return vehicle to bring the sample back to Earth, a Lander and an Ascent Vehicle that will collect the sample for rendezvous with the service module. The spacecraft is expected to have a launch weight of about 8 tonnes. A new launcher the CZ 5 will inject the Chang'e 5 spacecraft on a direct trajectory to the Moon.[13]

Most of the technologies needed for the mission have already been demonstrated by China. These include

Flight into orbit around the Moon;

Landing on the surface of the Moon;

The robotic acquisition of samples;

Rendezvous of the service module with a sample carrying ascent vehicle;

The deployment of a return capsule from the service module;

---

[11] http://spaceflight101.com/change/chinese-yutu-moon-rover-pronounced-dead/.

[12] http://spaceflight101.com/change/change-5-test-mission-completes-lunar-orbit-rendezvous-demo/.

[13] http://spaceflight101.com/change/change-5-test-mission-completes-lunar-orbit-rendezvous-demo/.

Recovery of the return capsule on the Earth.

The one component that had not been demonstrated was the deployment of the ascent module from the Lander and the takeoff and rendezvous of the ascent module with the service module in lunar orbit. The other component that had not yet been demonstrated is the transfer of material between the ascent module, the service module, and the return capsule.

It is likely that China may fly another mission not only to preserve the momentum for the Lunar Exploration Programme but also to address these gaps in capabilities.

Though there could be some issues in scaling up and modifying the demonstrated technologies for the sample return mission these seem to be minor. Once the CZ 5 launcher becomes available there appears to be no major gap in realizing a successful lunar sample return mission.

With four successful Lunar Exploration Missions, China has completed Phase 1 and Phase 2 of its Lunar Exploration Programme. It has developed and demonstrated most of the technologies for realizing the sample return mission. The CZ 5 launcher, which flew for the first time in November 2016, would be used to place the 8 tonne Chang'e 5 spacecraft on a direct trajectory to the Moon. The failure of its second flight delayed the mission originally slated for 2017.[14]

To keep the momentum going and to ensure that the various technical teams are kept busy, China went ahead with another Chang'e mission in July 2018. The landing of the Chang'e 4 on the dark side of the moon followed by the deployment of a rover marked a new first in the scientific exploration of the moon. The choice of the landing site (the oldest and deepest crater on the lunar surface), the choice of a novel set of experiments that include plant growth on the lunar surface and the relay of data from the dark side via a separate satellite are all novel achievements.

To relay data from the dark side of the moon, a special Queqiao satellite was launched in May 2018 just prior to the launch of the Chang'e 4 mission. This satellite with a large unfurlable antenna is located at the L2 (Earth-Moon Lagrange Point 2) and is used to relay the data from the lander and rover. Along with the Queqiao satellite two other radio astronomy satellites were also launched into lunar orbit These are called Discovering the Sky at Long Wavelengths Pathfinder (DSLWP) Satellites. On deployment from the rocket, they were expected to go into a 200 km by 9000 km lunar orbit on their own. One of the DSLWP satellites was not able to go into lunar orbit while the second one was successful. They carry radio astronomy payloads as well as imaging cameras. These successful feats have raised the stature of China's image as a major power in the international arena.

After the failure in 2017, the CZ 5 rocket returned to active flight at the end of 2019. On November 23, 2020, the Chang'e 5 was launched on a CZ 5 rocket from

---

[14] The second launch of the CZ 5 in 2017 failed immediately after takeoff. This delayed the launch of the Chang'e 5 mission which was completed only in 2020.

Wenchang. The composite Orbiter–Lander-Ascender-Returner Chang'e 5 modules entered lunar orbit on November 20, 2020. The Lander-Ascender combination then separated from the Orbiter-Returner module to land on the moon's surface on December 1, 2020. After collecting the samples, the Ascender separated from the Lander and docked with the Orbiter-Returner module on December 3, 2020. The samples were then transferred to the Returner. The Ascender then separated from the Orbiter-Returner module to crash-land on the lunar surface on December 8, 2020. The Orbiter-Returner combination then set out on its return trajectory to Earth. The Returner module separated from the Orbiter to land safely back on the Earth.

With the completion of the Chang'e 5 mission phase 3 of China's Lunar Exploration Programme has been completed. There are likely to be many more lunar missions in the coming years. The availability of bigger launchers that are on the anvil would make possible more ambitious missions. A human flight mission to the Moon could well be the next phase of China's Lunar Exploration Programme.

12.4 Organizational Challenges for the Lunar Programme

The Organization Structure for the CLEP is very similar to the structure that has been created for the management of the Human Space Flight Programme[15] (Besha 2010). It has a two-tier management structure. The top layer is responsible for providing guidance to the programme and is the final authority for approving all major decisions and changes in strategy. A Lunar Probe Project Office represented by high-level officials from all the major stakeholders directly reports to the Politburo and the State Council. It is likely that this is anchored by a high-level official from the GAD.

A Leading Small Group (LSG) headed by the SASTIND Director constitutes the second layer that is responsible for the actual implementation of the approved programme. The LSG has members that include representatives from the Ministry of Science & Technology (MOST), Ministry of Finance, GAD, CAS, CASC, and CNSA. A General Commander, a Chief Designer, and a Chief Scientist under the LSG have direct project implementation responsibilities. This troika provides the core technical leadership for the programme. A Lunar Exploration & Engineering Center, responsible for mission management, functions directly under the troika. A Lunar and Planetary Science Research Center has also been set up under the CAS for applications. This may be connected to a larger Space Science Center that functions under the CAS.

There are five major components of work under the programme. There are prime contractors identified for each of them. The five components are as follows:

- The development of the Lunar exploration spacecraft (CAST);
- The delivery of the launch vehicle (CALT & SAST);

---

[15] pp. 214–221.

The Organization Structure – Lunar Exploration Programme

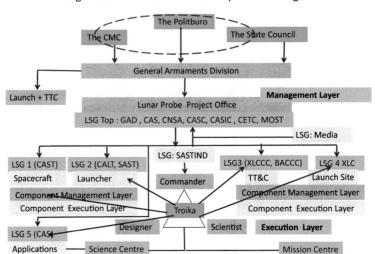

Fig. 12.1 The Organization Structure for the Lunar Exploration Programme

- The TT&C Network needed for the mission (X'ian Control Center & Beijing Aerospace Command and Control Center);
- Launch site related Activities (Xi Chang Launch Complex);
- The ground-based applications needed for using the data (CAS).
- Each of these components may also involve a troika with LSGs for each of them.

In 2007 just before the launch of Chang'e 1 another SLG was created to manage the Publicity and Media Coverage for the lunar mission. This reflects the political importance and prestige associated with the CLEP. The current organization structure used for the CLEP is shown in Fig. 12.1. The re-organization of the PLA military space programme and the elimination of the GAD in 2015 may result in some changes in the organizational arrangements for the LEP.

## 12.5	A Space Science and Technology Future for China—A CAS Perspective

The Box below provides a perspective on how Space Scientists in China view the future trends in Space Science and Technology and the priorities for China to become a major space science and technology power on the global stage (Kulacki 2013).

Case Study 4

Chinese Academy of Sciences (CAS) Strategic Vision 2020 Document

In April 2013 the CAS published a document entitled "Vision 2020: The Emerging Trends in Science & Technology and China's Strategic Options". A translation of the Space Science & Technology section of this report by the Union of Concerned Scientists is available in the public domain. A summary of this is provided below to provide a flavor of how the Chinese see the role of space science and technology in their larger national strategy.

The major science problems in which space could play a role are as follows:

- The Search for Life within the solar system;
- The form, structure and evolution of black holes and their effect on the structure of the universe;
- Understanding the characteristics of Dark Matter and the nature of Dark Energy and their role in the evolution and future of the universe;
- The Origin of the Universe and the Big Bang;
- The Origins of celestial bodies within the structure of the Universe;
- The Origins and evolution of life;
- Understanding the role of solar activity and its influence on the inter-planetary and Earth Systems;
- Use of Space Stations such as the ISS for investigations of fundamental phenomena related to the Life Sciences, Materials Science, Physics, Chemistry and for Astronomy;
- Use Space capabilities for monitoring the impact of human activities on the Earth system for resolving urgent global problems;
- The report then goes on to look at some of the common problems affecting all countries that use space.

The first of these problems relates to the need for an improved system for tracking and monitoring space debris; the report also mentions a need for monitoring the space environment.

The other issue identified deals with how space capabilities can be transformed into commercial products and services. Lower costs for accessing space and trends toward the use of smaller satellites are specially highlighted.

The report then makes the case that China should engage in these activities for making path breaking and revolutionary discoveries in the areas of space science and technology. The demands on new ways of thinking, new ways to approach design as well the new methods of manufacture that will arise out of such a progamme will be a driver of innovations that will transform China.

Apart from the above objectives the space programmme should also provide real-time information needed for development and national security purposes. An independent capability for acquiring high-resolution satellite imagery for meeting its development and strategic needs is identified as one important area.

The report then goes on to reiterate the need for China to significantly strengthen research and capabilities for Space Situational Awareness (SSA). This is required to take care of the increasing risk to operational satellites from space debris. The need for space capabilities to monitor space weather is also highlighted.

The Vision document also emphasizes the need for a series of Space Science satellites with instruments that provide state-of-art data to address the various research problems. Since the lead times for some of these efforts could stretch to over a decade the vision stresses the need to take a longer-term strategic view for sustaining such a programme.

The use of China's new space station for carrying out cutting edge work in Space Science and Applications Research also emerges as major area that requires sustained support.

The establishment of a global system for surveying changes in the environment arising from human activity is identified as another challenge.

Vision 2020 proposes the creation of a National Space Information Support and Services System. This includes satellites, global reception facilities, automation in data reception and data handling, acquiring daily, global data as well as research and model building for using the data.

Finally, the report makes the point that though China has made significant advances there are still several organizational and institutional problems that it needs to address for it to maximize the returns from the investments in the space programme. This requires a major restructuring of the space programme and the creation of a new space authority for coordinating, managing, and executing the programme.

12.6 The Tianwen Mission to Mars

An inter-planetary mission to Mars had always been on the agenda of Chinese scientists. A Yingho probe to Mars in 2011, as a part of the Russian Phobos Grant mission failed during launch. When the CZ 5 heavy lift launcher became available, such a mission could be realized. Though the mission could have been in planning for quite some time, a formal approval was given in 2016.

On December 9, 2020, China launched a Tianwen spacecraft on a mission to Mars. The spacecraft was placed on a Mars trajectory by the CZ 5 launch vehicle lifting off from the Wenchang launch site. After a 7-month journey through the solar system it entered Martian orbit on February 10, 2021.

The probe then spent the next three months surveying the Martian surface for a suitable landing site. On May 14, 2021, the lander-rover combination separated from the spacecraft to land on the surface of Mars. This made China the third country in the world to land an object on the Martian surface. On May 21, 2021, the Zhurong rover was deployed from the Lander on to the surface of Mars. A small camera was then deployed from the rover.

The orbiter and the rover carry a complement of sensors for the exploration of the Martian surface. Many stunning images from the rover and the deployed cameras have been put out in public domain by China. The rover and the orbiter are currently functioning well and providing valuable scientific data and images of Mars.

12.7 Other Major Achievements of China's Space Science Programme

On August 16, 2016, China launched the world's first Quantum Science Satellite (QSS). The satellite is being used to test the distribution of quantum keys via satellite and to investigate the phenomenon of entanglement.

On September 11, 2016, it also launched the world's first X-ray pulsar satellite. The aim was to investigate the possibility of using X-ray pulsar signals as a time standard for navigation in deep space.

In June 2017, as a part of its Sky Exploration programme China launched a Hard X-ray Modulation Telescope (HXMT) into a 550 km 43° inclination orbit. The satellite carries three instruments. All of them are collimated and co-aligned to explore the sky in the X' Ray region. This was part of the original Space Science Plan. With this success most of the programmes outlined in the earlier Space Science Plan have been realized.

In December 2020, two X-ray astronomy satellites Gecam A and Gecam B were launched by a CZ 11 rocket from Xichang. They are in a 29° inclination 600 km Low Earth Orbit (LEO).

An experimental 2 kg satellite was also orbited in 2017 using India's PSLV launcher. It carried an Ion / Neutral Mass Spectrometer.

12.8 Overall Assessment of China's Space Science Activities

Though there are reports that China had launched many satellites for various science missions especially under the Shijian name, a closer scrutiny of the available data suggests that barring a few early missions most launches were test satellites or dual use military satellites. Though there is not much information, the early recoverable satellite effort may have carried experimental payloads from space scientists functioning under the CAS umbrella.

It was only after the initiation of the 863 Programme that space science seems to have gotten importance within the political elite governing China. Highly visible missions such as human space flight or a mission to the Moon or to Mars are increasingly seen as programmes that project China's capabilities favorably in the international arena. The prestige they bring adds to the Comprehensive National Power of China. Chinese political leaders also perceive major space efforts as transformational interventions that can create advanced industrial ecosystems and promote innovation. The alignment between the Science elite and the political leaders at the highest level seem to be getting stronger. These appear to be the major drivers for the increased focus on space science in the Chinese Space Programme.

China has so far launched 23 satellites as a part of its space science and space exploration programmes. These constitute about 4% of all Chinese satellite launches.

As mentioned earlier, the pace of science launchings has picked up appreciably in recent years as China strives to catch up and excel in the creation of new knowledge. The China Academy of Sciences (CAS) has become a significant and more powerful player in China's space ecosystem.

Table 12.2 provides a listing of the major science missions that China has carried out after 2000.

The Doublestar programme to study the connections between solar activity and the Earth system in collaboration with ESA is identified in the CAS vision document as a major thrust area.

The Lunar Exploration Programme with the completion of lunar sample return mission appears to be the first step in Chinese plans for solar system and deep space exploration. There is a possibility that the human space flight programme and the lunar exploration programme will converge into a programme of sending humans to the Moon.

All these missions not only look at pushing the technology envelope but also address fundamental issues relating to the origin of the solar system, the universe and of course the origin of life. The launch of the Dark Matter Particle Explorer Satellite (DAMPE) in 2015 the HXMT in 2017 and the Gecam X-ray satellites in 2020 are consistent with the major thrust areas identified in the CAS Vision document.

There is also a significant space science and space applications effort that is being carried out under the auspices of the human space flight programme. The Shenzhou orbital modules as well as the Tiangong 1 and 2 Space Laboratories are being used both for space science as well as for carrying out experiments in physics, chemistry, astronomy, life sciences and materials sciences. When the 60-tonne Space Station becomes operational these activities are likely to get a further boost.

Though the Yinghuo mission to Mars launched as a co-passenger on a Russian Mars mission failed, the Tianwen 1 mission to Mars is a huge success story. Mars will be a priority area for future missions.

The availability of the more powerful CZ 5 launcher would enable Chinese scientists to significantly enhance their capacity to explore the solar system and venture into deep space. They are also looking at innovative ways to use their space capabilities to address cutting edge problems in strategically important areas.

Table 12.2 China's space science satellites

Satellite	Launch date	Launcher	Launch site	Mission
Double Star 1	12/29/2003	CZ 2C SM	Xichang	Sun-Earth coupling ESA collaboration
Double Star 2	7/25/2004	CZ 2C SM	Taiyuan	Sun-Earth coupling ESA collaboration
Chang'e-1	10/24/2007	CZ 3A	Xichang	Lunar exploration Orbiter Mission
Shijian 8	9/9/2006	CZ 2C	Jiuquan	Recoverable capsule—test bed
Chang'e-2	10/1/2010	CZ 3C	Xichang	Lunar Orbiter, L2, Deep Space probe
Yinghuo-1	11/8/2011	Zenit 3F	Baikonur	Mars Explorer Mission—Failed
Chang'e 3	12/1/2013	CZ 3B	Xichang	Lunar Lander, Lunar Rover Mission
Chang'e 5 T1	10/24/2014	CZ 3C	Xichang	Lunar Orbiter Capsule Return Mission
Wukong (DAMPE)	12/17/2015	CZ 2D	Jiuquan	Dark Matter Particle Explorer Satellite
Shenzhou missions	1999–2016	CZ 2F	Jiuquan	Multiple Areas—Experiments, Payloads
Tiangong 1 & 2	2013,2016	CZ 2F	Jiuquan	Multiple Areas—Experiments, Payloads
Shijian 10	5/4/2016	CZ 2D	Jiuquan	Recoverable capsule—test bed
Quantum Science Satellite (QSS)	8/16/2016	CZ 2D	Jiuquan	Quantum Communications test satellite—Quantum Key distribution
X-ray Pulsar Navsat	11/09/2016	CZ 11	Jiuquan	Uses X-ray Pulsar signals for time measurements for future navsats
Hard X-ray Modulation Telescope (HXMT)	15/06/2017	CZ 4B	Jiuquan	Three collimated and co-aligned telescopes used. Part of China's five-year space plan
Queqiao Lunar Data Relay Satellite	5/20/2018	CZ 7	Xichang	Data relay satellite in HALO Orbit Earth-Moon L2 point
DSWLP 1	5/20/2018	CZ 7	Xichang	X-ray Telescope 200 × 9000 km Lunar Orbit. May not have reached lunar orbit

(continued)

Table 12.2 (continued)

Satellite	Launch date	Launcher	Launch site	Mission
DSWLP 2	5/20/2018	CZ 7	Xichang	X-ray Telescope 200 × 9000 km Lunar Orbit. Successful lunar orbit
Chang'e 4	07/12/2018	CZ 3B	Xichang	Landing on the Dark side of the moon with deployment of a rover—data relay via lunar relay satellite
Tianwen 1	07/23/2020	CZ 5	Wenchang	Orbiter-Lander-Rover Mission to Mars
Chang'e 5	11/23/2020	CZ 5	Wenchang	Sample return mission from the Moon
Gecam A	12/09/2020	CZ 11	Xichang	X-ray Astronomy satellite
Gecam B	12/09/2020	CZ 11	Xichang	X-ray Astronomy satellite

The launch of a dedicated satellite to further quantum communications technology provides an illustration of this more confident China.

Another significant first is China's X-ray Pulsar mission for measuring time accurately. Though both these ideas have been talked about and researched in the US and Europe, China has been the first country to launch satellites specifically to advance the state of the art. The Box below provides a flavor of this new Chinese thrust toward cutting edge research.

Case Study 5

China Launches the World's First Quantum Science Satellite (QSS) and the World's first X-ray Pulsar Navigation Satellite(XPNAV)

On August 16, 2016, China launched the world's first Quantum Science Satellite (QSS) from the Jiuquan Launch Center. The satellite's main function will be to test the fundamentals of quantum communication over long distances. The main instrument on the satellite is an interferometer that is used to generate two entangled infrared photons by shining an ultraviolet laser on a non-linear optical crystal. The satellite will be used to demonstrate quantum key distribution between the satellite and two stations on the ground. The generation and use of such a key can ensure secure communications between the parties concerned. It will also reveal the presence of eavesdroppers.

It will also try to establish that entanglement between a pair of photons can exist over distances of as much as 1200 km and teleport a photon state from the ground to the satellite.

After these goals are completed by Chinese scientists there is also a plan to collaborate with scientists from the University of Vienna for establishing an inter-continental Quantum Key Distribution (QKD) channel between Beijing and Vienna.[16]

The QSS is managed by the Chinese Academy of Sciences (CAS). The payload for the mission was developed by the CAS's Shanghai Institute of Technical Physics and the University of Science and Technology in China. It is one of the four missions identified by the National Space Science Center's strategic space science programmes. Three of these identified four missions—the Dark Matter Particle Explorer (DAMPE) Satellite, the Shijian 10 recoverable satellite and the QSS have been completed (Raska 2013).

On November 9, 2016, China launched an experimental satellite that uses the X-ray emissions of Pulsars for improving the measurement of time. This could help in improving navigation especially for deep space missions. Though the idea has been around for some time with research being carried out in the US, China through the Chinese Academy of Sciences is the first country to launch a dedicated satellite for this purpose.

All Chinese activities in space science are consistent with the vision document put out by the CAS. These activities will continue to be supported by the political establishment. The importance attached to space science as a major instrument to shape and address China's status as a major power will create a climate for fundamental research both in space sciences as well as in other areas such as physics, chemistry, life sciences, materials science, and astronomy. Though military and practical applications will continue their dominance, there will be strong support for research programme in the identified areas of the CAS vision document.

The user community also appears to be getting more diversified. There are several entities under different agencies that can provide the required space hardware. The space ecosystem is becoming more competitive. This is a good sign for China and creates the necessary conditions for fostering innovation—a major objective identified in China's grand strategy.

The science community and the CAS which had taken a backseat in the earlier phases of the space programme are now once again well set to become major players in China's space ecosystem.

---

[16] http://physicsworld.com/cws/article/news/2016/aug/16/china-launches-world-s-first-quantum-science-satellite.

References

Aliberti M (2015) When China goes to the moon, studies in space policy Vol 11, European Space Policy Institute (ESPI), Springer

Besha P (2010) Policy making in China's space program: a history and analysis of the chang'e lunar orbiter Project. Space Policy 26, 2010

Kulacki, Jeffrey LG (2009) A place for one's mat: China's space program 1956–2003. Amer Acad Arts Sci

Kulacki G (2013) Strategic options for chinese space science and technology, a translation and analysis of the 2013 report from the Chinese Academy of Sciences, http://www.ucsusa.org/sites/default/files/legacy/assets/documents/nwgs/strategic-options-for-chinese-space-science-and-tec hnology-11-13.pdf

Liu ZX, Escoubet CP, Pu Z, Laakso H, Shi JK, Shen C, Hapgood M (2005) The double star misión. Ann Geophys 23, http://www.ann-geophys.net/23/2707/2005/angeo-23-2707-2005.pdf

Raska M (2016) China's quantum satellite experiments: strategic and military implications—analysis, RSIS (Rajaratnam School of International Studies) Commentary No. 223, 5 September, https://www.rsis.edu.sg/wp-content/uploads/2016/09/CO16223.pdf

Chapter 13
China's Military Satellites

13.1 Command, Control, Communications, and Computers (C4) Capabilities

Chapter 7 of the book that deals with communication satellites identifies 9 dedicated military communications satellites out of the 37 built and launched by China since 2000. It is quite evident that some of the domestic communication satellite capacity is also being used by the military—a common practice across the world. There are also five Tianlian TDRS satellites in GSO that relay data from orbiting satellites. These are also covered in the Communications chapter of the book. There are therefore at least 14 dedicated satellites that serve the C4 function.

13.2 Intelligence, Surveillance, and Reconnaissance (ISR) Capabilities

Early efforts for ISR included the launch and recovery of film capsules from cameras on board the FSW satellites. Twenty-five of these satellites have been launched with capsules being recovered. Some of the later launches involve microgravity experiments including some experiments from Germany, France, and Japan. Two satellites with the Shijian name are also recoverable satellites. Some on board testing for the human space flight programme could also have been a part of these launches. On balance all of them are classified as military satellites. (see Chap. 6).

The remote sensing chapter of the book (see Chap. 9) provides an overview of all satellites launched into LEO and Sun Synchronous Orbits (SSO). These include remote sensing, military test as well as some small satellites. ELINT and Missile warning satellites that have the Yaogan, Shijian, and other names are also included.

Out of a total of 139 successful satellites launched into SSO only 77 satellites can be considered as serving a civilian function. 62 of the 139 satellites (45%) serve a military function. In 2017 and 2018, China launched four LKW satellites that share

© National Institute of Advanced Studies 2022
S. Chandrashekar, *China's Space Programme*,
https://doi.org/10.1007/978-981-19-1504-8_13

very similar orbital and launch patterns to the Yaogan Electro-Optical (EO) satellites. They perform a direct military function. The Yaogan 32 A and 32 B satellites though called Yaogan share orbital and launch characteristics very similar to the Shijian 11 series of missile warning satellites. **Of the 116 remote sensing satellites in the overall grouping** (see Chap. 5.1) **32 satellites can be considered to serve a military ISR function** (see Sect. 9.17 **for details)**.

Details of the ISR operational Yaogan constellation that includes ELINT and other military functions are provided in Case Study 1.

13.3   Space-Based Navigation Satellites

China has over time created a global satellite-based navigation system made up of GSO, inclined Geosynchronous and orbiting satellites. 60 satellites have been launched as of the end of 2020. The Beidou navigation system is now operational. Though they are dual use, the creation of this constellation is primarily dictated by military needs. The details of this system are covered separately in Chap. 10 of this book. All of them are classified as military satellites.

13.4   Weather Satellites

China has launched a total of 24 satellites for monitoring the weather. Twelve of these satellites are polar orbiting satellites and nine of them are GSO-based satellites.

Though necessary for military operations these primarily serve a civilian public good function. They are therefore excluded from being accounted under the military head. The details of these satellites and China's relative position with respect to the other powers are in Chap. 8 of the book.

13.5   The Human Space Flight Programme

Chapter 11 of the book provides the details. There were 18 launches (as of end 2020) of the CZ 2F and the CZ 7 launchers. The CZ 2F placed 11 Shenzhou modules and 2 Tiangong Space Laboratories in orbit. The CZ 7 launched the first unmanned cargo vessel Tianzhou 1 that docked and undocked with the Tiangong 2 Space Laboratory. Two satellites were deployed from the Shenzhou spacecraft and one satellite from the Tianzhou 1. The third flight of the CZ 7 launcher failed in 2020. The CZ 5 also placed an uncrewed test capsule in LEO as a preparatory launch for the future space station. The human flight programme is military run and classified as a military programme.

13.6 Military Satellites

301 satellites out of 678 launched by China have a predominantly military function. After taking out the 73 commercial launches this accounts for nearly 50% of all satellites built and launched by China. This military part of the programme includes C4ISR, navigation, human space flight as well as early warning, ELINT, BMD, and ASAT components. China is now a full spectrum space military power with strength in all military components.

13.7 ELINT Military Satellites

The backbone of the space component of China's Anti-access and Area Denial (A2AD) strategy is their Electronic Intelligence (ELINT) capability. ELINT provides the cues for the Electro-Optical (EO) and Synthetic Aperture Radar (SAR) satellites to track ships in the oceans that could threaten China's core interests. China's ELINT programme has evolved considerably since the launch of its first satellite in 1970.

Immediately after launching its first satellite China devoted considerable efforts and resources for the development of ISR capabilities. The parallel development of two launchers, the Feng Bao and the CZ 2, and the launch of recoverable film capsules very early on in the space programme indicate that military needs (especially ISR needs) were the drivers of the early phase of the programme. Chinese efforts mirror US efforts albeit with some time lag during this period. Obviously Chinese space and military strategists were closely tracking and monitoring US developments in the military uses of space.

By the early 1970s the US had put in place a global Ocean Surveillance ELINT system. Each ELINT system consisted of three satellites flying in formation at an altitude of around 1100 km in a 63.4° inclination orbit. Each triplet was placed in orbit by a single launcher. The spacing between the three satellites in this orbit enabled the location of radio-emission sources with a fair degree of precision.

China launched two triplet constellations during the early phase of its programme. Both triplet constellations were launched by the now abandoned Feng Bao launcher. The first launch that took place in July 1979 was a failure. The second launch took place in September 1981 and succeeded in placing the three satellites in a 59.5° inclination orbit (Lan and Mark 2011). While China claims that these satellites were purely for scientific and development purposes the triangular formation flying satellite configuration in a near 63° inclination orbit suggests that this was possibly an early ELINT effort.[1] After the 1981 effort China moved away from military towards civilian uses of space under the Deng regime.

---

[1] https://directory.eoportal.org/web/eoportal/satellite-missions/s/shi-jian.

In the aftermath of the first Gulf War, Chinese leaders seemed to have realized the key role of space assets in war and war deterring strategies. The resumption of the human space flight programme in the early 1990s provided them with a new platform for experiments with ELINT payloads. The Shenzhou 2, 3 and 4 precursors to human flight all carried an ELINT payload on the orbital module that was left behind in orbit for experiments. These modules remained in orbit for about 8 months and would have provided valuable experience for the Chinese in dealing with operational aspects of a satellite-based ELINT system.[2] Detailed descriptions of the Shenzhou ELINT packages are available at the Astronautix website.[3]

Almost immediately after the last Shenzhou unmanned flight (2003) China launched a pair of Shijian satellites (SJ 6A and SJ 6B) on a CZ 4C vehicle from Taiyuan in 2004. These were launched into a 600 km 97.9° inclination Sun Synchronous Orbit. Though details on this launch are sparse, the paired launch suggests that this was also an ELINT mission like the two-satellite ELINT missions that succeeded the original three-satellite US Ocean Surveillance Missions.

Paired launches every 2 years followed with the SJ 6C and SJ 6D in 2006, SJ 6E and SJ 6F in 2008, and the SJ 6G and SJ 6H in 2010. This pattern and sequence suggest that these were ELINT missions with replacements approximately every 2 years though they were not placed in a 63.4° inclination orbit.

In March 2010, China placed its first triplet of ELINT satellites into an 1100 km, 63.4° inclination orbit. Its orbit characteristics make it very similar to the early US Ocean Surveillance System. Further launches in 2012, 2013, two launches in 2014, and the launch of the Yaogan 31 A, B, C triplet in 2018 indicate that these ELINT missions are an integral part of China's space-based ISR assets that are used for tracking US aircraft carriers in the Pacific Ocean.

In addition to the triplet ELINT satellites in a 63° orbit, China is also establishing a theatre ELINT constellation of satellites. Starting from 2017 onwards China has been launching these ELINT constellations three at a time into 35° inclination orbits. These satellites do not fly in a triangular formation. They are co-planar satellites that provide near-constant surveillance over the region between 35 North and 35 South latitudes. Their orbital characteristics closely resemble those of the Shijian six series of eight satellites that were launched between 2004 and 2010.

Nine of these Yaogan 30 ELINT satellites were launched in 2017, three in 2018, another 3 in 2019, and 6 in 2020. There are 21 of these satellites in orbit. They occupy six different 35° inclination planes separated by 60° which ensures near-constant surveillance over Taiwan and the oceans and seas that provide access to it. Further details are available in Case Study 1.

In 2018, China launched a pair of XJSW satellites into 35° inclination orbits from XiChang. These satellites are very similar to the Yaogan 30 ELINT series except that the altitude of these satellites was only 500 km. In 2020, six more XJS satellites were launched. From the orbital parameters, these are likely to be ELINT satellites.

---

[2] http://www.svengrahn.pp.se/histind/China12/sz3notes.html.

[3] http://www.astronautix.com/s/shenzhou.html.

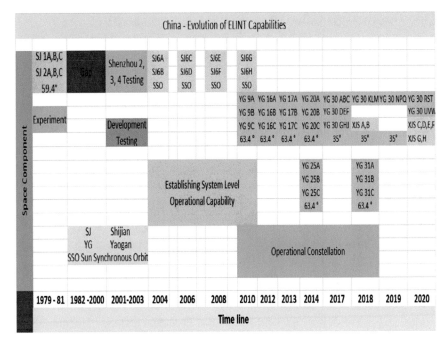

Fig. 13.1   China—evolution of ELINT capabilities

The orbits of the SJ 6 series suggest monitoring the northern parts of Canada and Alaska that house US installations for BMD-related activities. The orbits of the new Yaogan 30 series and the XJSW 1 and 2 ELINT satellites indicate surveillance of the Indo-Pacific region. Figure 13.1 provides a timeline for the evolution of China's space-based ELINT system.

China appears to have at least three operational Yaogan large area ELINT triplets (they have launched six triplets since 2010) at any given point in time. These ELINT triplets are now complemented by the theatre-level co-planar ELINT satellites that operate in lower 600 km, 35° inclination orbits. Preliminary analysis suggests that Taiwan and the waters and seas adjacent to it would be under electronic surveillance almost all the time. Along with the EO and SAR Yaogan satellites this should permit near-constant surveillance over the oceans and seas that provide access to China's eastern shoreline. An assessment of the ELINT coverage is available in Case Study 1. Though currently the ELINT capability is built around the need to monitor the western Pacific Ocean it is possible for China to monitor global developments if it chooses to do so.

From 1970 to 2020, China has launched 61 ELINT satellites. It is possible that China may also replace its three-satellite broad area Yaogan ELINT cluster with a two-satellite system like the US.

13.8   Ballistic Missile Defence (BMD), Anti-satellite (ASAT), and Space Situational Awareness (SSA)

13.8.1   US Developments

Traditionally, ballistic missile launches are detected by GSO-based satellites equipped with Infra-Red (IR) sensors that track the hot plumes of the launches. Based on a preliminary assessment of the trajectory this information is passed on to ground-based sensors for continued tracking of the missile launch and for the initiation of countermeasures.

The US has always had infrared sensors on board GSO platforms for the early detection and tracking of missile launches. Currently, it has at least four earlier generation Defence Support Programme (DSP) satellites as well as four later generation Space-Based Infra-Red System (SBIRS) satellites equipped with a suite of IR sensors for detecting and tracking missile launches (UCS 2021).

As a part of its GSO surveillance efforts, the US has also placed four satellites in GSO for providing Space Situational Awareness (SSA) in GSO. These Geosynchronous Space Situation Awareness Programme (GSSAP) satellites drift slightly above and slightly below the GSO and provide a real-time situation report of happenings in GSO.

More recent US responses to ballistic missile threats involve creating capabilities to intercept and deal with the threat during the boost and mid-course phases of the missile trajectory. To perform these functions, US is looking at a constellation of satellites in LEO and maybe in MEO to perform the mid-course and boost phase detection as early as possible after the missile is launched.

Discrimination of the warhead from the decoys and their continued tracking is a crucial requirement. Clouds, the sunlight, and the moon are sources of radiation that could corrupt the signal. These need to be addressed in the design of any space-based system.

The US launched the first of these experimental satellites called the Mid-Course Space Experiment in 1996. The fundamental purpose was to collect signatures of various reentry vehicles to build a data bank for the designing of suitable algorithms that help identify the RV from clutter, backgrounds, and chaff. The satellite after completing its mission also provided very useful inputs to astronomers by mapping several IR sources in the sky. The satellite functioned till 2008 (Stair and John 2002). Though the major objective was to track the missile RV it also built up a data set of Resident Space Objects (satellites and debris). In 2010, the US launched the Space-Based Surveillance Satellite (SBSS) into a 540 km 97.99° inclination SSO. This satellite has a broad field of view telescope that scans the skies for locating satellites including satellites in GSO.[4]

---

[4] https://directory.eoportal.org/web/eoportal/satellite-missions/s/sbss.

The US has the STSSATRR (Space Tracking and Surveillance System Advanced Tracking Risk Reduction Satellite) in an 870 km 98.93° inclination SSO. This satellite was launched in 2009. Along with the SBSS it provides inputs for the US Space Situation Network (SSN). The US has launched several experimental and demonstration satellites for space-based BMD and SSA capabilities. These include a pair of satellites called STSS Demo 1 and Demo 2. These were launched into a 58° inclination 1350 km orbit. The orbit suggests that this is a lookdown telescope. The pair of satellites will function together to provide a 3 D profile of the launch trajectory.

The US also launched an NFIRE (near-field infrared experiment) into a 440 km 48° inclination orbit for ballistic missile interception from space. Though this did not carry any kill vehicle it does signal US intentions. Another US satellite that provides real-time information on space debris is the STARE B (Space-Based Telescope for Actionable Refinement of Ephemeris) satellite. This satellite was launched into a 41° inclination 500 km orbit in 2013. It provides information on orbiting satellites and debris that could potentially affect functioning satellites.

The US is investing a considerable amount of effort into developing smaller satellites that perform the same functions that are currently performed by larger satellites. Many of them are oriented towards BMD and SSA functions. On August 26, 2017, the US launched another SSA kind of satellite the Operationally Responsive Space Satellite 5 (OROS 5) into a 600 km equatorial orbit. The purpose of the satellite is to keep track of satellites and other objects in the GSO.

Canada also contributed to the US Space Situation Network (SSN). A mini satellite was launched aboard a PSLV launcher in February 2013. The satellite is in an SSO orbit with a mean altitude of about 800 km. Its ECT of 06 00 h makes the orbit a dawn to dusk orbit. The satellite carries a visible sensor that could possibly detect missile launches in the night side of the orbit. Its orbit altitude could also enable imaging all the satellites orbiting below it.

13.8.2 China's Responses

The Shijian 7 was launched in July 2005 on a CZ 2D launcher from Jiuquan. It was placed in a dawn to dusk 06 00 h SSO with an altitude of about 555 km. Not much information is available on the functions of this satellite. From the orbit and altitude characteristics, this could have been a precursor test satellite for the detection and tracking of missile launches.

Starting from 2009 onwards China has launched eight Shijian 11 satellites (Shijian 11–01 to 11–08) into a 700 km, 98.1° inclination sun synchronous orbit. The first of them was launched in 2009 followed by three launches in 2011. Their ECTs were 10 00, 11 30, 15 00, and 16 30, respectively. In 2018, the Yaogan 32 A and 32 B satellites were launched together into a very similar orbit. Though it has the Yaogan nomenclature it may be performing a missile detection function for BMD.

The pattern of coverage suggests a possible near-continuous monitoring of Northern Canada and Alaska. A large field of view sensor (or a combination of smaller field of view overlapping sensors) may provide the desired large area coverage for detecting missile launches.

In December 2015, China also launched the Gaofen 4 satellite into GSO (106 E) using its heavy launcher the CZ 3BE. The satellite ostensibly is a civilian satellite used for the detection of hot spots in tropical storms. However, given its reported stop and stare optical and IR sensors it could be the precursor for China's GSO-based "Early Warning" satellite. China followed this up with another Gaofen 13 launch with a similar payload in December 2020.

China has also placed five TJSW satellites in geostationary orbit with launches in 2015, 2017, and 2018, and two launches in 2019. From publicly available images and information, they are likely to be Signals Intelligence (SIGINT) satellites.

In June 2010, China also launched the Shijian 12 satellite. The orbit achieved was very similar to SJ 6 series of ELINT satellites. After it was launched it carried out a series of maneuvers that took it close to several satellites of the SJ 6 series. Some of the SJ 6 series also carried out maneuvers in turn. It appears that these demonstrations of capabilities have surveillance, monitoring, and inspection roles possibly related to an ASAT function.

In August 2013, China launched the Shijian 15 satellite along with three other satellites. The main Shijian satellite also released a smaller sub-satellite followed by maneuvers. The Shiyan 7 that was co-launched had a remote manipulator arm whose use was also demonstrated. These operations signify Chinese capabilities for ASAT operations.

In October 2013, China also launched the Shijian 16 satellite into a 600 km by 616 km 75° inclination orbit. No Chinese satellite has used this inclination before this launch. From the inclination, one would assume that this covers the northern latitudes—Canada and Alaska particularly well. With the availability of the Yaogan constellation it is unlikely that this satellite is needed for an ISR function. It is probably related to a BMD and or an ELINT function. There has been a shift in the US pace of activities relating to mid-course BMD testing most of which involves real-life demonstrations with launchings from many sites in Alaska, Kwalejein Atoll in Hawaii, the new Reagan Test site, Vandenburg as well as other locations.

On June 29, 2016, China launched another Shijian 16 B with almost identical parameters to those of the Shijian 16. This was launched at 11 21 CST—a half hour earlier as compared to the earlier launch. The inclination suggests more intensive coverage over the Alaska region (launches from the Kodiak Launch complex) related to BMD testing for mid-course interception. The most likely mission of these two Shijian satellites maybe ELINT- or BMD-related functions.

The Shijian 17 satellite launched by the new CZ 5 Heavy Launcher on November 3, 2016 carried out a series of maneuvers in GEO close to Chinasat 5A. This has not been included as a military satellite.

In November 2018, the Shiyan 6 satellite was placed in a 500 km SSO by a CZ 2D launcher from Jiuquan. Though it is like the Yaogan EO recon satellites that are also launched from Jiuquan, its Equatorial Crossing Time (ECT) of 06 00 h suggest it maybe an early warning test satellite.

Table 13.1 provides an overview of the military satellites (Gaofen, Shiyan, Chuangxin) dealing with the ELINT, BMD, and ASAT functions. The table excludes the Yaogan ELINT and Yaogan 32A and 32B BMD satellites and the later ELINT

Table 13.1 China's military satellites (ELINT, BMD, and ASAT)

Satellite	Launch Date	Orbit	Inclination °	Launcher	Launch site	ECT	Function	Comments
Shijian 1	3/3/1971	266 x 1826	69.9	CZ 1	Jiaquan	NA	Test	Tech development
Shijian 1A	7/28/1979	Failure			Jiaquan	NA	ELINT	Early ELINT
Shijian 1B	7/28/1979	Failure			Jiaquan	NA	ELINT	Early ELINT
Shijian 1C	7/28/1979	Failure			Jiaquan	NA	ELINT	Early ELINT
Shijian 2	9/20/1981	232 x1598	59.4	FB 1	Jiaquan	NA	ELINT	Early ELINT
Shijian 2 A	9/20/1981	232 x 1615	59.4	FB 1	Jiaquan	NA	ELINT	Early ELINT
Shijian 2 B	9/20/1981	232 x 1608	59.4	FB 1	Jiaquan	NA	ELINT	Early ELINT
Shijian 4	2/8/1994	210 x 36125	28.6	CZ 3A	Xi Chang	NA	Test	Piggyback
Shijian 5	5/10/1999	849 x 869	98.8	CZ 4B	Taiyuan	08 45	Test	HY Ocean Test
Shijian 6A	9/8/2004	590 X 602	97.7	CZ 4B	Taiyuan	06 30	ELINT	Dual Launch
Shijian 6B	9/8/2004	590 X 602	97.7	CZ 4B	Taiyuan	06 30	ELINT	Dual Launch
Shijian 6 C	10/23/2006	595 x 600	97.7	CZ 4B	Taiyuan	06 45	ELINT	Dual Launch
Shijian 6 D	10/23/2006	595 x 600	97.7	CZ 4B	Taiyuan	06 45	ELINT	Dual Launch
Shijian 6 E	10/25/2008	580 x604	97.7	CZ 4B	Taiyuan	08 30	ELINT	Dual Launch
Shijian 6 F	10/25/2008	580 x604	97.7	CZ 4B	Taiyuan	08 30	ELINT	Dual Launch
Shijian 6 G	10/6/2010	588 x 604	97.8	CZ 4B	Taiyuan	08 00	ELINT	Dual Launch
Shijian 6 H	10/6/2010	588 x 604	97.8	CZ 4B	Taiyuan	08 00	ELINT	Dual Launch
Shijian 7	7/5/2005	550 x 569	97.6	CZ 2D	Jiaquan	06 00	Missile Detection	BMD warning
Shijian 8	9/9/2006	178 x 449	63	CZ 2C	Jiuquan	NA	Microgravity	Recoverable
Shijian 9 A	10/14/2012	623 x 650	98	CZ 2C	Taiyuan	10 45	Test	Electric Propulsion
Shijian 9 B	10/14/2012	623 x 650	98	CZ 2C	Taiyuan	10 45	Test	Tech development
Shijian 10	4/5/2016	220 x 482	43	CZ 2D	Jiuquan	NA	Microgravity	Recoverable
Shijian 11 01	11/12/2009	699 x 703	98.3	CZ 2C	Jiuquan	09 50	Missile Detection	BMD warning
Shijian 11 03	7/6/2011	690 x704	98.2	CZ 2C	Jiaquan	11 34	Missile Detection	BMD warning
Shijian 11 02	7/29/2011	689 x 704	98.1	CZ 2C	Jiaquan	14 50	Missile Detection	BMD warning
Shijian 11 04	8/18/2011	Failure		CZ 2C	Jiaquan	16 34	Missile Detection	BMD warning
Shijian 11 05	7/15/2013	680 x 703	98.1	CZ 2C	Jiaquan	16 34	Missile Detection	BMD warning
Shijian 11 06	3/31/2014	687 x 704	98.3	CZ 2C	Jiaquan	09 50	Missile Detection	BMD warning
Shijian 11 07	9/28/2014	686 x 705	98.1	CZ 2C	Jiaquan	12 20	Missile Detection	BMD warning
Shijian 11 08	10/27/2014	688 x 703	98.3	CZ 2C	Jiaquan	14 00	Missile Detection	BMD warning
Shijian 12	6/15/2010	580 x 606	97.8	CZ 2D	Jiaquan	08 45	ASAT	ASAT SSA
Shijian 15	7/19/2013	666 x 673	98.1	CZ 4C	Taiyuan	06 45	ASAT	ASAT SSA
Shijian 15 sub	7/19/2013	666 x 673	98.1	CZ 4C	Taiyuan	06 45	ASAT	ASAT SSA
Shiyan 7	7/19/2013	666 x 673	98.1	CZ 4C	Taiyuan	06 45	ASAT	ASAT SSA
Chuangxin 3	7/19/2013	666 x 673	98.1	CZ 4C	Taiyuan	06 45	ASAT	ASAT SSA
Shijian 16	10/25/2013	600 x 616	75	CZ 4B	Jiuquan	NA	Missile Detection	BMD warning
Shijian 16 B	6/29/2016	595 x 616	75.1	CZ 4B	Jiuquan	NA	Missile Detection	BMD warning
Shiyan 5	11/25/2013	739 x 754	97.8	CZ 2D	Jiuquan	09 30	Test	Tech development
TJSSW 1	9/12/2015	GSO	NA	CZ 3BE	Xi Chang	NA	SIGINT	ASAT SSA
Gaofen 4	12/28/2015	GSO	NA	CZ 3BE	Xi Chang	NA	Missile Detection	BMD warning
TJSSW 2	1/5/2017	GSO	NA	CZ 3BE	Xi Chang	NA	SIGINT	ASAT SSA
Shiyan 6	11/20/2018	495 x 512	97.4	CZ 2D	Jiuquan	06 00	Missile Detection	BMD warning
TJSW 3	12/24/2018	GSO	NA	CZ 3C	Xi Chang	NA	SIGINT	ASAT SSA
TJSW 4	10/17/2019	GSO	NA	CZ 3BE	Xi Chang	NA	SIGINT	ASAT SSA
TJSW 5	1/7/2019	GSO	NA	CZ 3BE	Xi Chang	NA	SIGINT	ASAT SSA
Shiyan 6-02	7/4/2020	666 x 799	98.2	CZ 2D	Jiuquan	06 00	Test	Tech development
Tiantuo 5	8/22/2020	490 x 510	97.5	CZ 2D	Jiuquan	08 45	Test	Tech development
Shiyan	8/22/2020	490 x 510	97.5	CZ 2D	Jiuquan	08 45	Test	Tech development
Gaofen 13	10/11/2020	GSO	NA	CZ 3BE	Xi Chang	NA	Missile Detection	BMD warning
Gaofen 14	12/6/2020	495 x 497	97.4	CZ 3BE	Xi Chang	11 00	Test	Tech development

launches. There are also 12 satellites in the table that seem to be tests for concept and subsystem testing. Several other military test satellites are not included in this table. These are separately identified and included in the other analyses, tables, and charts.

From the table, satellites with the generic names Shijian, Chuangxin, Gaofen, and Shiyan are related to the development of space weapon capabilities that include BMD, ASAT, and SSA functions. Some of them may also have ELINT capabilities for integrating these functions into a total weapons capability. China has launched 14 satellites for Missile detection (BMD) (excluding the Yaogan 32A and 32B) as well as 10 satellites for demonstrating ASAT and related SSA capabilities.

13.9 Small Satellite Constellations and Testing

Recent testing of small satellite constellations for military use and other development testing of military space systems account for 37 satellites during the period 1979 to 2020.

13.10 Organization Changes Lend Teeth to China's Military Strategy

The creation of these military space capabilities has also gone hand in hand with the restructuring of the organization and institutions of the PLA. The most recent of these periodic restructuring of the PLA took place in 2015. It represents a major shift in the distribution of power and resources within the PLA. These changes appear to be consistent with the PLA military mission of "fighting and winning local wars under conditions of informationization" (see Sect. 4.5 for an overview).

The first major change has been the elimination of the four General Departments that linked the Central Military Commission (CMC) with the various service arms. The General Staff Department, the General Political Department, the General Logistics Department (GLD), and the General Armaments Department (GAD) that function directly under the CMC with oversight functions over the Armed Forces were abolished. Fifteen smaller entities based on their functions have been created. These smaller mainly oversight, staff and support groups report directly to the CMC.

The seven military regions that were responsible for coordination with the individual services for operations have been replaced by five theatre commands in the recent re-organization.

Unlike the military regions who merely served as coordination nodes for operations by the individual services the theatre commands will be directly responsible for operations within their theatre. This is major shift in power away from the individual services commands to command of a joint force under a theatre commander.

In the earlier pre-reform structure, the PLA Navy, the PLA Air Force, and the Second Artillery Corps (called the PLA Rocket Force now) had a HQ that assumed overall responsibility. The Army however did not have such a HQ and was directly controlled by the four staff departments. A new Army HQ has been created that will come directly under the CMC. The HQ of the other services also report directly to the CMC as do the theatre commands.

The reforms make clear that the theatre commands are directly responsible for the operations in their theatres. They will exercise direct control over various service units in their theatre of operations. This is significantly different from the earlier system where the individual services were responsible directly for the operations in various regions under the coordination of the regional commands.

In the new dispensation, the Army, the Navy, the Air Force, and the Rocket Forces serve as suppliers to the theatre commands. Their main task is to organize, equip, and train the military units required by the theatre commands.

This is a major change that takes power away from the various services towards a more functional and integrated theatre command-based structure. Overall, it dilutes the power of the middle level General Department/Services HQ and puts it more directly under the CMC and the theatre commands.

One of the major highlights of the reforms is the importance assigned to the role of information in the war fighting and war deterring strategies of China. The CMC has created an entirely new entity that it calls the Strategic Support Force (SSF). The SSF operates directly under the CMC and is responsible for space, cyber, and electronic warfare. This arrangement seems to be directed towards meeting the need for real-time information dominance that is common to and cuts across all theatres of military operations.

Nuclear forces irrespective of where they are located come directly under CMC command only.

By eliminating a lot of unnecessary middle level military bureaucracy, transferring integrated military operations to the theatre commands, creating an SSF, and by redefining the role of the four traditional services to that of a supplier of equipment and trained military units, China is creating the organizational and institutional competencies for fighting, winning, and deterring local wars under conditions of informationization.

Through the creation of the SSF, space capabilities are given a central role in providing the basic shared battlefield and warfighting information that cuts across functional and geographic domains. Associated with space and closely related to it are the cyber and electronic warfare functions.

By linking all these closely coupled functions under the umbrella of the SSF, the restructuring exercise has put in place the necessary organization infrastructure for achieving information dominance that is so crucial for fighting and winning wars in the world of today.[5]

---

[5] There could be some ambiguities still. The Kinetic Kill ASAT capability, for example, could still vest with the PLA Rocket Forces whereas satellite-based counterforce capabilities and other ground-based ASAT assets may be with the SSF.

The US is also responding to these Chinese moves by creating a new space force that will be responsible for all space operations. This US-China dynamic will have spillover effects on the Indo-Pacific Region including India (Chandrashekar 2018).

13.11   Overview—Military Space

Table 13.2 summarizes all the satellites launched by China that predominantly perform a military function. Figures 13.2 and 13.3 provide a pictorial presentation by function.

Table 13.2   Overview of China's military satellites

Function	Satellites	Comments
C4	14	Includes 5 TDRS satellites—extensive civilian capacities used
Recoverable	25	Film capsule recovery—test bed military, microgravity, science
ISR	62	Military satellites used for civilian applications
Navigation	60	Operational 35 satellite dual-use Beidou constellation
ELINT	61	Essential part of Yaogan constellation—SSA, Space weapons
BMD SSA	14	Space part of mid-course BMD developments and GSO early warning
ASAT SSA	6	Reported numbers may be lower than the actual tests
Human flight	10	Political/Military Space Power Objective
Military tests	37	Number of other satellites during early period have military use
Total	301	Total number of satellites with a predominant military function

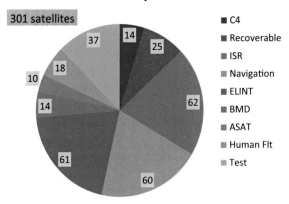

China's Military Satellites End 2020

Fig. 13.2   China's military satellites 2020—Nos

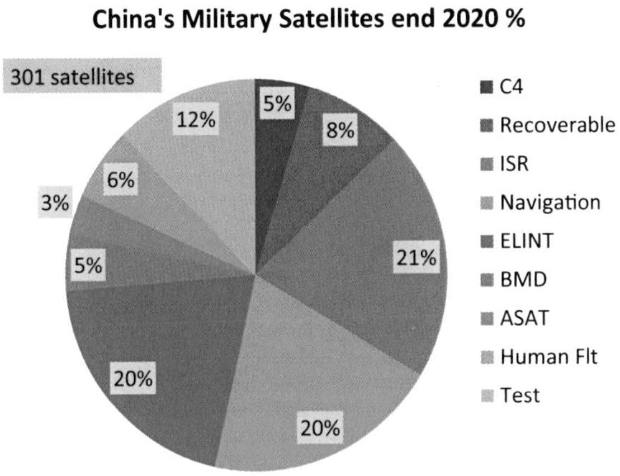

China's Military Satellites end 2020 %

Fig. 13.3 China's military satellites 2020 percentages

C4ISR functions account for 162 satellites or 54% of the 301 military satellites launched. There are 60 navigation satellites that constitute 20% of China's military satellites. The space weapons part that includes ASAT and BMD total 24 satellites or 8% of the total while the human space flight programme accounts for 6% of all satellites.

China launched a total of 678 satellites during the period 1970 to 2020. This includes 73 satellites for various countries as commercial payloads. When these are taken out China launched about 605 satellites for its own use which includes procured satellites as well. **From this assessment, 301 out of these 605 satellites or nearly 50% of all satellites have a military purpose. The Chinese space programme which originated as a military programme continues to be predominantly military driven.**

The data trends also suggest a shift in focus from a largely C4ISR function towards the development and demonstration of space weapons. As the tempo of the space competition between the US and China increases one would expect that China will enhance its military space efforts aimed particularly at SSA, ASAT, and BMD. In addition to this hard power demonstration China will also use its human space flight programme, its lunar exploration, and other deep space endeavors to enhance its status as a major soft power in the world stage.

13.12 Case Study 1

China's Space Power and Military Strategy—the Role of the Yaogan Satellites

13.12.1   Background

The Yaogan series of satellites launched by China from 2006 onwards are a constellation of operational military satellites that provide China with a global surveillance capability. In combination with its demonstrated ASAT capabilities, the architecture of this space constellation suggests that it is a vital component of its Anti-Access Area Denial (A2AD) strategy aimed at deterring US intervention over events in China's sphere of influence.

Till the end of the year 2016[6] the constellation consisted of three kinds of satellites:

- Electronic Intelligence (ELINT) satellites that pick up the electronic emissions and locate the object of interest with a relatively coarser spatial resolution.
- Synthetic Aperture Radar (SAR) carrying satellites that are cued by the ELINT satellites or by other satellites in the constellation that have located the object of interest.
- Electro-optical (EO) satellites that are cued by the ELINT satellites or by other satellites in the constellation that had located objects of interest earlier.

In the period 2017 to 2020, the Yaogan nomenclature has also been used for theatre-level ELINT satellites that provide additional surveillance capabilities over the Indo-Pacific region.

In 2018, another new type of Yaogan satellite was added with the launch of the paired Yaogan 32A and 32B satellites. The orbital characteristics of this launch resemble those of the Shijian 11 series of satellites used for missile detection and warning.

Till the end of 2020, China had launched a total of 68 Yaogan satellites. Apart from catering to specific missions such as the ASBM, this constellation provides the needed large area and theatre-level surveillance capability for its A2AD strategy.

Chapter 9 of this book that deals with remote sensing also covers the Yaogan broad area and theatre-level ELINT satellites in Sect. 9.2. Sections 9.3, 9.4, and 9.5 provide details of the Yaogan broad area Electro-Optical (EO), high-resolution EO, and SAR satellites, respectively. Section 9.6 provides details of the new Yaogan 32A and Yaogan 32B missile warning satellites that closely resemble the SJ 11 series of satellites. These chapters also cover the new military satellites such as the XJSW and LKW series of satellites that appear to be functionally similar to some Yaogan satellites.

The impact of the new theatre-level Yaogan ELINT satellites on China's ELINT surveillance capabilities is studied here. Specifically, this case study addresses how the addition of theatre-level ELINT satellites enhances the ISR coverage over the seas and oceans that provide access to China's shoreline.

---

[6] Though the Asia–Pacific now termed the Indo-Pacific is China's main region of concern the space capabilities it has created can provide it a global surveillance capability if needed.

13.12.2   *Shift in Focus Towards Theatre Surveillance and Military Operations*

From the second half of 2017, there are changes in the use of the Yaogan nomenclature for a new type of ELINT satellite. In September 2017, China launched a new triplet of Yaogan satellites. However, unlike the earlier triplets these did not fly in a triangular formation. Their altitude at 600 km and their inclination of 35° were also significantly different from the 1200 km altitude and 63.4° inclination of the earlier triplets. After launch the three satellites were maneuvered and spaced 120° apart in the same orbital plane. Unlike other ELINT satellites launched earlier these are all launched from Xi Chang the site from where launches to geostationary orbit are made.

The September 2017 launch was followed by further triplet launches in November, December 2017 that continued through 2018 to 2020. The later triplets were also maneuvered after launch to fly in the same orbital plane. A total of 21 satellites constitutes this theatre-level ELINT constellation. Table 13.3 provides details of the geometry of this constellation based on the open-source orbit information available as of September 4, 2021.[7]

The Right Ascension of the Ascending Node (RA)[8] in Table 13.3 is an important parameter to look at the architecture of this theatre-level ELINT constellation. From the table, the RA for the Yaogan 30 A, 30 B, and 30 C is 114°. They all cross the equator from south to north at a longitude of about 114°. They are all in the same plane and will cover the same area on the ground in succession one after the other.

Yaogan 30 D, 30 E, and 30 F are also co-planar and cross the equator at about 355° longitude. They are separated from the Yaogan 30 A, 30 B, and 30 C by 240° of longitude.

Yaogan satellites G, H, J, K, L, M are also nearly co-planar and cross the equator at a longitude of 233°. They are separated from the Yaogan A,B,C satellites by 120° and from the Yaogan D, E, F satellites also by 120° of longitude. The three-satellite planes are equally spaced around the equator separated by 120°. The satellites are in three different 35° inclination planes separated by 120° at the equator. Figure 13.4 captures the geometry of this theatre ELINT constellation.

Figure 13.5 depicts the spacing between the satellites in plane 3. The six satellites are spaced equally within the plane. This means that as one satellite goes out of visibility of a target another successor satellite would take over surveillance.

As the earth rotates the first plane will move out and the second plane of satellites will take over. This will then be followed by the third plane.

---

[7] This listing only includes the Yaogan 30 launches to end 2020.

[8] In simple terms, this is the longitude at which the satellite crosses the earth's equatorial plane from south to north. It is provided in an inertial reference frame and needs to be corrected for the rotation of the earth to get the longitude in a rotating earth framework.

Table 13.3 Yaogan 30 ELINT theatre constellation details

Satellite	Launch site	Launcher	Launch date	Apogee Km	Perigee Km	Inclination (deg)	Period (Min)	RA (deg)	Plane
30 A	Xi Chang	CZ 2C	Sep 29 2017	609	603	35	96.6	113.72	1
30 B	Xi Chang	CZ 2C	Sep 29 2017	608	603	35	96.6	114.37	1
30 C	Xi Chang	CZ 2C	Sep 29 2017	610	601	35	96.6	114.52	1
30 D	Xi Chang	CZ 2C	Nov 24 2017	607	603	35	96.6	355.85	2
30 E	Xi Chang	CZ 2C	Nov 24 2017	609	602	35	96.6	355.59	2
30 F	Xi Chang	CZ 2C	Nov 24 2017	608	603	35	96.6	353.93	2
30 G	Xi Chang	CZ 2C	Dec 25 2017	608	601	35	96.6	233.07	3
30 H	Xi Chang	CZ 2C	Dec 25 2017	609	602	35	96.6	233.55	3
30 J	Xi Chang	CZ 2C	Dec 25 2017	608	601	35	96.6	232.98	3
30 K	Xi Chang	CZ 2C	Jan 25 2017	610	601	35	96.6	234.23	3
30 L	Xi Chang	CZ 2C	Jan 25 2017	608	603	35	96.6	233.71	3
30 M	Xi Chang	CZ 2C	Jan 25 2017	608	603	35	96.6	233.82	3
30 N	Xi Chang	CZ 2C	Jul 26 2019	609	601	35	96.6	52.88	4
30 P	Xi Chang	CZ 2C	Jul 26 2019	612	599	35	96.6	52.66	4
30 Q	Xi Chang	CZ 2C	Jul 26 2019	606	605	35	96.6	53.74	4
30 R	Xi Chang	CZ 2C	Mar 24 2020	609	601	35	96.6	293.15	5
30 S	Xi Chang	CZ 2C	Mar 24 2020	608	603	35	96.6	293.76	5
30 T	Xi Chang	CZ 2C	Mar 24 2020	607	604	35	96.6	291.77	5
30 U	Xi Chang	CZ 2C	Oct 26 2020	606	605	35	96.6	173.92	6
30 V	Xi Chang	CZ 2C	Oct 26 2020	607	603	35	96.6	174.00	6
30 W	Xi Chang	CZ 2C	Oct 26 2020	610	601	35	96.6	174.80	6

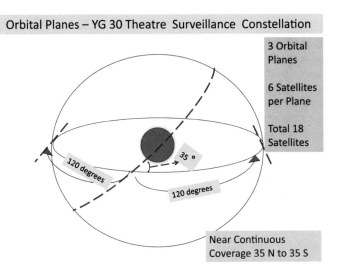

Fig. 13.4 Orbital plane geometry yaogan 30 theatre ELINT constellation

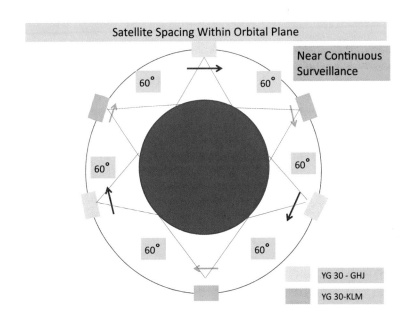

Fig. 13.5 Yaogan 30 ELINT satellite spacing in orbital plane

The architecture of this satellite configuration indicates a near-constant ELINT surveillance over the land and ocean areas of the world that lie between 35° north latitude and 35° south latitude.[9]

---

[9] These would provide a constant vigil over Taiwan, the Korean peninsula, as well as the southern islands and seas of Japan. The seas and waters off China's eastern coastline including the South

Based on the logic of this approach one would have expected that the next triplet (Yaogan N, P, Q) would be in the same plane as the first or second triplet. However, with a RA of 53° the fifth triplet occupies a plane that is midway between plane 1 and plane 2. Yaogan R, S, T, with a RA of 293 occupy a position midway between Plane 2 and Plane 3. The last triplet with an RA of 174 occupies a plane between Plane 1 and Plane 3. The 21-satellite constellation is thus distributed in 6 different planes spaced equally along the equator.

13.12.3   Studying ELINT Coverage of Taiwan—Approach

To evaluate the improved coverage that is likely with the addition of the new theatre-level ELINT constellation, a three-step procedure was followed in a study conducted at the National Institute of Advanced Studies (NIAS) in 2018.

The first step involved studying the ELINT coverage of the Large Area Yaogan ELINT constellation that involves a satellite triplet flying in a triangular configuration in a 1200 km 63.4° inclination orbit. The coverage of a given target area (Taiwan) was studied over a 26-h period using the NORAD Two-Line Element orbital data and propagating the orbits of the satellites for a 26-h period.

The second step involved repeating the same exercise over Taiwan with the two-line element data for the theatre Yaogan constellation.

The third step combined both the large area and theatre-level satellites for studying the ELINT satellite coverage over Taiwan.

13.12.4   Large Area ELINT Coverage over Taiwan

For a satellite surveillance system to work, the coverage of the ELINT part of the system is most important since it provides the initial cues for the other satellites. To understand this coverage, the study used the two-line element orbital data put out by NORAD[10] with the in-house developed Veni Vidi Vici orbit propagation software. Five of the broad area ELINT constellations were taken and their coverage of Taiwan analyzed for a period of 26 h.[11] The ELINT satellites chosen were the Yaogan 16

---

and East China seas as well as the Sea of Japan would also be covered. Since they do not fly in a triangular formation, they will not be able to locate the position of the emission source. However, since they cover the area closer to China's shoreline more intensely there will be other ways to locate their position. New electronically switched array-type antennae may also help location.

[10] North American Aerospace Defence Command.

[11] The choice of Taiwan was based on the view that it is the most likely venue for a future conflict given China's vulnerabilities and its security priorities. Taiwan provides only a static target whereas the threat is posed by a mobile Aircraft Carrier Group (ACG). A similar analysis is needed for a moving target. This will be addressed in a subsequent paper.

ABC, Yaogan 17 ABC, Yaogan 20 ABC, Yaogan 25 ABC, and the Yaogan 31 ABC satellites.[12] The ability of these satellites to monitor electronic emissions emanating from Taiwan was studied.

There are 29 passes during the 26-h period (a little over one pass every hour) that can monitor electronic emissions over Taiwan. This location information can then cue the SAR and EO satellites for a more precise location of the object of interest such as an Aircraft Carrier Group (ACG).[13]

There are also about 30 periods between the passes during which emissions cannot be detected. Of these 30 gap periods there are 6 periods where the gaps between passes exceed 60 min. The average gap between passes is about 36 min with a maximum gap of 92 min and a minimum gap of 5 min.

The target is visible for 495 min out of the total time period of 1560 min which translates into a 32% emission monitoring efficiency.

While this monitoring capability may be good enough for long-range surveillance of the ocean areas around China's eastern coastline extending into the northern Pacific Ocean it may not be adequate as the adversary comes closer to its coastline. A continuing surveillance capability may be needed to deal with such eventualities.

The launches of a new series of Yaogan 30 theatre ELINT satellites promise to provide this near-constant surveillance capability.

13.12.5 The Yaogan Theatre ELINT Coverage

To investigate this improved coverage achieved with the theatre-level ELINT system, all the Yaogan 30 satellites available at that time (Yaogan A to Yaogan M a total of 12 satellites) were chosen to study the constellation's ability to monitor electronic emissions over Taiwan.

The same 26-h period used for the earlier broad area ELINT surveillance study was used along with the in-house developed Veni Vidi Vici software.[14]

The major findings are summarized below. There are a total of 103 orbital passes during a period of 26 h during which electronic emissions from Taiwan can be monitored by the Yaogan theatre-level ELINT constellation.

Out of a total duration of 1560 min (26 h) electronic emissions emanating from Taiwan can be monitored for 1091 min for an overall efficiency of 70%.

There are a total of 58 gaps during this period. The maximum gap between emission monitoring passes is 22 min with the minimum being 1 min. The average gap is 8 min. The surveillance has improved considerably.

---

[12] The Yaogan 9 triplet was launched in 2010. This has not been included in the exercise since it may have reached its end of life.

[13] Ideally this exercise must be carried out for a moving object like an Aircraft Carrier. Such an exercise is currently underway using the Veni Vidi Vici software tool developed by ISSSP, NIAS.

[14] The orbit propagation was for a period of 26 h starting from June 17, 2018, 11:00 h to June 18, 2018, 13:00 h.

There are also overlaps in coverage among the satellites in the constellation. This means that electronic emissions can be monitored at the same time by more than one satellite.

The total coverage during the 26-h period is 1340 min though because of overlaps in coverage this translates only into 1091 min of actual coverage. This overlap provides resilience to the system in case of satellite failures or because of some other operational problems.[15]

It is also possible that with improved orbit management practices the coverage efficiency can also be improved beyond the current level of 70%.

The analysis so far has treated the large area ELINT surveillance system and the theatre-level ELINT surveillance system as two separate systems. In practice, they are complements to each other with one looking farther away and the other becoming more dominant as the threat comes closer to the Chinese mainland. How does the coverage change if the two systems function in tandem?

13.12.6 Combining Large Area and Theatre ELINT Surveillance Systems

The coverage of both the large area and the theatre-level ELINT surveillance systems over Taiwan for the same 26-h period was then analyzed.[16] Table 13.4 provides a summary overview of the combined coverage of the two Yaogan ELINT constellations.

Over the 1560 min of the orbit propagation involving both types of ELINT satellites Taiwan was covered for a total duration of 1238 min. This translates into a coverage efficiency of 79% over the 26-h period. The number of gaps has reduced from the theatre ELINT only configuration from 58 to 48 gaps though the maximum gap periods for both cases remain the same at about 20 min.

There is a reduction in the average gap period from 8 to 6.7 min. There is also overlap in the coverage providing some system-level resilience in case of satellite failures. Overall system efficiency can be improved with better orbit maintenance practices. As expected, the combination of the two systems provides significant improvement in the overall coverage efficiency from 70 to 79%.

Figure 13.6 provides one snapshot of such coverage. As we can see from the figure as, one satellite moves away another satellite takes over. Figure 13.6 is just one part of a 100-min continuous surveillance period.

As the theatre ELINT constellation gets completed with the addition of nine more satellites to the end of 2020, to make a constellation of 21 satellites, the gaps in

---

[15] The resilience factor can be estimated as 1.23 which is 1340 divided by 1091. There is nearly a 25% redundancy in the achieved coverage.

[16] The coverage is from June 17, 2018, 11:00 h, to June 18, 2018, 13:00 h.

Table 13.4 Yaogan ELINT coverage—large area and theatre surveillance constellations

| Start time 17 June 2018 1100 h | | | | | | End time 18 June 2018 1300 h | | | | |
Time (Min)	Interval (Min)	Cum time (Min)	Gaps No	Cum gaps No	Gap time (Min)	Cum gap (Min)	Max gap (Min)	Min gap (Min)	Interval Efficiency (%)	Dynamic Efficiency (%)
1 to 66	66	66	1	1	3	3	3	0	95	95
67–120	54	120	3	4	5	8	3	1	91	93
121–184	64	184	1	5	6	14	6	0	91	92
185–239	55	239	3	8	33	47	20	2	40	80
240–299	60	299	2	10	21	68	12	9	65	77
300–358	59	358	2	12	22	90	20	2	63	75
359–439	81	439	2	14	10	100	6	4	88	77
440–508	69	508	4	18	11	111	5	1	84	78
509–581	73	581	3	21	11	122	5	2	85	79
582–644	63	644	3	24	13	135	6	3	79	79
645–713	69	713	2	26	28	163	21	7	59	77
714–782	69	782	2	28	38	201	21	17	45	74
783–837	55	837	3	31	12	213	10	1	78	75
838–900	63	900	0	31	0	213	0	0	100	76
901–1002	102	1002	1	32	2	215	2	0	98	79
1003–1056	54	1056	3	35	12	227	6	2	78	79
1057–1112	56	1112	1	36	7	234	7	0	88	79
1113–1208	96	1208	1	37	20	254	20	0	79	79
1209–1311	103	1311	2	39	33	287	20	13	68	78

(continued)

Table 13.4 (continued)

Start time 17 June 2018 1100 h						End time 18 June 2018 1300 h				Interval	Dynamic
Time (Min)	Interval (Min)	Cum time (Min)	Gaps No	Cum gaps No	Gap time (Min)	Cum gap (Min)	Max gap (Min)	Min gap (Min)	Efficiency (%)	Dynamic Efficiency (%)
1312–1414	102	1414	4	43	14	301	6	1	86	79
1415–1515	101	1515	3	46	7	308	3	1	93	80
1516–1560	45	1560	2	48	14	322	12	2	69	79

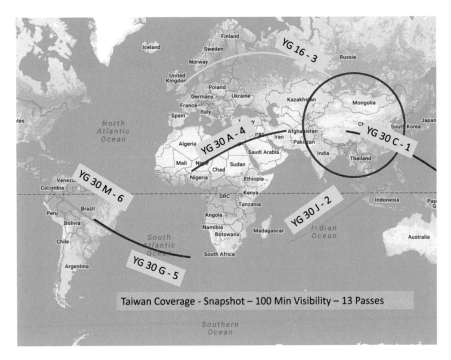

Fig. 13.6 Snapshot of Yaogan ELINT coverage over Taiwan

coverage are likely to get further reduced. The large values for the resilience or redundancy factor imply that there is some scope for improving the overall coverage efficiency.[17]

13.12.7 Complementary EO and SAR Military Satellites for Surveillance

Figure 13.7 is an overview of the current Yaogan and other military ISR satellites without the ELINT The altitude of the satellite is plotted against its Equatorial Crossing Time (ECT). This figure provides an overview of all the EO and SAR satellites that use the Yaogan and LKW nomenclature.

Out of a total of 31 satellites, 24 are high-resolution SAR and EO satellites launched as a part of the Yaogan and LKW military series of satellites. Eight of

---

[17] The addition of eight XJSW ELINT satellites should improve the coverage a bit more. When the ELINT information is used to cue other EO and SAR satellites the efficiency of coverage is likely to improve further.

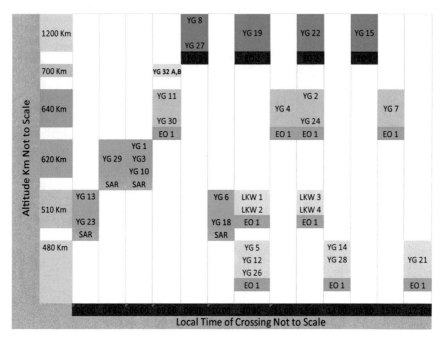

Fig. 13.7 EO and SAR military satellites—China End 2018

them are SAR satellites and 16 of them are high-resolution optical imaging satellites. As mentioned earlier, the Yaogan 32 A and 32 B appear to be Missile Warning Satellites based on their orbital characteristics.

Out of these about five to six SAR and about eight high-resolution EO satellites may be currently operational. Four broad area EO satellites out of the five orbiting at an altitude of 1200 km may also be operational.

Apart from the Yaogan and LKW satellites covered here China also has several high-resolution SAR and EO satellites that ostensibly support a civilian function. These dual-use satellites could provide additional SAR or optical coverage.

Chapter 9 of this book provides details of the remote sensing component of China's Space Programme. Section 9.17 specifically addresses the issue of dual-use satellites.

The data also suggests that China is now into its third generation of Yaogan SAR satellites. The first-generation SAR satellites were heavy (launched by the CZ 4C) and operated at altitudes of around 620 km. They continue to be used even as recently as 2015 as is evidenced by the launch of Yaogan 29.

The second-generation SAR satellites appear to be lighter and launched by the CZ 2C into a lower 500 km SSO. The third-generation SAR satellites are placed in orbit by the CZ 4B which can place a slightly heavier payload than the CZ 2C in SSO.

As far as the high-resolution EO satellites are concerned the first-generation satellites which are in a 630 km orbit are being replaced by the second-generation satellites which are in a 500 km SSO. Both generations use the CZ 2D launcher to achieve orbit.

These high-resolution SAR and EO capabilities are complemented by at least a pair of medium-resolution EO satellites that are in a 1200 km SSO.

At least three to four triplets of ELINT satellites flying in a triangular formation provide the coarse fix to the SAR and EO satellites for locating and continuously tracking the objects of interest. The constellation build-up started in 2006 and entered the initial operational phase with the launch of the Yaogan 9A, 9B, 9C ELINT triplet in 2010. It has been in operation since then. These are now being complemented by at least 21 theatre-level satellites that operate in six different planes with an inclination of 35° and an altitude of 600 km. It is thus well positioned to monitor the Indo-Pacific region on a near-continuous basis.

Different launch sites (Taiyuan, Jiuquan, Xichang) as well as different launchers with different capacities have been used for placing ISR satellites in orbit. The achieved orbits may reflect both improvements in capabilities as well operational improvements deriving from the experience of the earlier missions.

China also seems to have in place the capacities for both satellite and launcher manufacture. There are also multiple entities that can supply the required numbers of satellites and rockets. Details of these capabilities are available in Chap. 15 of this book.

13.12.8 Space—A Major Constituent of China's Military Power

The critical role of the space and cyberdomains along with the nuclear and more traditional modes of war is a reality today. China has for a very long time recognized the importance of military space capabilities to prevent or delay US intervention over the military actions it may take in its region of interest. The formulation of its A2AD strategy is an asymmetric response to a dominant adversary[18] (Chandrashekar 2018). While the strengthening of its other more conventional military assets is also a high priority, Chinese investments in space and cyberassets are clearly on a steep upward trend.

The pace of launching military satellites by China has significantly increased over the last several years. The programme covers the full spectrum of space capabilities from human flight and space stations to advanced quantum communications and X-ray pulsar observation satellites that are oriented towards the advancement of basic science and technology knowledge. While China is striving to use its space assets for both economic and political ends, the fundamental driver of the programme has always been dictated by military needs.

---

[18] pp. 381–392.

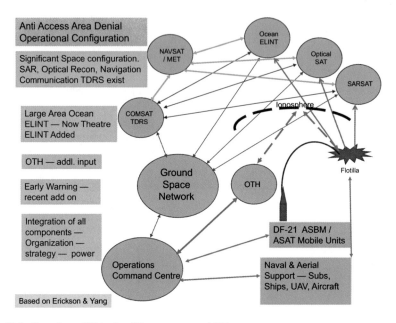

Fig. 13.8 Overview of China's military space capabilities

Figure 13.8 provides an overview of Chinese military space capabilities.

Apart from the C4ISR capabilities that include communications and data relay satellites in GSO, China has also invested in both ground-based and space-based ASAT capabilities that can extend to the GSO and maybe even beyond it. It has also demonstrated BMD capabilities. More recently it has launched two Gaofen and five TJS satellites located in the GSO that are Early Warning and SIGINT Satellites for ASAT and BMD.

The creation of the new Yaogan 30 theatre constellation provides a significantly improved surveillance capability.

In combination with the large area Yaogan ELINT constellation China has in place a robust ELINT capability that can be effectively used along with other C4ISR and conventional military assets to provide the real-time data needed to "deter, fight and win local wars under informationized conditions".

China is now clearly poised to use its space capabilities not only as a key component of its military strategy but also as an instrument to build up its Comprehensive National Power (CNP) that includes the economic, political, and knowledge dimensions. It has clearly identified space as a critical component of its geopolitical power architecture.

Whether it is hard power revealed through its increasingly sophisticated military space systems or commercial communications satellites or soft power as shown by the diffusion of its emerging Beidou navigation system into global commercial products and services China is emerging as the only country that can challenge the dominant position of the US as the most powerful space faring nation in the world.

What we are witnessing in the space domain is one facet of the larger geopolitical rivalry between the US and China. Though the Indo-Pacific region represents the current focus area of this rivalry one can expect it to expand into the larger global stage.

References

Chandrashekar S (2018) The China US space rivalry & the new world order what should India Do? ISSSP Report No. 03–2018. Bangalore, National Institute of Advanced Studies, September 2018

Chandrashekar S (2018) China's Anti-Access Area Denial (A2AD) strategy, India's national security annual review 2016–17. In: Kumar Satish (ed) Foundation for national security research. New Delhi, Routledge Taylor & Francis Group

Lan E, Mark SA (2011) China's Electronic Intelligence (ELINT) satellite developments–implications for US Air and Naval Operations, Project 2049 Institute, Arlington Virginia, February 23 https://project2049.net/documents/china_electronic_intelligence_elint_satellite_developments_easton_stokes.pdf

Stair AT Jr, John DMD (2002) The Midcourse Space Experiment (MSX). https://ieeexplore.ieee.org/document/577637

Union of Concerned Scientists (2021) UCS Satellite Data Base, May 2021. https://ucsusa.org/resources/satellite-database

Chapter 14
China's Launch Vehicle Programme

14.1 Background and Trends

As outlined in the first part all of China's current stable of operational launchers have their origins in its military missile programme. The DF 3 and DF 4 missiles were the direct predecessors of the CZ 1 responsible for placing the first Chinese satellite in orbit in 1970. The CZ 2 series of launchers (including the CZ 2D, the CZ 3 series as well as the CZ 4 series) and the defunct Feng Bao Launcher were all derived from the DF 5 ICBM. In all these developments, the space programme has used the power plants of their missiles as the base around which they have configured different architectures for realizing the payload requirements for different space missions. By adapting the missile through the stretching of the stages and through the addition of suitable upper stages and strap-on boosters China has created launchers that are built around the DF 5/CZ 2 basic architecture.

This trend has also been extended to their solid-propellant-based missiles. The DF 21 missile has been used for the development of several launchers. These include the earlier Kaitouzhe (KT) series and the more recent four-stage Kuiazhou (KZ) launch vehicle family that are based on the 1.4 m diameter DF 21 technology. The KZ series of launchers are not being developed by CALT or SAST both of which come under the purview of CASC. They are being developed by the Sanjiang Aerospace Group which is the 9th Academy of CASIC the other major aerospace corporation in the Chinese missile and space ecosystem.

In a similar vein, the DF 31 ICBM has given rise to the CZ 11 which uses the same 2 m diameter solid rocket motors of the DF 31 for powering the launch vehicle. This launcher comes under CALT.

The KZ and the CZ 11 both address the emerging market for small satellites. They are aimed at providing a quick launch capability for meeting military space needs as well. They are competitive products that can launch satellites singly or in multiples from a Transporter Erector Launcher (TEL) platform used for launching mobile missiles.

© National Institute of Advanced Studies 2022
S. Chandrashekar, *China's Space Programme*,
https://doi.org/10.1007/978-981-19-1504-8_14

In 2018, a small rocket, the Zhuque, developed by a private start-up, the Landspace Technology Corporation was launched. This was derived from the DF 21/DF 26 solid rocket stages. The first- and second-stage firing as well as the fairing separation of this three-stage rocket took place as expected. However, there was a problem with the third stage and the Weila satellite failed to reach orbit.

One Space Tech a start-up company failed to reach orbit with its OSM Chonqing launcher on March 27, 2019. Another start-up iSpace successfully placed four small satellites in Low Earth Orbit (LEO) on July 25, 2019, after an earlier failure. However, a third attempt on August 3, 2019 did not succeed.

A third successful flight of a company-built rocket took place on August 17, 2019 when the Jielong launcher built by China Rocket Company placed three small satellites in 500 km Sun Synchronous Orbit (SSO). China Rocket Company is a subsidiary of the state-owned China Academy of Launch Vehicle Technology (CALT).

From 2018, China does not officially report on failed private sector launches. Successful launches, however, are officially reported.

As a part of its new initiative related to Civil Military Integration (CMI) the Chinese government is opening-up some sensitive areas for private sector investment. There are other companies such as OneSpace, LinkSpace, and ExPace that also have plans to develop and launch new vehicles.

14.2   Greener Dedicated Satellite Launch Vehicles—A New Direction

While most of the earlier generation operational launchers and some of the new generation launchers are derived from their missiles, China has also embarked upon the development of a new series of launchers that are not derived from their missile programme. They are not only more capable but also involve greener technologies. The new series of launchers that include the CZ 5, the CZ 6, the CZ 7, and the CZ 8 launchers use stages that are powered by liquid engines. However, unlike the earlier engines that used toxic Unsymmetrical Di Methyl Hydrazine (UDMH) and Nitrogen Tetroxide (N_2O_4) as fuel and oxidizer the new engines either use cryogenic propellants (liquid hydrogen and liquid oxygen) or semi-cryogenics (kerosene and liquid oxygen) to provide more environmentally friendly launchers. The original plan involved the development of several common engines and stages that were modular in nature. Through an appropriate choice of the modules a series of launchers catering to a very large number of payloads could have been configured. However, the Chinese have chosen to move away from this approach and instead focused on a narrower set of choices built around the CZ 5, the CZ 6, the CZ 7, and the CZ 8.

The CZ 5 is China's Heavy lift launcher that is needed for the establishment of the Space Station, the Lunar Sample return mission as well as for the launching of multiple large or very large communication satellites. The first flight of this launcher took place in November 2016 from the new Wenchang Base on Hainan Island. A second launch in 2017 failed. The CZ 5 returned to flight at the end of 2019 with the successful launching of a Shijian test satellite. Three further successful flights in 2020 that included test launches for the space station as well as the sample return from the moon and a Mars rover mission indicate it is now operational. The launcher has been developed by CALT.

The launch of the CZ 5B on April 29, 2021 for placing the first module of the Chinese space station Tianhe in orbit created a crisis. After the module reached orbit and separated from the booster, the Chinese lost control of the spent stage. Concerns were raised about the damage that the debris from the 22-tonne rocket body would cause when it reentered the earth's atmosphere. Fortunately, the spent body made a safe reentry over the Southern Indian Ocean on May 9, 2021. This was the second major incident involving uncontrolled reentries of Chinese spacecraft following the reentry of the Tiangong 1 space laboratory in April 2018.

The CZ 6 is the Chinese launcher in the small lift category. The responsibility for the development of the launcher has been assigned to SAST. The first flight of the CZ 6 took place in September 2015 from the Taiyuan base. Fourteen small satellites were injected into a 530 km SSO. Further successful flights took place in 2017, 2019, and 2020.

The medium lift launcher that has been developed by China is the CZ 7. This can cater to either dedicated or multiple communications satellites to GTO as well as cargo and other missions to the space station. The first launch of this launcher took place from Wenchang in June 2016. Multiple satellites including a prototype of a crew return capsule were put into orbit using a special YZ 1 upper stage. A successful flight in 2017 was followed by a failure in 2020.

The first launch of the CZ 8 took place in December 2020 from the Wenchang Launch Complex. A military SAR satellite along with three smaller satellites was placed in a 500 km Sun Synchronous Orbit (SSO).

The CZ 5, the CZ 6, the CZ 7, and the CZ 8 are not derived directly from the missile programme. They are dedicated space vehicles meant only for the delivery of spacecraft, satellites, and other payloads in orbit. Some of them can launch spacecraft that can be maneuvered in orbit for rendezvous, docking and can also inject multiple payloads into orbit. The CZ 5 is powerful enough to place large payloads into trajectories to the moon and other planets.

These newer rockets can cater to both military and civilian application needs. Over time the current workhorse CZ 2, CZ 3, and CZ 4 series of launchers will be retired with the new generation taking over. This will happen over the next 2–3 years.

The heavy and medium lift launches the CZ 5, and the CZ 7 will be the workhorse vehicles for the human flight programme that will need the ferrying of crew and cargo for establishing and then operating the space station. The CZ 8 may be the vehicle for launches to Sun Synchronous Orbits. The CZ 6 will launch smaller satellites while the CZ 11 and the KZ launcher families will cater to the small and very small satellite markets. Though a few of the launchers developed by private companies have failed, they are likely to become operational soon.[1] China therefore has a spectrum of launch capabilities that can deliver payloads from the very small to the very large into any desired orbit.

14.3   Competition for Launch Services—Domestic and International

These launch capabilities are not under one organization. Several organizations under the two major space corporations CASC and CASIC will compete for a share of both the domestic and international markets. This competitive dimension goes back to the Cultural Revolution. The Beijing faction of the space ecosystem had to accommodate the interests of the Shanghai Group during the heydays of the Cultural Revolution to win political acceptability. This led to parallel development of the CZ 2 and Feng Bao space launchers by the Beijing-based First Academy (CALT) as well as the Shanghai-based Eighth Academy (SAST). Since that time there have always been at least two competing entities for the development of any missile or launcher variant.[2]

The resulting competition has had the benefit of providing different alternatives for evaluation and assessment as is evident in the way the Chinese have gone about finalizing their design of the launcher and the space craft needed for their human space flight programme.

The rationalization of work between CALT and SAST has also resulted in a bifurcation of responsibilities as far as different launchers are concerned. The responsibility for the CZ 2C series, the CZ 3 series as well as the CZ 2F launcher used in the Shenzhou missions vests with CALT. SAST is responsible for the CZ 2D as well as the CZ 4 series. Both are involved in the design of the Shenzhou spacecraft as well as in the Tiangong space station projects.

This practice of competitive procurement of launchers is likely to continue even as China moves into a new generation of launchers. CALT is responsible for the CZ 5, the CZ 7, and the CZ 11 launchers. SAST is responsible for the CZ 6 small lift launcher.

Recently, there is third entrant into the launch vehicle business. The Kuaizhou (KZ) launcher will also enter the field of competitive offerings through the Sanjiang Aerospace Group (the 9th Academy under CASIC). The two major aerospace corporations are not only in competition with each other but also within themselves for domestic and international market share. This competition dynamic is likely to foster and drive innovation in both the domestic and international launcher markets.

---

[1] The US Private Sector has now become dominant in the launch industry due to companies like Space X, Blue Origin, and Virgin Galactic.

[2] Both entities can also integrate the power plant with spacecraft that have to do a lot of maneuvers needed for docking as well as for reentry. The standard terminology used for these kinds of operations is spacecraft as opposed to satellites.

Unlike the US which has a dominant private sector presence private companies in China are just beginning to emerge as potential suppliers of launch services. As of now the launch vehicle market is dominated by state-run entities and this is likely to continue over the next few years. The stage and vehicle details for all the launchers are provided in Annexure 14.1. Annexure 14.2 provides details of the annual number of launches from 1970 to 2020 along with reliability estimates. A more detailed review of the Chinese launch vehicles covering the period 1970 to 2020 is provided below. This builds upon the earlier overview provided in Sect. 5.2 of the first part of this book.

14.4 Overview of the Launch Vehicle Programme

Figure 14.1 provides a year-wise timeline of all the launches carried out by China.

China used a total of 396 launchers to place 678 satellites into various orbits during the period 1970–2020. Ten launches were procured launches. Out of these 396 launches there were 22 failures providing an overall reliability of 93%. Many of the failures were during the early phases of the launcher development efforts. The launch record since 1995 has been significantly better. There were 352 launches from 1995 to 2020 with 20 failures giving an overall launch reliability of 94%.

This is comparable with the reliability record of other spacefaring nations. In 2018, China had a total of 39 space launches. In comparison, the US had only 34 launches. In 2019, China had 34 launches, the US had 27, and Russia had 25 launches. The

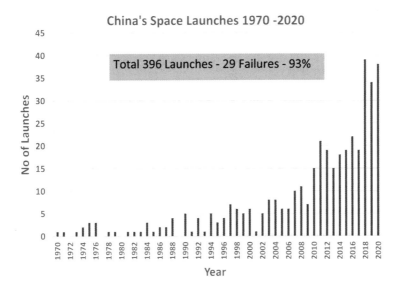

Fig. 14.1 China's space launches 1970–2020

US took the lead in 2020 with 44 space launches. China had 38 launches and Russia had 17. Elon Musk's US-based company set new records for private space flight with several other private players like Blue Origin and Virgin Galactic developing new rockets.

While the pace of Chinese launchings has increased severalfold as its competition with the US intensifies, the US is now poised to reclaim its position as the dominant player in space.

Figure 14.2 provides a chronology of the various launchers introduced by China over the period 1970–2020. Except for a gap of 10 years from 1974 to 1984 (corresponding to the period of the Cultural Revolution and the Deng Reforms) there has been steady and continuous progress. New variants as well as new configurations have been introduced on a continuous basis over the 1984–2020 period.

Figures 14.3 and 14.4 provide details of launches from different sites in number as well as percentage terms.

Most of the satellites launched by China use Chinese launchers both for meeting domestic requirements as well as for international customers. Because of International Trade Arms Regulation (ITAR) restrictions China has been forced to use foreign launch services for some of its communication satellite needs.

Jiuquan was the first of the Chinese launch complexes and has been used for most of the development flights. Xi Chang is used for launches to the GTO and Taiyuan for launches into SSO. The three launch sites provide China with the capacity to launch payloads into different kinds of orbits.

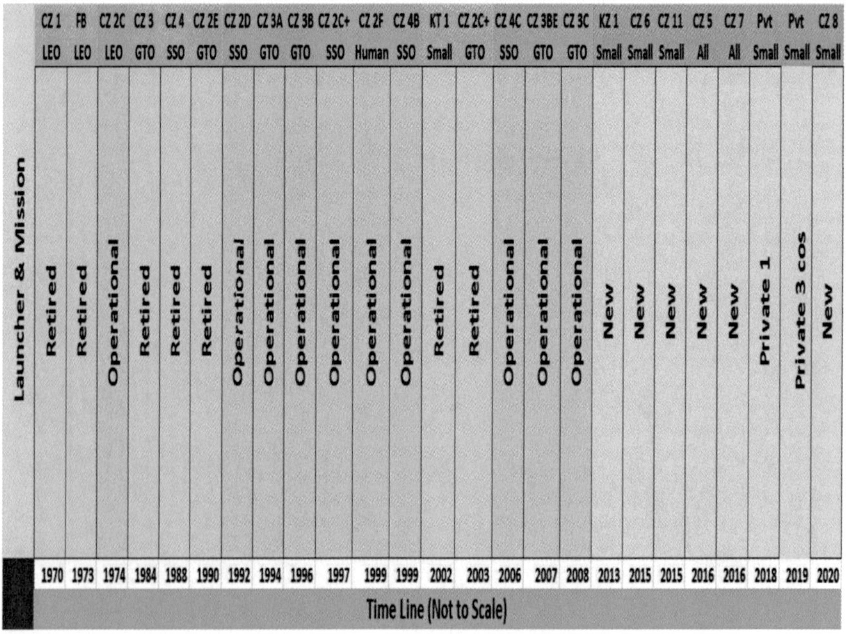

Fig. 14.2 A chronology of China's launch vehicle programme

Fig. 14.3 Launches from different sites

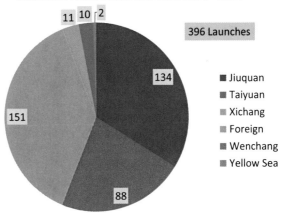

Launches From Different Sites 1970 -2020

396 Launches

- Jiuquan
- Taiyuan
- Xichang
- Foreign
- Wenchang
- Yellow Sea

11 10 2

134

151

88

Fig. 14.4 Launches from different site percentages

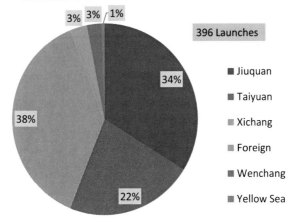

Launches from Different sites % 1970 -2020

396 Launches

- Jiuquan
- Taiyuan
- Xichang
- Foreign
- Wenchang
- Yellow Sea

3% 3% 1%

34%

38%

22%

A new launch complex at Wenchang in Hainan Island commenced operations from 2016. Multiple launch locations and a broad range of launchers provide the flexibility and capacity needed for China's emergence as a major space power.

Launches from offshore platforms in the Yellow Sea Area commenced from 2019. In 2019 and 2020, China had launches from different launch sites within the same day as well as multiple launches from the same launch site on the same day. Some small satellites are also launched from mobile launchers.

Figure 14.5 provides details of the various orbits reached by the 324 launches realized during the period 1970–2020.

Figure 14.6 provides the same details in percentage terms.

Chinese launchers have been used for many different space missions. The 42° as well as the 57–63° inclination orbit missions have been mainly used for the human

Fig. 14.5 Launcher orbits
China 1970–2020

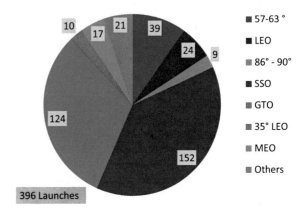

Fig. 14.6 Launcher orbits
China 1970–2020%

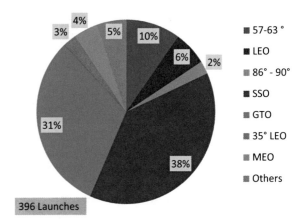

spaceflight and recoverable satellite programmes. More recently, the 63° launch inclination LEO orbits have been used for the Yaogan satellite ELINT missions. The 87° missions were used for placing the US Iridium satellites in the required LEO.[3]

GTO launches (31%) and SSO launches (38%) together account for nearly 70% of all missions. Apart from communications satellites, weather, navigation as well as Tracking and Data Relay Satellites have been placed in GSO and inclined GSO via the GTO route. Most of the SSO satellites are Earth Observation Satellites. More recently, these may include other types of satellites launched for testing out other functions such as ASAT BMD and SSA applications.

---

[3] The launch site location, the safety clearance zones needed for safe launch operations as well as sovereignty considerations constrain the inclinations that can be realized from a given launch site.

Fig. 14.7 CZ 2, 3, 4
launches 1970–2020

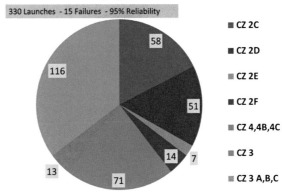

China CZ 2, 3, 4 Launches 1970 - 2020

330 Launches - 15 Failures - 95% Reliability

- CZ 2C
- CZ 2D
- CZ 2E
- CZ 2F
- CZ 4,4B,4C
- CZ 3
- CZ 3 A,B,C

The MEO orbits (4% of all launches) represent the beginnings of China's plans to place in orbit a complete constellation of orbiting satellites to complement and improve the GSO and IGSO navigation system that they had established earlier.

As mentioned earlier, about 301 satellites (out of a total of 678 satellites launched by China) serve military purposes. These 301 satellites have used 231 launchers or about 58% of all launches. Military functions do have a major role to play in China's space programme.

Apart from these missions into orbits for meeting major application needs, China has used its launchers for space science missions which require a more customized trajectory. These account for 5% of all missions.[4]

14.5   China's Workhorse Launchers—The CZ 2, CZ 3, and CZ 4 Series

Of the 678 satellites put into orbit by China 530 satellites were launched using 330 launch vehicles belonging to the CZ 2, CZ 3, and CZ 4 families. China has formally retired some launchers used earlier. These include the CZ 1 used for launching the first two Chinese satellites, the Feng Bao launcher developed by the Shanghai Group during the Cultural Revolution and the CZ 2E launch vehicle developed and used in the 1990s for commercial launchings into the GSO. Figure 14.7 provides an overview of the various CZ 2, CZ 3, and CZ 4 series of launch vehicles used by China.

The CZ 2C and the CZ 3 were the earlier generation vehicles. The CZ 2D and the CZ 4 series were later developments. Further improvements via the addition of strap-on boosters resulted in the extension of the 2C/2D series into the CZ 2E series which with the addition of a third solid stage tried to address the needs of the burgeoning

---

[4] These included two Shijian missions to a 75° inclination orbit that may have a military purpose.

Fig. 14.8 China CZ 2, 3, 4
launches 1970–2020%

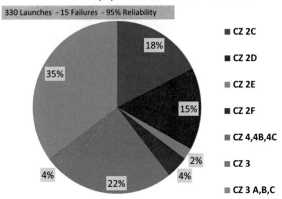

market for launch services into GTO. The CZ 2E with improvements became the CZ 2F launcher for the Shenzhou human space flight spacecraft.

Figure 14.8 presents the same data as in Fig. 14.7 in percentages.

An improved cryogenic upper stage to the CZ 3 led to the creation of the CZ 3A. With the additions of strap-on boosters more powerful launchers such as the CZ 3B, 3BE, and CZ 3 launcher families have been created. These not only cater to the different communications satellite market segments but also provide the launchers for its lunar exploration programme. The CZ 4 series (originally a non-cryogenic route for meeting the launch services requirements for communications satellites) has become the workhorse vehicle for launch into polar and sun synchronous orbits.

The CZ 2E, CZ 3 and the CZ 3A, 3B, 3BE, and 3C launchers used mainly for placing communications satellites in GTO account for 41% of the launches. The CZ 2C, CZ 2D, CZ 4, CZ 4B, and CZ4C that are mainly used for LEO and SSO missions account for 55% of the launchings. The CZ 2F launcher used for the manned Shenzhou missions account for the remaining 4%.

Annexure 14.3 provides an overview of the connections between the power plants and the various stages that together create the configurations necessary for the CZ 2, CZ 3, and CZ 4 families of launch vehicles. A new generation of launch vehicles that include the CZ 5, the CZ 7, the CZ 6, the CZ 8, the CZ 11, and the Kuaizhou (KZ) launchers have completed development and entering operational status. The earlier generation launchers will be phased out and replaced by these new launchers over the next few years.

14.6   Assessment of Capabilities

LEO Capabilities

Figure 14.9 provides a timeline of how Chinese launch capabilities for Low Earth

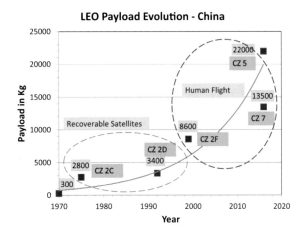

Fig. 14.9 The evolution of LEO payload capability—China

Orbits (LEO) have evolved over the years. The two major requirements were the recoverable satellite programme in the early phases and the needs of the human flight programme that officially started in 1992.

The first-generation launchers powered by the same basic CZ 2C engines have been modified progressively to go from recovering capsules from LEO to placing humans in orbit around the earth. This transition from the CZ 2C of 1975 vintage to the 1999 CZ 2F represents the maximal stretching of the limits of this technology. The first generation of Chinese launch vehicles are now being replaced by the more powerful CZ 5 and CZ 7 launchers powered by cryogenic and semi-cryogenic engines. These have significantly enhanced China's capabilities for performing LEO missions.

Figure 14.10 provides a comparison between Chinese launch capabilities into LEO with other launch vehicles that are currently operational. This does not reflect

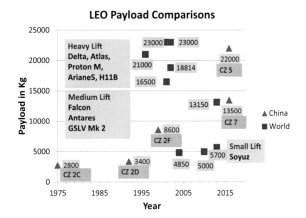

Fig. 14.10 LEO payload comparisons

the launch of the Falcon Heavy by the US company SpaceX that placed a Tesla car on an inter-planetary trajectory to Mars.[5]

The new CZ 5 launcher that was successfully launched in 2016 can place 22,000 kg in LEO. This is almost on par with the Delta IV and Proton M both of which can place about 23,000 kg in LEO. China is ahead of Ariane 5 (21,000 kg), the Atlas 5 (18,800 kg), and the HIIB (16,500 kg).

The CZ 7 launcher (13,500 kg in LEO) that was also launched in 2016 is slightly superior to the US Falcon 9 (13,150 kg). It is a Medium Lift Launcher to reach LEO. China also has the currently operational CZ 2F launcher (8600 kg) that straddles the gap between the medium and small lift categories. The small lift category of launchers (about 5000 kg) forms a third segment.

The main requirements of payloads for LEO are for the human space flight programme. The 5000 kg segment represents the lower end of any possible human space flight programme. China with two big launchers in the heavy and medium lift categories is well placed for establishing their space station in the next few years. They also have the CZ 2F as a backup.

These launcher capabilities are also adequate to cater to the needs of a Lunar Sample Return Mission and any other unmanned exploration missions of the solar system. It may not however be sufficient for meeting the needs of a human mission to the moon. China does have ambitious plans for a human flight to the moon. It has also revealed plans to build a much bigger launch vehicle for realizing these goals.

China is well placed to meet the needs of its human space flight as well as its lunar and solar system exploration programmes. It matches and in certain cases surpasses the lift capabilities of Russia and Europe. It is clearly ahead of Japan. However, the US with the launch of the Falcon Heavy in 2018 is ahead and likely to remain so for some time.

GTO Capabilities

Figure 14.11 provides a schematic of how China's launch capabilities have evolved over the years. The CZ 3 and the CZ 2E have been retired. The CZ 3A, the CZ 3B, the CZ 3BE, and the CZ 3C have provided launch services for meeting both domestic and international customer needs.

Figure 14.12 provides China's relative position in the GSO launch services market.

The global trend in communications satellites suggests a move away from the medium and intermediate class segments towards large and very large satellites. Though the use of ion propulsion may reduce the mass required the trend towards heavier satellites may continue. Figure 14.12 does not include the Elon Musk's SpaceX launcher which made its debut in 2018. This can deliver 21,000 kg in GTO making it the standard to beat.

In 2016, China launched the CZ 5 which can place 14,000 kg in GTO. However, a second launch in 2017 exploded after takeoff creating delays in some critical space

---

[5] Three launches of the Falcon Heavy (64 tonnes in LEO) have taken place since the first launch in February 2018.

Fig. 14.11 GTO payload China—evolution

Fig. 14.12 GTO payload comparisons

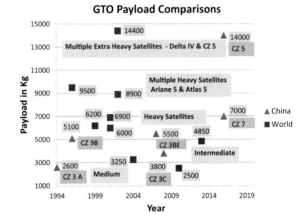

station and moon missions. Four successful launches in 2019 and 2020 have now made it operational.

It has successfully launched the CZ 7 medium class launcher twice in 2016 and 2017. This can place 7000 kg in GTO. Its new launch site at Wenchang is operational and can be used routinely to place heavier payloads in GTO. The CZ 5 and the CZ 7 can be configured to deliver multiple payloads into GTO. In principle, such capabilities can meet the needs of customers in the different segments of the communications satellite market. In addition, the CZ 3B and the CZ 3BE can be used to launch single large or very large satellites into GTO. The CZ 3C and the CZ 3A can also meet the needs of the few customers needing intermediate and medium class communications satellites. China is well placed with a product offering in all the segments of the communication satellite launch services market. Coupled with the ability to build ITAR free satellites China can counter any US move to prevent its emergence

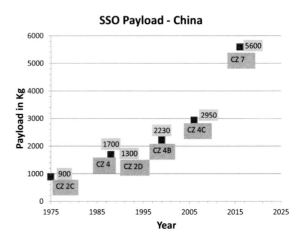

as a major player that combines launch services with delivery of communication satellites in GSO.

SSO Capabilities

Unlike the trends in the communications satellite, market remote sensing satellites show a trend towards smaller satellites. Constellations of such smaller satellites are also being planned to provide round the clock coverage services for various applications. Chinese launch vehicle developments have largely kept pace with these trends. Figure 14.13 provides an overview of various launch vehicles that China has developed to meet the needs of domestic and international customers for satellites in sun synchronous orbits.

Apart from the reliable CZ 2C, CZ 2D, CZ 4B, and CZ 4C launchers that have been used for placing civilian and military payloads, the CZ 7 can also place a 5600 kg payload in SSO if such missions are needed. What is not shown in figure is China's development of a new class of launch vehicles that can cater to the emerging markets for small satellites that include remote sensing satellite constellations in SSO. The KZ 1 launcher (derived from the DF 21 solid propellant missile), the CZ 11 (derived from the DF 31 solid propellant ICBM), and the CZ 6 (liquid propellant) space launcher provide payloads of 430 kg, 350 kg, and 1080 kg, respectively.

In 2020, China also launched a CZ 8 launcher that uses engines and stages derived from the CZ 7 and the CZ 3 series. The CZ 8 can deliver a payload of 4500 kg to a 700 km Sun Synchronous Orbit (SSO). This provides an offering between the CZ 4C (3000 kg) and the CZ 7 (5600 kg). Figure 14.14 provides a comparison of Chinese launch capabilities in SSO with those of other countries.

The CZ 4C can place nearly 3000 kg in SSO. Along with the CZ 4B (2300 kg in SSO) and the three-stage CZ 2C (1800 kg in SSO) China can cater to the needs of customized remote sensing satellites in the large and very large categories. If needed these vehicles can also provide launch services for constellations of smaller satellites suitably separated in any SSO orbit. Direct competitors to China in these segments

Fig. 14.14 SSO payload comparisons

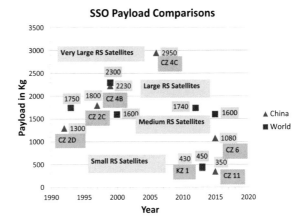

are Dnepr (2300 kg), PSLV (1750 kg), Vega (1740 kg), Minotaur (1600 kg), and Rocket (1600 kg). The successful CZ 8 launch in 2020 provides China a leadership position in this segment.

In the emerging market for small satellites, China has three product offerings. These are the CZ 6 (1080 kg), the KZ 1 (430 kg), and the CZ 11 (350 kg). All of them can be used for launching smaller satellites going from a few kilograms to a few hundred kilograms.

Chinese product offerings for missions to SSO can provide the entire range of services needed for domestic as well as international customers. It is strong not only in the more traditional markets for heavier customized satellites but also in the new emerging markets for small satellites and constellations of small satellites. With its lower price strategy, it is well placed for missions to SSO.

14.7   Overall Assessment

China has in place a wide spectrum of launch capabilities that stretches from human space flight to very small satellites weighing only a few kilograms. This large product range provides China with capabilities to launch any kind of mission needed for meeting both domestic and international needs. It is strong in all the major markets (LEO, GTO, and SSO). It is self-sufficient in meeting the needs of its military and space exploration programmes. With its ability to build ITAR free satellites that can be delivered in orbit using its own launchers China is well positioned to take on the major space powers including the US in direct headlong competition.

Annexure 14.1: Details of China's Satellite Launch Vehicles

The CZ 1 Launcher[6]

When the decision to launch a satellite was made in 1965 the Chinese needed a launch vehicle that could deliver the required payload as quickly as possible. The most realistic option was for them to use the two stages of the DF 4 missile that had by then reached an advanced stage of development. Since the two stages did not provide the required velocity to place a satellite in orbit a third solid stage (to be developed by the 4th Academy) was added.

The Cultural Revolution started almost immediately after the decision to modify the DF 4 for performing the satellite launch was taken. The development of the first Chang Zheng 1 (CZ 1) launch vehicle was transferred from the First Academy (CALT) to the Shanghai bureau (8th Academy) because of the politics of the Cultural Revolution. After it ran into problems the development was transferred back to the First Academy. The CZ 1 launcher placed the first Chinese satellite (Dongfanghong 1) weighing 173 kg in a 68.5° inclination orbit on April 24, 1970. It was again used to place a 221 kg satellite in a 70° inclination orbit the following year. Both flights were launched from the Jiuquan Launch Centre.

Well before the first satellite launch, the Chinese had initiated an effort to use the DF 5 ICBM stages to develop a new class of launchers that could place heavier payloads in orbit. The CZ 1 was effectively retired after two flights though there were attempts to offer an improved CZ 1D variant as a launcher for small satellites in the 1990s.

The first two stages of the CZ 1 used liquid engines for providing the required thrust. They used Unsymmetrical Dimethyl Hydrazine (UDMH) as the fuel and Red Fuming Nitric Acid (RFNA) as the oxidizer. The first stage was powered by an YF 2 that consisted of a cluster of four parallel YF 1 engines. It had a length of 17.84 m with a diameter of 2.25 m. It carried about 60 tonnes of propellant. It developed a sea level thrust of 1101 kN with a burn time of 130 s.

The second stage consisted of a cluster of two YF 3 engines (a high-altitude version of the YF 1). It had a length of 5.35 m, a diameter of 2.25 m, and carried 12.13 tonnes of propellant with a dry mass of 2.65 tonnes. The stage developed a thrust of 294 kN with a burn time of about 126 s.

The third stage was powered by 770 mm solid rocket motor (GF-02) that was developed by the 4th Academy. It had a length of 2 m and carried a propellant mass of about 1800 kg with a thrust of 29 kN.

Jet vanes in the exhaust of the first two stages provided the necessary forces for control. An additional flight control system powered by gaseous nitrogen was added to provide control during the coasting phase. The third stage was spin stabilized

---

[6] http://www.b14643.de/Spacerockets_1/China/CZ-1/Description/Frame.htm.

Table 14.1 CZ 1
specifications

Launcher	CZ 1
Length (m)	30.45 m
Diameter (m)	2.25 m
No. of stages	3 stages
Propellant (fuel)	UDMH
Oxidizer	RFNA
3rd stage	Solid propellant
Launch mass (Kg)	81,600 kg
Payload	300 kg, 70° 440 km orbit
First launch (Year)	1970
Status	Retired

with timing signals being used for initiating spin and for separating the satellite. The launcher could place about 300 kg in LEO.

The overall specifications for the CZ 1 are provided in Table 14.1.[7] Figure 14.15 provides an image of the CZ 1 on the launch pad in March–April 1970.

The CZ 2 C Series

In parallel with the development of the CZ 1 launcher, China had also initiated work on the more powerful CZ 2 launcher derived from the stages of the DF 5 ICBM. In addition to the CZ 2 which came under the ambit of the First Academy, a second launcher that also used the DF 5 stages called the Feng Bao (FB) was developed in parallel by the Shanghai-based 8th Academy. In the initial period, both the launchers were used for the recoverable satellite programme as well as for testing some other military payloads. After a few successes and many failures, the Feng Bao launcher programme was cancelled. The CALT-led CZ 2 development became the mainstay of the space programme. Variants and extensions built around the CZ 2 have created an entire family of launchers geared towards meeting the various mission requirements needed by an emerging and ambitious space programme. However, because of the Feng Bao programme, the Shanghai-based 8th Academy also became a parallel development center for launchers. The launcher development is divided between CALT (1st Academy) and the Shanghai Academy for Space Technology (SAST also known as the 8th Academy).

The CZ 2C launcher was configured around the two stages of the DF 5 missile. Unlike the CZ 1 this launcher uses a more advanced fuel oxidizer combination— unsymmetrical dimethyl hydrazine (UDMH) and nitrogen tetroxide (N_2O_4). The launcher was first flown in 1975. It has been in continuous use since then. It has

---

[7] http://www.braeunig.us/space/specs/lgmarch.htm.

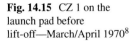

Fig. 14.15 CZ 1 on the
launch pad before
lift-off—March/April 1970[8]

flown 58 times till the end of 2020 with only two failures for an overall reliability of
97%.

Four YF 20 thrust chambers clustered together constitute the YF 21 engine that
powers the first stage of the CZ 2 C. Each of the four YF 20 engines can be gimbaled
for achieving control. The YF 21 engine along with other stage components make
up the L 140 stage 1 of the CZ 2C. The stage has a length of 20.52 m with a diameter
of 3.35 m. It carries 142 tonnes of propellant and has a dry mass of 9 tonnes. The
sea level thrust that the stage generates is 2785 kN with a burn time of 132 s.

A single YF 22 engine derived from the YF 20 optimized for high-altitude performance powers the second L35 stage of the CZ 2C. The YF 20 is combined with
four YF 23 Vernier engines to provide the YF 24 power plant for the second stage.
The four Vernier engines that can continue to operate after the main engine shuts off
provide the required control for the second-stage burn and coasting phases. The stage
has a length of 7.5 m with a diameter of 3.35 m. It can carry 35 tonnes of propellant
and the stage has a dry mass of 4 tonnes. The burn time for the main engine is 110 s

---

[8] https://news.cgtn.com/news/3d3d674e78516a4d33457a6333566d54/index.html.

Table 14.2 Specifications of the CZ 2 C launcher

Launcher	CZ 2C/SD/SM/SMA
Length (m)	43 m
Diameter (m)	3.35 m
No. of stages	2 normal + upper stages
Propellant (fuel)	UDMH
Oxidizer	N_2O_4
Lift-off weight (LOW)	245 tonnes
Payload (2 stage)	3850 kg LEO/900 kg to SSO
First launch (year)	1975
Status	Operational

while the verniers continue to burn till 190 s.[9] Several types of fairings are available for catering to different kinds of payloads. Payload injection takes place in the three-axis stabilized mode.

Though originally developed for the recovery satellite missions the CZ 2 has also been modified with the addition of third stage to meet other mission requirements.

A third-stage smart dispenser built around an FG 47 (542 mm diameter) solid rocket motor was used for the Iridium launches.

The CZ 2C was further modified to cater to the requirements of the Doublestar missions in 2003. A spin-stabilized 1.4 m diameter solid propellant motor was the base around which a new third stage was configured to cater to the requirements of this mission.

A further extension of the CZ 2C capabilities took place in 2008. Two Huanjing 1 A and I B remote sensing satellites were placed in SSO using a third stage configured around another three-axis stabilized solid rocket motor. Several extensions and modifications to the launcher have also been carried out to improve the performance of the launcher.[10]

The three-stage versions can place 2100 kg in SSO and 1250 kg in GTO. The overall specifications of the CZ 2C are provided in Table 14.2[11] (LM 3A Series 2011).

The CZ 2D Launcher

In the late 1980s, a new launcher was needed to meet the military requirements for a larger recoverable satellite. There were two competing proposals on offer. One proposal was from CALT (1st Academy) based on the CZ 2C launcher. The

---

[9] http://www.b14643.de/Spacerockets_1/China/CZ-1/Description/Frame.htm.

[10] https://chinaspacereport.com/launch-vehicles/cz2/.

[11] Over its development and use period the launcher has seen several extensions and changes. This is the most recent publicly available data.

other proposal was from SAST (8th Academy) that was based on the CZ 4 launcher under development there. The Ministry of Astronautics decided in favor of the SAST proposal in 1990. The first launch of the CZ 2D took place in 1992. The SAST designed CZ 2D is a two-stage version of the CZ 4 with some minor modifications. It is also derived from the DF 5 ICBM design.

The first stage of this launcher called the L 180 stage (carries about 180 tonnes of propellant) is the same stage as the one used for the CZ 4 launcher. The stage is powered by a cluster of 4 YF 20B engines called the YF 21B that uses UDMH as fuel and N_2O_4 as oxidizer. The four individual YF 21B engines can be gimbaled for control purposes. The stage has a length of 24.66 m and a diameter of 3.35 m. It carries 183.2 tonnes of propellant and has a dry mass of 10 tonnes. The stage produces a sea level thrust of 2961 kN at lift-off and has a burn time of 153 s.

The second stage of the CZ 2D (designated as the L35 stage) is also based on the second stage of the CZ 4. It consists of a single-stage YF 22B engine along with four YF 22 B Vernier engines that are used for control purposes. Together they constitute the YF 24 B power plant for the second stage. This stage also uses UDMH and N_2O_4 as fuel and oxidizer. The stage has a length of 10.41 m with a diameter of 3.35 m. The YF 24 B assembly provides a vacuum level thrust of 788.4 kN and burns for 115 s. The verniers continue to provide thrust after burnout of the main engine and cut off after 145 s of burning. The stage carries a propellant mass of 35.5 tonnes with a dry mass of 4 tonnes.

The CZ 2D has been launched 51 times from the time of its first launch in 1992 till 2020. There have been no launch failures giving it a reliability of 100%. Figure 14.16

The CZ 2D Launching the Venezuelan Remote Sensing Satellite Miranda

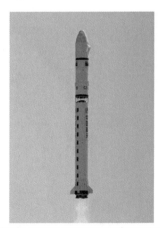

Fig. 14.16   CZ 2D launch of Venezuela's Remote Sensing Satellite October 12, 2012

Table 14.3 Specifications of the CZ 2D launcher

Launcher	CZ 2D
Length (m)	41 m
Diameter (m)	3.35 m
No. of stages	2
Propellant (fuel)	UDMH
Oxidizer	N_2O_4
Lift-off weight (LOW)	250 tonnes
Payload	4000 kg LEO/1150 kg SSO
First launch (year)	1992
Status	Operational

shows the CZ 2D launching the Venezuelan Remote Sensing Satellite Miranda.[12]

Table 14.3 provides the overall specifications of the CZ 2D (LM 3A Series 2011).

CZ 2E Launcher

As a part of the economic reform process, China made a formal announcement in 1985 that it would provide commercial launch vehicle services for international customers. The most lucrative market for launch services at that time was the market for placing telecommunication satellites in GSO. Launchers usually placed the telecom satellites into a Geostationary Transfer Orbit (GTO) from where a rocket motor onboard the satellite took it to GSO[13] (CALT 1999).

At the time of the announcement China had only one launcher the CZ 3 that could place about 1400 kg in GTO. This meant that the maximum mass of the telecom satellite when it began its life in GSO was likely to be between 600 and 700 kg.[14] However, this capability was not sufficient to cater to the needs of the 2000–2500 kg telecom satellites that had become common at that time.

In early 1986, CALT (the First Academy) came up with a proposal to modify the CZ 2C launcher by adding four liquid strap-on boosters to the first stage of the CZ 2C and an additional third solid fuel Perigee Kick Motor (PKM) cum stage to the CZ 2C for increasing the payload to GTO.

In September 1987 the Chinese space industry won a contract for launching two Hughes-built satellites for the Australian telecom company Optus. CALT was given the mandate to build and deliver the CZ 2E in 18 months for fulfilling the contract. The first launch of the CZ 2E launcher took place on July 16, 1990. It carried the

---

[12] https://en.wikipedia.org/wiki/Long_March_2D#/media/File:Long_March_2D_launching_VRSS-1.jpg.

[13] http://www.b14643.de/Spacerockets_1/China/CZ-1/Description/Frame.htm.

[14] As a rule of thumb about half the mass of the satellite in GTO consists of fuel that is required to take it from GTO to GSO.

Pakistani Badr satellite and a dummy Aussat satellite. Though the two satellites were successfully placed in Low Earth Orbit (LEO) the Chinese built Perigee Kick Motor (PKM) that was to take the dummy Aussat from LEO to GTO fired in the wrong direction resulting in a failure. For the first operational launch, the Chinese PKM was replaced by a US-built Star F 63 motor. Though the launch successfully placed an Aussat satellite in GTO, further launches ran into a string of problems with many of them failing because of the coupling between the CZ lower stages and the Star F 63 motor.

The design was such that it could be used as a two-stage launcher or as a three-stage launcher. Equipped with an ETS solid rocket motor powered upper stage, an Orbital Maneuvering System along with a smart dispenser, it could place multiple payloads in SSO and other orbits though the launch record shows that it has never been used for this purpose.

The first stage consists of a cluster of four YF 20B engines that together make up the YF 21B. Each of the four engines can be swiveled for control purposes. The stage has a length of 28.5 m with a diameter of 3.35 m. The stage carries a propellant mass of 186 tonnes, provides a sea level thrust of 2961 kN, and burns for a duration of 159 s.[15]

Four strap-on boosters are added to the first stage to increase the payload. The four boosters use the same YF 20B engine but have a diameter of 2.25 m. Each booster has a length of 15.3 m, carries 37.8 tonnes of propellant, and burns for a duration of 126 s. Each booster provides a sea level thrust of 740 kN.

The second stage of the CZ 2E (also called the L90 stage) consists of one main YF 22B along with 4 YF 23B Vernier engines that together make up the YF 24 B engine assembly. The stage has a length of 15.2 m with a diameter of 3.35 m. It carries about 85 tonnes of propellant and has a thrust of 788 kN with a burn time of 300 s. The verniers continue to burn after the main engine is shut down till about 413 s after second-stage ignition.

The CZ 2E launcher can place a 9500 kg into a 200 km 28.5° inclination orbit from Xichang. The payload from Jiuquan into a higher 53° inclination orbit is 8400 kg.

The CZ 2E has never been used as a two-stage launcher. It has always carried a solid upper stage for injection of communication satellites into GTO. Since it was originally designed to carry Hughes spinning communication satellites the Perigee Kick Motor (PKM) spins up the third-stage spacecraft stack before ignition and satellite separation.

The 4th Academy designed PKM has a diameter of 1.7 m with a length of 2.94 m. It carries 5444 kg of propellant with a total mass of about 6000 kg. The motor burns for 87 s and it has a vacuum-specific impulse of 292 s.

The three-stage PKM version of the CZ 2E can place a payload of 3500 kg into a 200 km by 35,786 km, 28.5° inclination Geostationary Transfer Orbit (GTO).

Out of a total of 7 launches (all of them to GTO) there were 3 failures for a reliability of 57%. The poor reliability record led to an early retirement of the launcher.

---

[15] This is very similar to the L180 stage of the CZ 2D—possibly the CZ 2E is closer to the CZ 2D than the CZ 2C.

Table 14.4 Specifications of the CZ 2E launcher

Launcher	CZ 2E
Length (m)	49.7 m
Diameter (m)	3.35 m
No. of stages	2 + solid upper stages
Propellant (fuel)	UDMH
Oxidizer	N_2O_4
Upper stage	Solid propellants
Lift-off weight (LOW) 2 stages	460 Tonnes
Payload LEO 2 stg./GTO 3 stg	9500 kg LEO/3500 kg GTO
First launch (year)	1990
Status	Retired

However, the CZ 2E capability to deliver a payload of over 8 tonnes (with strap-on additions) into LEO led to its selection as the preferred launcher for the Shenzhou Human Space Flight Programme. Table 14.4 provides the overall specifications of the CZ 2E.[16]

The CZ 2F

Though the CZ 2E was retired after seven flights it got a new lease of life when the Human Space Flight Programme was approved in 1992. Since the CZ 2E remained the most powerful rocket that was available the Chinese decided to use it to power their initial forays into establishing a space station. A major project was initiated to transform the CZ 2E into a human flight vehicle.

Many improvements had to be made and a major project was initiated as a part of the human space flight programme. One major change that had to be made was the addition of an escape system for a possible flight abort during the early lift-off phase of the flight. China turned to Russia for help with this system but ultimately built its own. The CZ 2F vehicle has so far flown 14 times and has a perfect record of 14 successful flights.

It can place a payload of 8100–8600 kg into LEO from the Jiuquan Launch Complex. Figure 14.17 provides a visual overview of the CZ 2F[17] (see also Chap. 11 on the Human Space Flight Programme).

---

[16] https://chinaspacereport.com/launch-vehicles/cz2/.

[17] https://chinaspacereport.com/launch-vehicles/cz2/.

Fig. 14.17   The CZ 2F on the launch pad with the Shenzhou 13 spacecraft

The CZ 4 Series

The CZ 4

There were two competing launcher proposals from CALT and SAST for the design of the launcher that would place China's first communication satellite in GSO. The SAST proposal involved the addition of a conventional liquid third stage to the CZ 2C for providing the necessary thrust to place the satellite in the required GTO. The CALT proposal involved the use of a cryogenic engine-powered third stage on the CZ 2C to provide the required capability.

After the selection of the CALT proposal, Chinese decision-makers did not abandon the SAST proposal. They wanted the same design to be tailor made for placing satellites in Sun Synchronous Orbits (SSO) from a new launch center in Taiyuan. The development of the rocket began in 1985.

The first launch of the CZ 4 took place on September 6, 1988, when the Feng Yun 1A weather satellite was placed into a near-polar sun synchronous orbit from Taiyuan. This was followed by a second successful launch of the Feng Yun 1B weather satellite along with two weather balloon satellites in September 1990. About a month later the third stage which was also in orbit exploded creating a large debris field. Subsequent launches were designed to take care of this problem by using a residual fuel venting system on the third stage.

The first stage of the CZ 4 launcher (also called the L 180 stage) was stretched from the CZ 2C version by 4 m to accommodate another 40 tonnes of propellant. It

Table 14.5 Specifications of the CZ 4 launcher

Launcher	CZ 4
Length (m)	41.9 m
Diameter (m)	3.35 m
No. of stages	3
Propellant (fuel)	UDMH
Oxidizer	N_2O_4
Lift-off weight (LOW)	249 tonnes
Payload SSO	1700 kg
First launch (year)	1988
Status	Replaced by CZ 4B after 2 flights

also had an improved YF 20 B engine cluster called the YF 21B that increased its thrust to 2942 kN. The stage has a length of 24.67 m with a diameter of 3.35 m. It carries about 183 tonnes of propellant and has a dry mass of 10 tonnes. The stage develops a sea level thrust of 2961 kN at lift-off and has a burn time of 151 s.

The second-stage main engine is a high-altitude version of the YF 20 B. This along with four FY 23 B Vernier engines constitutes the YF 24 B (L 35) second stage. Both the first- and second-stage engines are improved versions of the engines used for the original CZ 2C. The stage has a length of 10.41 m with a diameter of 3.35 m. It has a dry mass of 4 tonnes and can carry 35.5 tonnes of propellant. The YF 24 B generates a vacuum level thrust of 788 kN and burns for 127 s.

The third stage was powered by a new YF 40 liquid engine consisting of a pair of motors that could be swiveled. It also had an improved computerized guidance system and an onboard propellant management system. It has a length of about 4.9 m with a diameter of 2.90 m. The propellant mass is 14150 kg. The engine generates a thrust of 98.1 kN with a single burn time of 321 s. Fourteen hydrazine thrusters on the third stage provide attitude control during the coasting and orbit insertion phases. The CZ 4 had two kinds of fairings to take care of single and multiple satellite launchings.

This configuration was used for only the first two flights for placing two Feng Yun weather satellites in polar orbit. An improved version called the CZ 4B was developed for placing the CBERS satellite in SSO.

Table 14.5 provides details of the CZ 4 launcher.[18]

The CZ 4B

The CZ 4B was very similar to the CZ 4 in terms of design. It was flown for the first time in May 1999 for placing the FengYun 1C weather satellite in a near-polar sun synchronous orbit. A few months later it was also used for placing the Sino-Brazilian remote sensing satellite CBERS in a 730 km SSO.

The major addition to the CZ 4 configuration was the incorporation of a new payload fairing of length 8.48 m to accommodate bigger remote sensing satellites.

---

[18] http://www.braeunig.us/space/specs/lgmarch.htm.

Fig. 14.18 The CZ 4B on the launch pad at Taiyuan

Other improvements included an all-digital electronic control system, an improved TT&C, and self-destruct systems.

The nozzle of the second stage was also modified to provide better high-altitude performance. An improved propellant management system was also incorporated in the second stage to reduce the spare propellant loading requirements and increase the payload. The length of the CZ 4B is 45.8 m.

After the third stage of the CZ 4 launcher carrying the FengYun 1B weather satellite had exploded in orbit creating a debris problem, the designers of the CZ 4 incorporated a propellant venting system in the third stage to prevent this problem from recurring. This modification had however not been incorporated into the design of the CZ 4 B launcher that launched the CBERS due to objections from the satellite designers.

In May 2000, this upper stage again exploded creating debris. However, all CZ 4 series launches after that have incorporated a residual fuel venting system. No explosions of the third stage have taken place after this event.[19]

Figure 14.18 shows the CZ 4B on the launch pad at Taiyuan preparing for takeoff. Table 14.6 provides details of the CZ 4B launcher.

From its debut in 1999 till 2020, there have been 40 flights of the CZ 4B launcher. Only one of these launches failed in 2013 for an overall reliability of 98%. It can deliver a payload of 2230 kg in SSO (LM 3A Series 2011).

---

[19] https://chinaspacereport.com/launch-vehicles/cz4/.

Table 14.6 Specifications of the CZ 4B launcher

Launcher	CZ 4 B
Length (m)	45.8 m
Diameter (m)	3.35 m
No. of stages	3
Propellant (fuel)	UDMH
Oxidizer	N_2O_4
Lift-off weight (LOW)	249 tonnes
Payload SSO	2230 kg
First launch (year)	1999
Status	Operational

The CZ 4C Launcher

The CZ 4C launcher was introduced in 2006 to launch the Yaogan 1 SAR satellite. While the CZ 4B launcher continues to be the mainstay launcher for SSO missions the CZ 4C is used for launching heavier satellites as well launching multiple satellites on the same launcher. All launches of the CZ 4 series have taken place from the Taiyuan Launch Complex. A significant deviation from this pattern occurred on March 5, 2010, when a CZ 4C launcher carrying a constellation of 3 ELINT satellites were put into a 63° inclination orbit from the Jiuquan Launch Complex. Since then, several such triplet Yaogan satellites have been launched from Jiuquan using the CZ 4C.

The major difference between the CZ 4B and the CZ 4C is in the third-stage engine. An improved YF 40A engine in the third stage provides the CZ 4C with a multiple restart capability. Improvements in the avionics and third-stage propellant management system have raised the payload delivered in SSO to 2950 kg (LM 3A Series 2011). Apart from these minor changes, the CZ 4B and CZ 4C have very similar configurations.

The CZ 4C has been launched 29 times since its first launch in 2006 including 7 launches from Jiuquan. It has had two failures for a reliability rating of 93%.

The CZ 4 series have had a total of 71 launches with only three failures for a reliability of 96%. This is on par if not superior to the launch record of other major players in the space game.

The CZ 3 Series of Launch Vehicles

The CZ 3

When China decided to initiate the first communication satellite project it had two competing proposals for the development of the required launch vehicle. Both involved the use of first two stages of the flight-tested CZ 2C launcher. The CALT (First Academy) proposal involved the development of new cryogenic third stage using liquid hydrogen and liquid oxygen as fuel and oxidizer for placing the required mass in GTO. The SAST (8th Academy) proposal involved development of a conventional UDMH/N_2O_4 third stage for the satellite mission. Both proposals were supported in parallel.

The development of the YF 73 cryogenic engine faced many problems including an explosion during a ground test and a fire. However, it continued to be supported by the Director of the First Academy Ren Xinmin who managed to convince the authorities that this was the way to go. Finally, in 1980, the FY 73 engine successfully passed 800 s of ground testing to become the preferred choice for powering the third stage of the CZ 3 launch vehicle.

On January 29, 1984, the CZ 3 launcher lifted off from the new Xi Chang launch complex carrying an indigenously built DFH 2 communication satellite. The first and second stages functioned normally and the third-stage cryogenic was also fired for the first time to place the satellite in a 200 km LEO.

However, the cryogenic stage failed to fire the second time to raise the orbit from LEO to GTO. The satellite was left stranded in LEO and though the Chinese used the satellite's apogee boost motor to raise the orbit the mission could not be termed a complete success.

Two months later a second CZ 3 launcher lifted off from Xi Chang carrying another indigenous communications satellite. This time around the launch took place flawlessly. The third stage fired a second time to take the satellite from a 450 km by 170 km 27° inclination orbit to the designated GTO. After the separation of the satellite from the 3rd stage, the Apogee Boost Motor (ABM) on the satellite was fired to place it in GSO. From there it was successfully moved to its GSO slot at 125°. The DFH 2 communications satellite started providing telecom and TV services to domestic users.[20]

The first two stages of the CZ 3 launcher are based on the L140 and L35 stages of the CZ 2C launcher. The third stage is powered by an YF 73 cryogenic (liquid hydrogen liquid oxygen) specifically developed for the CZ 3.

The YF 73 engine that powers the third stage consists of four chambers with fuel being supplied by a single turboprop. The stage has a length of 7.48 m with a diameter of 2.25 m. It carries a propellant mass of 8500 kg with a dry mass of 2000 kg. The vacuum thrust developed by the stage is 44.15 kN. The YF 73 can burn for a total of 800 s with a restart capability. A set of hydrazine thrusters provides attitude control during the coasting and orbit insertion phases.

There have been 13 launches of the CZ 3 with 3 failures for a reliability of 77%. The CZ 3 was retired in 2000 after the CZ 3A with a larger GTO payload mass became available.

Table 14.7 provides an overview of the CZ 3[21] (CALT 1999). Apart from placing the early weather satellites in GSO the CZ 3 also placed some bought satellites of

[20] https://chinaspacereport.com/launch-vehicles/cz3/.

[21] http://www.braeunig.us/space/specs/lgmarch.htm.

Table 14.7 Specifications of the CZ 3 launch vehicle

Launcher	CZ 3
Length (m)	44.6 m
Diameter (m)	3.35 m
No. of stages	3
Propellant (fuel)	UDMH Stg 1and 2 liquid H_2 stg 3
Oxidizer	N_2O_4 Stg 1 and 2 liquid oxygen stg 3
Lift-off weight (LOW)	204 tonnes
Thrust	2785 kN
Payload (GTO)	1450 kg
First launch (year)	1984
Status	Retired

the Asiasat and Apstar series in GSO for domestic use in China. For a good image of the CZ 3 launcher, see the attached link.[22]

The CZ 3A

The CZ 3A launcher was developed by China to capture some of the market for GSO communication satellites. The CZ 3 payload capacity was not sufficient for launching the heavier satellites that were beginning to appear in the middle to the late 1980s. Though like the CZ 3 it was also derived from the CZ 2C basic design it differed from it in several ways. A major difference is the addition of a new YF 75 cryogenic engine-powered third stage that enhances the GTO payload to 2600 kg.

The CZ 3A was designed as a Chinese launcher to cater to the emerging market for heavier (medium class) geostationary communication Satellite. Development was initiated in 1986 with the first launch scheduled for 1992. Because of CALT's preoccupation with the CZ 2E, the development of the CZ 3A was delayed with the first launch taking place only in 1994.[23]

The CZ 3A was the first of the three variants that China wanted to develop to cater to the requirements of the emerging markets for launches into GTO. Plans for enhancing payload capacity of the CZ 3A through the addition of strap-on boosters were included in the original design.

The first stage of the CZ 3A is based on the L180 stage of the later CZ 2C variants. It has a length of 23.27 m and a diameter of 3.35 m. It carries 171.8 tonnes of propellant and generates a sea level thrust of 2962 kN. The four nozzles of the YF 21 C engine can be swiveled for control purposes. The specific impulse is 2556.5 N seconds per Kg.

The second stage is also a derivative of the L35 stage of the CZ 2C. It has one main engine along with four Vernier engines for control purposes. It has a length of 11.28 m with a diameter of 3.35 m. It carries 32.6 tonnes of propellant. The total thrust

---

[22] https://chinaspacereport.com/launch-vehicles/cz3/.

[23] https://chinaspacereport.com/launch-vehicles/cz3/.

Table 14.8 Specifications of
the CZ 3A launch vehicle

Launcher	CZ 3 A
Length (m)	52.5 m
Diameter (m)	3.35 m
No. of stages	3
Propellant (fuel)	UDMH stg 1 & 2 Liquid H_2 stg 3
Oxidizer	N_2O_4 stg 1 & 2 Liquid O_2 stg 3
Lift-off weight (LOW)	241 tonnes
Lift-off thrust	2962 kN sea level
Payload (GTO)	2600 kg
First launch (year)	1994
Status	Operational

(vacuum) of the stage is 790 kN. The main engine has a vacuum-specific impulse of 2922.57 N seconds per Kg while the Vernier engine has a vacuum-specific impulse of 2910.5 N seconds per Kg.

The third stage is powered by the new advanced YF 75 cryogenic engine (Zhang 2013). The YF 75 has a pair of thrust chambers linked together to provide the required power plant. It has a length of 12.38 m with a diameter of 3 m. It carries 18,200 kg of propellant. The stage provides a vacuum level thrust of 167.77 kN and has a vacuum level specific impulse of 4295 N seconds per Kg (LM 3A Series 2011).

The CZ 3A has been launched 28 times till the end of 2016 and it has never failed for a reliability of 100%. Table 14.8 provides the specifications of the CZ 3A. Figure 14.19 shows a lift-off of the launcher.

The CZ 3B

The CZ 3B launcher can place one heavy class or two medium class communication satellite in GTO. It uses the same three-stage configuration as the CZ 3A. However, to increase the payload to 5100 kg in GTO four liquid strap-on boosters are added. Its development began in 1989 with the first flight taking place in 1996. Its maiden flight in February 1996 carrying an Intelsat 708 satellite resulted in a major launch vehicle disaster. Immediately after lift-off the launcher veered away and hit a hill very close to the launch pad. The resulting explosion and fire killed several people injured others and destroyed many buildings. The problem was traced to a soldering defect in the outputs of one of the gyro servo-loops. Despite this problem the CZ 3B successfully orbited an Agila Comsat for the Philippines in 1997.[24] There were no further problems with the CZ 3B till 2008. In 2008, the Indonesian Palapa 1D satellite was not able to reach its designated GTO orbit because of problems in the second firing of the third-stage cryogenic engine. The mission was retrieved by using the satellite propulsion system to raise it to GSO though the satellite life was shortened.

---

[24] https://chinaspacereport.com/launch-vehicles/cz3/.

A CZ 3A Lift-off

Credit: CGWIC

Fig. 14.19 A CZ 3A lift-off

There have been no failures since then till 2018. In 2019 and 2020, however, two CZ 3B launchers failed.

The three main stages of the CZ 3B are identical to those of the CZ 3A. Four liquid strap-on boosters are added to the first stage to increase the payload. The strap-on booster uses UDMH and N_2O_4 as fuel and oxidizer, respectively. They are derived from the boosters developed for the CZ 2E and the CZ 2F launch vehicles. Each booster has a length of 15.33 m with a diameter of 2.25 m. It carries 37,700 kg of propellant and develops a sea level thrust of 740.4 kN. The boosters have a sea level specific impulse of 2556.2 N seconds per kg (LM 3A Series 2011). Table 14.9 provides the basic specifications of the launcher. Figure 14.20 gives a visual overview of the CZ 3B.

The CZ 3BE Launcher

An enhanced variant of the CZ 3B called the CZ 3BE was introduced in 2007 for launching Nigeria's Nigicomsat. The first stage of the CZ 3B was extended by 1.5 m and the four strap-on boosters were stretched by 0.8 m each to accommodate more propellant.[25] This enabled the CZ 3BE to place 5500 kg in GTO.

The CZ 3B/CZ 3BE rockets have been launched 72 times till the end of 2020 with four failures for a reliability of 94%. Table 14.10 provides the basic specifications of the CZ 3BE (LM 3A Series 2011).

---

[25] https://chinaspacereport.com/launch-vehicles/cz3/.

Table 14.9 Specifications of the CZ 3B launch vehicle

Launcher	CZ 3 B
Length (m)	54.8 m
Diameter (m)	3.35 m
No. of stages	3
Propellant (fuel)	UDMH stg 1 and 2 liquid H_2 stg 3
Oxidizer	N_2O_4 stg 1 and 2 liquid O_2 stg 3
Lift-off weight (LOW)	425.8 tonnes
Lift-off thrust	5923 kN
Payload (GTO)	5100
First launch (year)	1996
Status	Operational

A CZ 3 B Lift-off from Xi Chang

Fig. 14.20 A CZ 3B lift-off from the Xi Chang launch complex

The CZ 3C Launcher

The CZ 3C is the third variant built around the basic CZ 3A configuration. Its main purpose is to fill the gap in payload delivery to GTO that lies between 2600 kg (medium class comsats) and 5100 kg (heavy class comsats). Through the addition of two liquid strap-on boosters to the CZ 3A three core stages the payload capacity to GTO is raised to 3800 kg which can cater to the intermediate class of communication satellites. One major difference between the CZ 3C and the CZ 3A and CZ 3B

Table 14.10 Specifications of the CZ 3BE launch vehicle

Launcher	CZ 3 BE
Length (m)	56.3 m
Diameter (m)	3.35 m
No. of stages	3
Propellant (fuel)	UDMH stg 1 & 2 Liquid H_2 stg 3
Oxidizer	N_2O_4 stg 1 & 2 Liquid O_2 stg 3
Lift-off weight (LOW)	456 tonnes
Lift-off thrust	5923 kN
Payload (GTO)	5500
First launch (year)	2007
Status	Operational

Table 14.11 Specifications of the CZ 3C launch vehicle

Launcher	CZ 3 C
Length (m)	54.8 m
Diameter (m)	3.35 m
No. of stages	3
Propellant (fuel)	UDMH stg 1 & 2 Liquid H_2 stg 3
Oxidizer	N_2O_4 stg 1 & 2 Liquid O_2 stg 3
Lift-off weight (LOW)	345 tonnes
Lift-off thrust	4443 kN
Payload (GTO)	3800 kg
First launch (year)	2008
Status	Operational

launchers is the absence of fins on the CZ 3C.[26] The CZ 3C made its maiden flight in 2008 when it launched the first Chinese Tracking and Data Relay Satellite (Tianlian 1). The overall specifications of the CZ 3C are given in Table 14.11 (LM 3A Series 2011). The CZ 3C has been launched 16 times till the end of 2020 with no failures for a reliability record of 100%.

The New Heavy Lift Launcher CZ 5[27]

As China's ambitions in space soared, it was no longer possible to realize all of them through the proven CZ 2 derived stable of launch vehicles.

---

[26] https://chinaspacereport.com/launch-vehicles/cz3/.

[27] https://chinaspacereport.com/launch-vehicles/cz5/.

Around the year 2000 the Chinese government approved the development of a new series of advanced rocket engines to power the new launchers required for meeting its needs. One of the engines was the 120-tonne thrust YF 100 that used kerosene and liquid oxygen as fuel and oxidizer. A smaller version of this engine called the YF 115 with a thrust of 18 tonnes was also cleared for development. The second major engine development that was approved was the 50-tonne thrust YF 77 cryogenic engine that used liquid hydrogen and liquid oxygen as fuel and oxidizer.

The programme also included several conventional hypergolic as well as cryogenic engines for powering the upper stages of the new launch vehicles. The original plan was to develop stage modules of 2.25, 3.35, and 5 m built around the use of the YF 100 and YF 77 engines. These modular stages could then be assembled into several configurations based on common engines and stage designs.

The modular approach was eventually abandoned. Instead, the Chinese opted to build four new launchers—the small payload CZ 6, the medium payload CZ 7, the heavy payload CZ 5, and a small to medium launcher the CZ 8. The go ahead given to the new projects especially the heavy payload CZ 5 also involved a major revamping of the manufacturing and transportation infrastructure. The CZ 2 series of launchers and their derivatives were constrained because China's railway network could not handle sizes that had a diameter more than 3.35 m and a length of 14 m.[28] A new rocket and spacecraft manufacturing and test facility has been built at Tianjin on China's east coast to overcome this constraint. A new launch center has also been built at Wenchang on Hainan Island with two launch complexes for the launch of the larger vehicles.[29] The rocket and stage components are transferred by sea to the launch site. Two specially built Yuanwang ships are available for this operation.

Development activities began in 2004 with the main contract being awarded to CALT. SAST however also had responsibilities for the development of some parts. The YF 100 engine was successfully tested for 200 s of operation in May 2012. With this China became only the second country in the world to have mastered the high-pressure staged combustion rocket engine technology.

Though the original CZ 5 design had provisions for six variants using different modules only two versions have been developed so far. The basic two-stage CZ 5 variant is to be used for GTO missions while the single-stage CZ 5B will be used for missions to LEO. Both variants will have four strap-on 3.35 m boosters to provide the required thrust during lift-off. In its different avatars, it can deliver a 14-tonne payload to GTO and a 22-tonne payload to LEO.

The first stage of the CZ 5 has a length of 31.02 m with a diameter of 5 m. It is powered by a pair of YF 77 cryogenic engines that can be independently swiveled for steering the launcher. Each YF 77 engine generates a thrust of 50 tonnes at lift-off (70 tonnes vacuum level thrust). It carries 158 tonnes of propellant with a stage dry mass of 17.8 tonnes. It burns for a total of 520 s. The YF 77 does not use the staged combustion cycle but the less efficient gas generator cycle. This results in some loss

---

[28] The tunnels apparently could not accommodate larger sizes.

[29] The location also provides a range of trajectory options including a more optimum trajectory to the GTO that increases the payload delivered in orbit.

Table 14.12 Specifications of the CZ 5 launch vehicle

Launcher	CZ 5 basic configuration
Length (m)	56.97 m
Diameter (m)	5 m
No. of stages	2 + 4 strap-on boosters
Propellant (fuel)	LH2 stg 1 and 2 kerosene strapons
Oxidizer	LO2
Lift-off weight (LOW)	869 tonnes
Lift-off thrust	10,573 kN
Payload (GTO)	14,000 kg
First launch (year)	2016
Status	Initial phase of development

of performance as compared to other engines developed by some of the major space powers.

In addition to the cryogenic first stage, the CZ 5 also has four strap-on boosters powered by LOX Kerosene YF 100 for providing the initial thrust during lift-off. Each strap-on has a length of 26.28 m with a diameter of 3.35 m. Each of the boosters is powered by a pair of LOX kerosene YF 100 engines that generates 2680 kN of thrust during lift-off. Each booster carries 135 tonnes of propellant with a stage dry mass of 12 tonnes. The burn time is 155 s.

The second stage also has a diameter of 5 m with a length of 12 m. It is powered by a pair of YF 75D cryogenic engines derived from the YF 75 that has been extensively used for the CZ 3A, 3B, 3BE, and 3C launchers. The engines can be swiveled for control purposes. The gross mass of the stage is 26 tonnes with an empty mass of 3.1 tonnes. The stage burns for 780 s.

CALT has also developed a new upper stage the YZ 2 (Yuan Zheng 2) for use with the CZ 5. Its purpose is to function as a space tug for delivering spacecraft and satellites directly into their orbits. It can perform multiple burns and performs a series of orbital maneuvers for carrying out a complete mission. The diameter of this stage is 5.2 m. The payload fairing for the CZ 5 has a 5.2 m diameter with a length of 12.5 m.

The first launch of the CZ 5 took place on November 3, 2016, from the new Wenchang Launch complex. The configuration flown was the basic two-stage core vehicle with four strapons. The CZ 5 placed the YZ 2 upper stage carrying the Shijian satellite in orbit. After the Shijian was injected into the required orbit by the YZ 2 it was moved to the GSO using its ion propulsion thrusters.[30] Table 14.12 provides the specifications of the CZ 5 launcher. The second launch in 2017 however failed. The CZ 5 returned to service with a successful flight in 2019. It has since then been launched thrice that include missions to the Moon and Mars. Along with the CZ 7 it will be used for the construction of China's Space Station. Figure 14.21 provides a visual overview of the launcher.

---

[30] https://www.nasaspaceflight.com/2016/11/china-long-march-5-maiden-launch/.

A CZ 5 being transported to the Launch Pad

Credit: 篁竹水声

Fig. 14.21   The CZ 5 being transported to the launch pad at Wenchang

The CZ 6 Launcher

The CZ 6 is the small lift launcher of the new generation of satellite launchers that China has had under development for quite some time now. It is a three-stage launcher with a length of 29.24 m. It has a lift-off weight of 103.217 tonnes with an empty mass of about 9 tonnes. It can place a payload mass of 1080 kg in a 700 km SSO.

The CZ 6 project was approved in 2009. It uses the same engines that had been developed to power the first and second stages of other launchers including the CZ 5. Its development was first contracted to CALT but was later transferred to SAST.

The first stage of the CZ 6 has a length of about 15 m with a diameter of 3.35 m. The stage carries 76 tonnes of propellant. It is powered by the YF 100 engine that uses kerosene and liquid oxygen as fuel and oxidizer. It develops a sea level thrust of 1180 kN.

The diameter of the second stage is 2.25 m with a length of about 7.3 m. It is powered by the YF 115 engine which is a scaled-down version of the YF 100 that uses the same propellant combination of kerosene and liquid oxygen. It carries about 15 tonnes of propellant and generates a vacuum level thrust of 175 kN. Thrust vectoring along with small thrusters provides control for the launcher.

The stage can be stopped and started again for several burns. During the time the vehicle coasts the third-stage control takes over to guide the vehicle.

The third stage of the CZ 6 has a length of 1.8 m with a diameter of 2.25 m. It is powered by four 1000 N thrusters that allow for trimming and insertion into the

Table 14.13 Specifications of the CZ 6 launch vehicle

Launcher	CZ 6
Length (m)	29.24 m
Diameter (m)	3.35 m stg 1, 2.25 m stg 2 and 3
No. of stages	3
Propellant (fuel)	Kerosene (all stages)
Oxidizer	LO2 (all stages)
Lift-off weight (LOW)	103.22 tonnes
Lift-off thrust	1180 kN
Payload (700 km SSO)	1080 kg
First launch (year)	2015
Status	Initial phase of development

required orbit. It has another eight additional thrusters (100 N thrust) that are used for attitude control.[31]

A new launch pad for the CZ 6 has been built at Taiyuan. The launcher can also be launched from the new launch complex at Wenchang.

The small size of the launcher makes it possible for the whole launcher to be assembled at the factory and then transported to the launch site. Arrangements to check out the launcher and the conduct the fueling operations have also been simplified and automated to a great extent. The launch campaigns are typically expected to be about 7 days as compared to the earlier campaigns that could take as long as 30 to 40 days. Table 14.13 provides the overall specifications of the CZ 6 and the reference provides a visual of the lift-off.

There is some information that suggests that the CZ 6 will be modified to make it on par with the CZ 7 by adding strap-on boosters and improved first and second stages.

It was launched for the first time on September 19, 2015, placing 14 small satellites in sun synchronous orbits of about 520 km. A second launch in 2017 placed 3 Jilin commercial satellites in SSO. Two successful launches followed in 2019 and 2020. It can be used with two types of fairings with diameters of 2.25 or 2.6 m.[32]

The CZ 7 Launcher

The CZ 7 is the medium lift launcher in China's new generation of launch vehicles.

The origins for the CZ 7 go back to ideas put forth in the mid-2000s that were related to upgrades of the CZ 2F human flight launcher. By replacing the hypergolic propellants used in the CZ 2F with greener propellants such as kerosene and liquid

---

[31] http://spaceflight101.com/spacerockets/long-march-6/.

[32] https://chinaspacereport.com/launch-vehicles/cz6/.

oxygen a more environmentally friendly launcher could be realized. This concept was merged with the plans for development of a new generation of launchers. The CZ 7 thus became the medium lift launcher of this new generation. The development was entrusted to CALT.

Actual development of the launcher was initiated in 2010. The first flight of the launcher took place on June 25, 2016 from the new Wenchang launch complex on Hainan Island. A second launch took place from Wenchang in 2017 for testing the unmanned Tianzhou ferry for the space station. A third launch in March 2020 failed.

According to Chinese sources, over 80% of the satellites to be launched within the next 20 years would require the services of a medium lift launcher. This would comprise heavy telecom satellites to GSO, large earth observation payloads to SSO as well as crewed and cargo flights.

The CZ 7 will be initially used to launch automated Tianzhou cargo vessels to the Tiangong space laboratory and to the "soon-to-be-established" Chinese space station. A test flight in 2017 successfully demonstrated this capability.

The core CZ 7 vehicle consists of two stages both of which have a diameter of 3.35 m. Four liquid strap-on boosters are added to the core first stage to provide the required thrust during lift-off. All engines on the first second and booster stages use kerosene as fuel and liquid oxygen as oxidizer. The thrust at lift-off is 7200 kN.

The launcher can place a payload of 13,500 kg in 200 by 400 km 42° inclined orbit or 5500 kg in a 700 km SSO. It can place a mass of 7000 kg in GTO. An additional upper stage (YZ 1) option is also available to increase the payload delivered in orbit or improve the orbit insertion accuracy.[33]

The first stage of the CZ 7 is 26 m long and is powered by a pair of YF 100 engines. It has a dry mass of 12.5 tonnes with propellant mass 174 tonnes. The engines can be swiveled for achieving control. The stage (two YF 100 engines) develops a sea level thrust of 2360 kN and burns for about 215 s. Each of the four strap-on boosters is also powered by a single YF 100 engine generating a sea level thrust of 1080 kN. The stage has a length of 26.5 m. Each booster can carry about 75.5 tonnes of propellant with a dry mass of about 6000 kg. The burn time for each booster is about 185 s.

The second stage has a diameter of 3.35 m (the same as the first stage) with a length of 11.5 m. It carries about 65 tonnes of propellant and has a dry mass of 5.5 tonnes. It is powered by four YF 115 engines that provide a total vacuum level thrust of 180 kN with a specific impulse of about 341.5 s.

The CZ 7 fairing has a diameter of 4.32 m with a length of 12.4 m. Other fairing options are also available depending on the user requirements. As of now it can only be launched from Wenchang. However, in the future, it could be launched from any one of the other three sites as well. Figure 14.22 provides a full-scale view of the launcher. Table 14.14 provides the broad specifications of the CZ 7.

The first launch (June 25, 2016) which was also the first launch from the new Wenchang complex carried a prototype version of the next-generation crew return vehicle. After injection into orbit, it was successfully recovered. Three other small

---

[33] https://chinaspacereport.com/launch-vehicles/cz7/.

Model of the CZ 7 Rocket – Paris Air Show

Credit: Pline

Fig. 14.22 A mock-up model of the CZ 7 at the Paris air show 2015

Table 14.14 Specifications of the CZ 7 launch vehicle

Launcher	CZ 7
Length (m)	53.1 m
Diameter (m)	3.35 m
No. of stages	2 + 4 strap-on boosters
Propellant (fuel)	Kerosene all stages
Oxidizer	Liquid oxygen all stages
Lift-off weight (LOW)	595 tonnes
Lift-off thrust	7080 kN
Payload LEO	13,500 kg 200 × 400 km 42°orbit
Payload GTO	7000 kg
Payload (700 km SSO)	5500 kg
First launch (year)	2016
Status	Initial phase

satellites were also carried on this first flight.[34] A second flight in 2017 demonstrated the cargo capability to the future space station by docking and undocking with the Tiangong 2 Space Laboratory. A third flight to the GSO in 2020 ended in a failure.

---

[34] https://www.nasaspaceflight.com/2016/06/china-debuts-long-march-7-rocket/.

The CZ 8 Launcher

In December 2020, the maiden flight of the CZ 8 launch vehicle took place from the Wenchang Launch center. Five satellites including a classified SAR payload were placed in a 500 km Sun Synchronous Orbit (SSO). With this launch all the new green launchers have made their debut. The first stage is powered by the same liquid oxygen kerosene engines used in the CZ 7. The upper stage uses a cryogenic (liquid hydrogen liquid oxygen) engine derived from the CZ 3 series of launch vehicles. Two to four strap-on motors can be added to the first-stage core to increase the payload. The CZ 8 can deliver 3 to 4.5 tonnes to a 700 km SSO, 2.5 tonnes to GTO, and about 8000 kg to LEO. In keeping with current trends, the first stage will be recovered through a vertical landing arrangement and re-used. The reference provides an image of the CZ 8 on the launch pad ready for takeoff.[35]

The CZ 11 Launcher

The CZ 11 is a four-stage solid propellant rocket designed to place single or multiple small satellites in orbit. This launcher is a CALT development and is derived from the DF 31 and DF 31A ICBM. A small fourth upper stage has been added to provide the necessary velocity for injection into orbit. The development of the rocket began in 2010.

The first launch of the CZ 11 took place from the Jiuquan Satellite Launch Complex on September 25, 2015. The launch took place with the missile being carried to the launch site on a mobile transporter erector vehicle. The missile which is inside the canister was launched from this mobile platform.

This ability to launch from any location with minimal preparation provides a flexible and quick turnaround capability that may be needed for military applications.

A second CZ 11 launch took place on November 9, 2016. This launch was also from a canister carried on a mobile transporter erector vehicle. The launch placed a 240 kg X-ray Pulsar Navigation Satellite in a 500 km SSO.

Three successful launches of the CZ 11 took place in 2018, 2019, and 2020, respectively. These demonstrate operational capabilities to launch on demand mainly to meet military needs.

Flights from a mobile platform in the Yellow Sea Launch Area (YLSA) also took place in 2019 and 2020.

The CZ 11 has a length of 20.8 m with the diameter of the first three stages being 2 m. The fourth stage has a smaller diameter which is most probably 1.4 m. The lift-off weight of the CZ 11 is 58 tonnes with a thrust of 120 tonnes. It can place a 700 kg satellite in LEO or a 350 kg satellite in a 700 km SSO.

---

[35] http://www.xinhuanet.com/english/2020-12/22/c_139610000.htm.

The first stage of the CZ 11 has a diameter of 2 m and a length of 9 m. The second and third stages also have diameters of 2 m with lengths of 3 m and 1 m, respectively. The fourth stage has a diameter of 1.4 m with a length of 1 m. The payload fairing has a diameter of 1.6 m with a length of 5.7 m.[36] There is little open-source information available on the design of the CZ 11.

The references provide a visual overview of the canister, the launch of the missile from the canister, and the TEL used for transporting and launching the CZ 11.[37,38]

The Kuaizhou Family of Launchers

The Kuaizhou series of launch vehicles are developed by the China Aerospace and Industry Corporation (CASIC). It is directly derived from China's first Kinetic Kill Vehicle or ASAT weapon that is based on the 1.4 m DF 21 missile.

Work on the ASAT weapon started in 2002 with a contract being awarded to the Sanjiang Aerospace Group (also known as CASIC's 9th Academy or the 066 Base) for the development of the KT 409 (also called the SC 19) missile. The KT 409 was flight tested between 2005 and 2007. The last test was used to directly intercept and destroy the defunct FY 1C weather satellite in a 700 km SSO.

The KZ 1 launcher is directly derived from the KT 409. It has four stages. The first three are all fueled with solid propellants while the fourth last stage uses a small liquid stage. The overall length of the KZ 1 is about 20 m. The lift-off weight of the KZ 1 is about 30 tonnes. The first two stages have a diameter of 1.4 m with the third and fourth stages having a diameter of 1.2 m. The fourth stage of the launcher and the satellite functions are all integrated into one stage. This reduces operational complexity making the launcher ready very quickly. The satellite along with the launcher is integrated into one unit at the factory itself. It is then transported to the launch site by a TEL from where it can be launched in a matter of hours rather than days. The KZ 1 can place a 430 kg satellite in a 500 km SSO.

Two launches of the KZ 1 have taken place in 2013 and 2014. Two satellites were placed in near-polar 300 km orbits. The satellites were remote sensing satellites testing out optical CCD-based camera systems. Launches in 2017 and 2018 placed a Jilin constellation and a test Centispace navigation satellite in SSO. Five launches of the KZ took place in 2019 followed by four in 2020. The last two launches in 2020 ended in failure.

There is a second variant of the KZ 1 which is referred as the KZ 1A. In this variant, the satellite and the fourth stage are not integrated into one but are separate entities. Two fairing options with diameters of 1.2 and 1.4 m are available for use. Because of these changes the payload to a 500 km SSO is only 250 kg.

---

[36] http://www.spacelaunchreport.com/cz11.html.

[37] http://www.orbiter-forum.com/showthread.php?p=515896.

[38] http://defence-blog.com/army/china-unveils-model-of-new-special-vehicle-to-transport-cz-11-rockets.html.

In addition to the KZ 1 CASIC also plans to introduce two more launchers the 2.2 m diameter KZ 11 and the 3 m diameter KZ 21, into the commercial launch services market. Though there is not a great deal of information on the Kuaizhou family the KZ 11 will most probably be a three-stage launcher. It will have a lift-off weight of about 78 tonnes with the core having a diameter of about 2.2 m. It could possibly deliver about 1000 kg to a 700 km SSO.

In February 2016, the CASIC Launch Vehicle Ltd. Company (Expace) was registered in Wushan Hubei Province. This company is expected to serve as the commercial front for the marketing of Kuaizhou launchers. A new production facility at Wuhan is also being established with a capacity to produce 50 Kuaizhou launchers and 140 satellites.[39]

The Kuaizhou series is CASIC's first offering of launch vehicle services for use by global customers. It will directly compete with its sister company CASC that is also offering a similar solid missile derived CZ 11 "quick to orbit" launch vehicle for use by commercial customers. A Kuaizhou 1 being readied for launch from a TEL can be accessed through the reference link.

The Kaituozhe Launcher

In March 2017, a new Kaituozhe launcher placed a small Tiankun satellite in a 400 km Sun Synchronous Orbit (SSO). Two earlier Kaituozhe launches were reported in 2002 and 2003. Both failed to place satellites in orbit. The 2017 launcher called the Kaituozhe 2 (KT 2) appears to be a three-stage solid rocket with the stages being derived from the DF 21 missile. Like the Kuiazhou and the CZ 11 it appears to be able to place payloads in orbit using a mobile launch platform. Open-source data suggest it can place 250 kg in a 700 km SSO or 350 kg in LEO. The link provides a visual of the KT 2 flight of March 2017.[40]

Private Sector Small Satellite Launch Vehicles

In October 2018, a Zhuque privately developed rocket failed to place a small Weilai satellite in orbit.[41]

---

[39] https://chinaspacereport.com/launch-vehicles/kuaizhou/.

[40] http://spaceflight101.com/china-conducts-secretive-debut-launch-of-kaituozhe-2-rocket/.

[41] https://www.nasaspaceflight.com/2018/10/chinese-landspace-launches-weilai-1-zhuque-1-rocket/.

In 2019, three private companies launched three different rockets. The OSM launcher launched in March 2019 failed. The Hyperbola launched in July 2019 placed four satellites in a Low Earth Orbit (LEO). This was followed by the launch of the Jielong which placed three small satellites in SSO.

As a part of the China's Civil Military Integration Policy, military technologies and capabilities are now being commercialized via industry. These private launchers are most probably derived from China's missile programme. The launchers may be result of technology transfer arrangements from the military to the commercial sector. China will try to mimic the US in trying to create viable launch capabilities in the private sector.

References

China Academy of Launch Vehicle Technology (CALT) (1999) LM 2C User's Manual 1999
LM 3A Series Launch Vehicle User's Manual Issue (2011)
Zhang (2013) The development of LOX/LH2 Engine in China. In: 64th international astronautical congress, Beijing, China, IAC-13, C4-1,1x18525

Annexure 14.2: The Launcher Record

See Table 14.15.

Annexure 14.3: Connections—Engines, Stages, and Launchers in the CZ 2, CZ 3, and the CZ 4 Family of Launch Vehicles

See Fig. 14.23.

Table 14.15 The Launcher Record

Year	CZ 1	CZ 2C	CZ 2D	CZ 2E	CZ 2F	CZ 4	CZ 4B	CZ 4C	CZ 3	CZ 3A	CZ 3B	CZ 3C	FB	KT	KZ	CZ 6	CZ 11	CZ 5	CZ 7	CZ 8	Private	Bought	Total
1970	1																						1
1971	1																						1
1972																							0
1973													1										1
1974		1											1										2
1975		1											2										3
1976		1											2										3
1977		0											0										0
1978		1											0										1
1979		0											1										1
1980		0											0										0
1981		1											1										1
1982		1																					1
1983		1							2														3
1984		1							0														1
1985		1							1														2
1986		1							0														2
1987		2							2														4
1988		1				1			0														0
1989		0				0			2														5
1990		1		1		1			1														1
1991		0	1	0					0														4
1992		1	0	2					0														1
1993		1		0					1	2													5
1994		0		1					0	0													3

(continued)

Table 14.15 (continued)

Year	CZ 1	CZ 2C	CZ 2D	CZ 2E	CZ 2F	CZ 4	CZ 4B	CZ 4C	CZ 3	CZ 3A	CZ 3B	CZ 3C	FB	KT	KZ	CZ 6	CZ 11	CZ 5	CZ 7	CZ 8	Private Bought	Total
																	Launcher Record					
1996		0	1						2	0	1											4
1997		2	0						1	1	2										1	7
1998		4	0						0	0	2											6
1999		1	0				2		0	0	0										1	5
2000		0	0	1			1	1	1	3	0										1	6
2001		0	0				0			0	0											1
2002		0	1	1			2			0	0			1								5
2003		1	1	1			1			2	0			1							1	8
2004		4	1	0			2		1	1	0											8
2005		1	2	1			0			0	1										1	6
2006		1	0	0			1	1		2	1											6
2007		1	1	1			1	1		4	2											10
2008		1	2	0			2	1		1	2	1									1	11
2009		2	3	0			0	1		1	1	0										7
2010		0	1	2			3	3		3	5	4										15
2011		4	1	2			3	0		3	5	1									2	21
2012		2	3	1			2	4		1	5	3										19
2013		2	2	1			2	3		0	3	0			1							15
2014		4	2	2			4	3		1	0	1			1						2	18
2015		0	4	0			2	2			8	1				1	1					19
2016			6	2			2	2			3	3				1	1	1	1			22
2017		3	3				1	1		2	5	1		1	1	1	1	1	1		1	19
2018		6	8				2	4			11	1			1		3	1				39
2019			1	1			4	3			11	1			5	1	3	3	1	1	3	34
2020		3	7	1			5	1			8				4	1	3	3	1			38
Total	2	58	51	7	14	2	40	29	13	28	72	16	8	3	13	4	11	6	3	1	11	396
Failures	0	2	3	0	0	0	1	2	3	0	4	0	4	2	2	0	0	1	1	0	2	29
Reliability	100%	97%	100%	57%	100%	100%	98%	93%	77%	100%	94%	100%	50%	33%	85%	100%	100%	83%	67%	100%	82%	93%

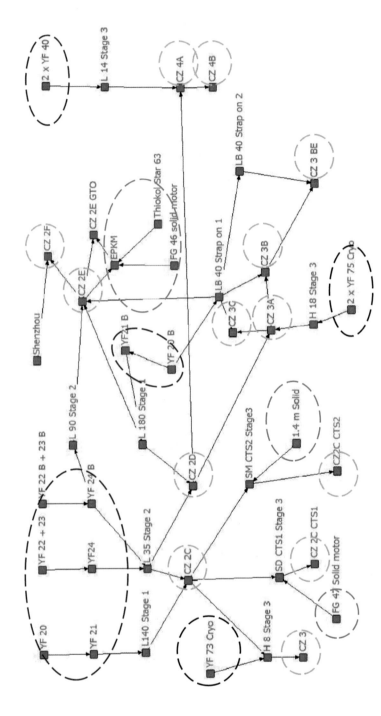

Fig. 14.23 Connections—Engines, Stages and Launchers in the CZ 2, CZ 3, and CZ 4 Family of Launch Vehicles

Chapter 15
Space Infrastructure in China

15.1 Background

Over the years China has built up an extensive and comprehensive infrastructure to cater to the needs of an increasingly complex space programme. The advent of the human spaceflight programme in the early 1990s provided a fillip to the creation of new state-of-the-art facilities (see Chap. 11 for more details). China is now on par with both the US and Russia in satellite and launch capacities. The space industry ecosystem of China encompasses the entire value chain from satellite building and launch to applications and use. There are many entities within China who perform a wide spectrum of space-based value addition functions.

Apart from the basic facilities, China also has in place an organizational and institutional infrastructure that can carry out the complex coordination and resource allocation functions needed for creating and running a dynamically evolving technology-driven high-tech programme. These have been covered in detail in the first part of this book. In this chapter, the major components of China's space infrastructure are covered to complete the picture.

15.2 Launch Complexes

Figure 15.1 provides an overview of the location of China's launch complexes.

China currently has four operational launch complexes. The Wenchang launch complex (located on Hainan Island) that is optimized for launches to LEO and GTO became operational with the launches of two new launch vehicles the CZ 5 and the CZ 7 in 2016. With the commissioning of this launch complex China may be able to substantially increase the number of launches from the current level of 18 to 20 launches per year to a much higher number. This may place it on par or even ahead of both the US and Russia for launch capacity.

© National Institute of Advanced Studies 2022
S. Chandrashekar, *China's Space Programme*,
https://doi.org/10.1007/978-981-19-1504-8_15

Fig. 15.1 Location of China's launch complexes

The **Jiuquan** Launch Centre (40.96 N 100.29E) is the oldest launch complex in China. It was originally set up to test ballistic missiles.[1] Between 1950 and 1970, the facility was used for testing missile launches only.[2] In 1970, China's first satellite, the Dong Fang Hong 1, was launched from here. The facility operates under the supervision of the PLA. The Center although located in Gansu Province is referred to as Jiuquan Launch Centre. It is also referred to as the PLA 20th Test and Training Base.

The currently active launch complex in Jiuquan has two launch pads—Pad 921 and Pad 623. This active launch complex is referred to as the South Launch Site (SLS) or Launch Complex 43. Pad 921 (or SLS—1) is used to launch the Shenzhou crew vehicle and space laboratory modules, using the CZ-2F launch vehicle. Pad 603 or SLS-2, (also known as "Jianbing Pad"), supports launches to low earth orbit using the CZ-2C, CZ-2D, and CZ-4B launch vehicles.

The China Great Wall Industry Corporation (CGWIC), a subsidiary of the China Aerospace and Technology Corporation (CASC), markets Jiuquan Commercial launch services to the international customers.

The **Taiyuan** Launch facility (38.8491 N 111.608 E) located in Sichuan Province started operations in the late 1960s for testing ICBMs and Submarine Launch Ballistic Missiles (SLBM). Space launches began in the late 80 s. The site was selected to optimize payloads for Sun Synchronous Orbits (SSO) using the CZ-4 family of launch

---

[1] https://chinaspacereport.com/facilities/.

[2] The first full range test of China's DF-5 used a launch from Jiuquan to an impact point in the Pacific Ocean to signal to the world that China had achieved ICBM capabilities.

vehicles. The first China-Brazil Earth Resource Satellite (CBERS-1) was launched from here. This facility is also operated by the PLA and is known as the 25th Test and Training Base (or Base 25).

Taiyuan has three launch complexes—LC 7, LC 9, and LC 10—a technical area for vehicle and spacecraft receiving and checkout, a communication center, a launch control center, and a TT&C center. Of these currently LC 9 and LC 10 are used for satellite launches. LC 10 is a dedicated launch facility for the new-generation CZ6 small launcher and is relatively new (2014). LC 9 is the primary space launch pad at Taiyuan. LC 7 the earliest pad supported SSO missions using the CZ-4A/B/C and LEO missions using the CZ-2C till 2008. Currently, it is used to test China's missiles and other space weapons. The WU-14 (DF-ZF) hypersonic glide vehicle (HGV) flight tests using the CZ-2 booster was conducted from here.

The **Xichang** Satellite Launch Centre (28.2459 N 102.0272 E) operates under the PLA 27th Training and Test Base and is dedicated for satellite launches to the GTO. It was specially established for launching China's first geostationary communication satellite in 1984. Since then, it has been China's main spaceport for geostationary and lunar exploration missions. The launch complex has two launch pads. China's first anti-satellite (ASAT) weapon test was carried out from this site. It has also been featured in a major launch mishap when a CZ 3B malfunctioned immediately after launch resulting in several fatalities in adjoining villages. With the commissioning of the Wenchang Complex GTO missions may be partially shifted there.

The Wenchang Space Launch Centre (WSLC) is the latest launch center from where the first launch took place in June 2016. There are two launch complexes for the CZ 5 and CZ 7 launchers with associated assembly buildings. The WSLC will support the launch missions of China's new-generation launch vehicles CZ 5 and CZ 7.

Wenchang is closer to the equator than China's three other launch sites. This makes it a better location for Geostationary Transfer Orbit (GTO) missions. Rockets can also now be transported from the Tanjin production plant by sea, eliminating any limits on the size of rocket stages dictated by road or rail transport (Spaceflight 2016). This is also the first launch site close to the sea which enables launches in a southeast direction into the South Pacific (see Sect. 14.4 dealing with launch vehicles for details of launches from these sites).

In 2019, China also started satellite launchings from offshore platforms in the Yellow Sea. In the launch log used in this book, these launches are designated as YLSA for Yellow Sea Launch Area.

One of the requirements for space-related military operations is a country's ability to launch repeatedly and launch quickly from different locations. China has demonstrated launches from land-based mobile platforms as well as offshore platforms. In 2019, it has shown that it can launch satellites on the same day from several sites. It has also been able to demonstrate multiple launches from mobile launchers on the same day.

15.3   China's Tracking Telemetry and Command (TT&C) Centres

Each of the launch complexes has facilities which usually track the launch vehicle from the time of takeoff to about 500 s.

The Dashuli tracking station located 36 km southwest of the Jiuquan launch center serves as the Jiuquan's TT&C Centre. This facility tracks the launch vehicle from takeoff until around 500 s into the flight. The tracking station consists of radar, optical tracking facilities, communications network, computers, meteorological weather stations as well as technical and logistic support systems. The Dongfeng Space City located 6.5 km west of the South Launch Site is the main administrative headquarters of the Jiuquan Launch Centre and its TT&C.

The Tracking Centre associated with the **Taiyuan** Launch facility is known as Lüliang Command Post with its headquarters in the city of Taiyuan. It has four subordinate radar tracking stations in Yangqu (Shanxi), Lishi (Shanxi), Yulin (Shaanxi), and Hancheng (Shaanxi).

The Xichang TT&C system includes three tracking stations: Xichang, Yibin, and Guiyang. In a typical launch campaign, they are supplemented by the Weinan Tracking Station, Ximen Tracking Station, and two Yuanwang space tracking ships stationed in the South Pacific.

The Wenchang Launch Centre has two tracking stations located at Tongguling in Wenchang and another at the Xisha Islands.

China has a comprehensive satellite tracking infrastructure for monitoring and controlling all their space assets. There are three Aerospace Command and Control Centres—Xian Control Centre (XSCC or Base 26), Beijing Aerospace Control Centre (BACC), and Dongfeng Mission Command and Control Centre (MCCC, located within Jiuquan Satellite Launch Centre). XSCC or Base 26 is the administrative headquarters of the China's TT&C. The BACC largely focuses on the human space flight missions.

Apart from the TT&C facilities associated with the launch stations, there are seven fixed stations—Weinan, Changchun, Kashi, Minxi, Xiamen, Naning, and Qingdao. All of these serve as TT&C stations for various missions. Figure 15.2 shows the locations of these TT&C stations. China also has two mobile TT&C stations that can be moved around to cater to the specific needs of any new missions.

Some of these stations have been equipped with larger antennae to provide the deep space TT&C facilities needed for the lunar and inter-planetary missions. A deep space tracking station has also been set up in the Neuquen Province of Argentina that became operational in 2017.

In addition to the above land-based stations, China has **seven** Yuanwang TT&C ships (Xinhua 2016) and five Tianlian Tracking and Data Relay (TDRS) satellites that provide additional support for critical missions. The Tianlian TDRS satellites were launched specifically for the human space flight missions.

The Yuanwang ships have carried out some 70 expeditions and traveled more than 1.5 million nautical miles in the Pacific, Atlantic, and Indian oceans. The first of these

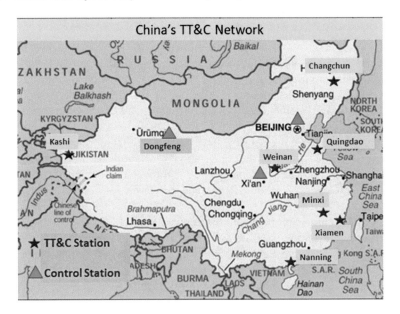

Fig. 15.2 China's tracking, telemetry, and command ground network

ships came into operation in 1979. The new generation ship, the Yuanwang-7, is used for tracking of the Shenzhou spacecraft and for other space missions as well. They are used for tracking both missiles and satellites and carry instrumentation both for missile range monitoring and intelligence collection.[3] They carry a large parabolic tracking antenna, two log-periodic HF antennae (fore and aft) shaped like fish spines, two small missile-tracking radars, and several precision theodolite optical tracking director stations.

The fixed and mobile ground tracking stations in China can only communicate with spacecraft visible from the Chinese landmass. The Tracking and Data Relay Tianlian satellites positioned in the GEO orbit provide low-, medium-, and high-speed data communications between the satellites and the ground stations in K- and S-bands. The first relay satellite was launched in 2008, the second in 2011, and the third in 2012. Two more relay satellites were launched in 2016 and 2019 (see Chap. 7 for details of the Tianlian satellites).

Since meteorological conditions can constrain optical tracking during the April to October months, China has built overseas stations in Swakopmund in Namibia, Karachi in Pakistan, and Malindi in Kenya. In addition to optical tracking, they also provide conventional radio tracking data for improved orbit estimation and mission management. China also uses additional TT&C stations in Dongara (Western Australia), Santiago (Chile), Alcantara (Spain), Aussaguel (France), and Kerguelan Islands (Australia) for managing their Tiangong Space Laboratory.

---

[3] http://www.military-today.com/navy/yuan_wang_class.htm.

15.4   Remote Sensing Satellite Ground Station (RSGS)

Figure 15.3 shows the location of the stations that receive data from remote sensing satellites. There are four satellite remote sensing data receiving stations located in Miyun, Kashi, Sanya, and Kunming, respectively. Together they have the capacity to receive real-time satellite data that not only covers all Chinese territory but also 70% of the land area of Asia.

The Institute of Remote Sensing and Digital Earth (RADI) under the Chinese Academy of Sciences (CAS) is the organization that controls the Satellite Data reception centers. It was created in September 2012 through the merging of the Institute of Remote Sensing Applications (IRSA) and the Center for Earth Observation and Digital Earth (CEODE).

The China Remote Sensing Satellite Ground Station (RSGS) established in 1986 to receive remote sensing satellite data from SPOT and Landsat was also brought under RADI. The Miyun Station located in the outskirts of Beijing is the oldest which began operation in 1986. It has seven antennas and covers Central China, NE China, and adjacent foreign countries. The Kashi Station in Xinjiang Province began operation in 2008. It has five 12 m antennae and covers Western China and Central Asian countries. Sanya in Hainan Province started functioning in 2010. It has five 12 m antennae and covers South China, and neighboring countries in SE Asia. The data reception station at Kunming in Yunnan Province started functioning in 2015 and has one 7.3 m antenna. It covers SW China and neighboring regions.

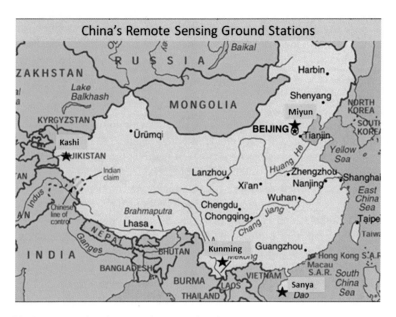

Fig. 15.3 Remote sensing data reception ground stations

The Headquarters of RADI is in Beijing. A high-speed network ensures that all the data received by the four stations are transmitted within 20 min to the main center in Beijing.

Recently (December15, 2016), RADI opened the China Remote Sensing Satellite North Polar Ground Station (CNPGS) near Kiruna, Sweden to receive data transmitted from high-resolution Earth Observation satellites. The station was constructed by RADI and is also operated by RADI. This is the first overseas satellite data receiving station for China. The station has three antennae capable of receiving data from all types of remote sensing satellites.

RADI also has an Airborne Remote Sensing Center that conducts all-weather flight operations with different sensors ranging from aviation cameras, scanners, and imaging spectrometers to imaging radar. It is a key national S&T infrastructure project equipped with two airplanes, a dozen advanced remote sensing facilities, and high-performance ground data processing systems. Chap. 9 on Remote Sensing Satellites and Chap. 16 on International Cooperation provide additional details on the scope and scale of China's global reach efforts.

15.5 National Satellite Meteorological Centre

The National Satellite Meteorological Center (NSMC) was set up in 1971 and is one of the operational centers of China Meteorological Administration. It provides nation-wide weather forecasts, climate research, and natural disaster monitoring with space-based EO data and derived products. Towards achieving this goal, the center operates equipment to receive, process, and disseminate meteorological satellite data. There are four satellite meteorological data receiving stations located at Beijing, Guangzhou, Wulumuqi, and Jaimusi within China. Figure 15.4 shows the location of these stations. The stations receive process and disseminate data from both geostationary and orbiting weather satellites (see Chap. 8 for details). China also provides weather stations and data to many countries (see Chap. 16 for details).

15.6 Satellite Communication Infrastructure

Immediately after the launch of its first communications satellite China initiated moves to create domestic capabilities for utilizing communication satellite capacities. It has created three satellite companies. These were Asiasat, Sinosat, and Chinasatcom.

Each of them either bought foreign satellites or used indigenously built satellites to build up the required communications and TV networks to realize domestic coverage. In the earlier pre-ITAR period, many of them were launched with Chinese launchers.

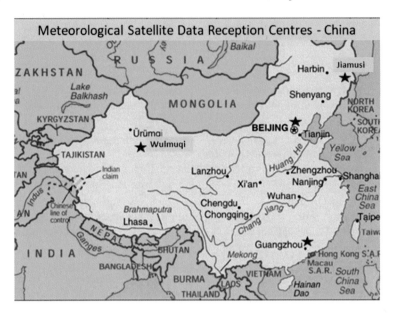

Fig. 15.4 Meteorological satellite data reception centres—China

After ITAR they were free to buy both satellites and launch services from the international market to provide cost-effective and economically viable services. As Chinese capabilities in satellite manufacture improved, these satellites and launch services are all being sourced domestically (see Chap. 7 on Communications satellites for more details).

Over the years, these companies have established the basic infrastructure to cater to the large domestic base of consumers for a variety of communications and TV services. Most of the equipment used for these applications are domestically built and serviced creating a large industry ecosystem (Weimin 2013).

Many of these were started as joint ventures with Hong Kong and foreign-based companies. Over time they have been subjected to the ups and downs of market forces. The Chinese government has however managed to retain control over them.

More recently in keeping with global trends some degree of consolidation is going on within this industry. Chinasat which is a company owned by the China Aerospace Corporation (CASC) took over APT the company that owned the APSTAR satellites. It also took over Sinosat another major player in the domestic market to create a very large conglomerate that provides the range of satellite services[4] (Harvey 2013). These moves are to make sure that Chinese companies can compete on equal terms with other major global players in the COMSAT markets.

Many of these communication satellite companies used their spare capacities to cater to the needs of customers from other countries. This trend is likely to continue

---

[4] pp. 162–163.

and gain momentum as China leverages its space capabilities for achieving commercial returns. Chinasat 12, for example, that was launched on November 27, 2012 has leased out capacity for use by Sri Lanka where it is called Supremsat.[5] Many transponders on Asiasat and Apstar are also leased out to many neighboring countries including India. As China expands its space capabilities, it will leverage its strong domestic strengths to address global market needs.[6]

15.7 Organizations

The Space assets in China are not necessarily owned by the PLA as pointed out by Dutton et al.[7] Over time the division of work and coordination of work within the space and missile ecosystem has changed and assumed different forms depending on the specific challenges faced at that time. The most recent of these changes took place in 2015. Many of these organizational changes have been addressed in different chapters of the book. It is sufficient to state here that the overall Chinese space ecosystem has displayed an ability to adapt and change to the increasingly complex demands placed on it. Chapters 1, 2, 3, 4 and 5 provide specific details of the complex and unique structures adopted by China for meeting the demands placed on it. These hybrid structures have evolved and assumed their present form because of the need to adapt to technological, political, and military complexity. Details on how the various organizational, political, and control issues have been resolved are provided in Chap. 11 that deal with the Human Space Flight Programme. The increasing influence of the scientific elite is covered in the discussions on the Lunar Exploration Programme in Chap. 12 of the book[8] (Aliberti 2015). The major restructuring of the military operations of the PLA including the creation of a new Strategic Support Force (SSF) is covered in Sects. 4.5 and 13.10.

Many organizations operate and control the various satellites that have been launched so far. Prior to the 2015 restructuring, the General Armaments Division (GAD) was responsible for launch facilities and on-orbit command and control; the General Staff Department (GSD) was responsible for their military operations. After the 2015 reform that resulted in the elimination of the various General Staff Departments, space activities come directly under the Central Military Commission (CMC). The Strategic Support Force (SSF), agencies such as the State Oceanic Administration, the China Meteorological Agency, the China National Navigation Project Centre, the National Committee for Disaster Reduction, the Ministry for Land and Resources, the Ministry of Science and Technology (MOST), and a few state-owned enterprises and universities also contribute significantly to the space programme.

---

[5] http://www.supremesat.com/about-us/.

[6] China has signed a contract to launch a dedicated satellite for Sri Lanka in 2013. It will be delivered in orbit. http://www.cgwic.com/news/2013/0528_Telecommunication_Satellite.html.

[7] pp. 87–132.

[8] pp. 8–23.

Except for a brief period during the Deng regime, overall control over the allocation of resources to the space programme has always been with the PLA. After the creation of the GAD, it had become the big player responsible for overall direction. GAD was also responsible for the operations and performance of the various launch complexes and the TT&C network through a special corporation for interfacing with foreign users of China's launch vehicle services. After creation of the new SSF, the responsibility for space activities in the military domain will come under the PLA (see Sects. 4.6 and 13.10 for more details).

The GAD depended upon state-owned defense industrial establishments for research, development, and manufacturing of space systems. With the elimination of the various staff groups, the CMC via the SSF directly controls most of the organizations dealing with space within the space industrial ecosystem.

There are two huge state-owned industrial groups, China Aerospace Science & Technology Corporation (CASC) and China Aerospace Science & Industry Corporation (CASIC), that are actively involved in the administration and execution of China's space programme.

The various Academies and Industrial groups and research institutes originally under the Seventh Ministry of Machine Building were regrouped into these two corporations in 1999 along with the creation of the GAD.

From a very broad reading of the functions of these two corporations, it appears that CASC has a greater role to play in the civilian and industry domain of missiles and space while CASIC deals more with meeting the specific needs of the military. Both corporations can produce launchers, satellites, and many of the components that are needed for manufacturing them.

CASC

Table 15.1 provides details of the various academies that come under CASC.

CASC has three kinds of organizations—R & D and production complexes, specialized companies (or commercial entities), and directly subordinated units which carry out the PR functions in addition to training. The CASC S & T Committee serves as the advisory body for the PLA and the State Council.

According to their website, CASC is mainly engaged in the research, design, manufacture, and launch of space systems such as launch vehicles, satellites, and manned spaceships as well as strategic and tactical missiles. They also provide international commercial satellite launch services.

It has the capability and experience needed to perform large system engineering management. Its R&D and industrial bases are distributed over Beijing, Shanghai, Tianjin, Xi'an, Chengdu, Inner Mongolia, Hong Kong (Shenzhen), and Hainan. The China Great Wall Industries Corporation and the Chinasatcom Company come under CASC.

CASIC

According to its website, China Aerospace Science and Industry Corporation, CASIC, is a large state-owned high-tech enterprise under direct administration of central government and is the backbone of China's national defense science and

Table 15.1 The organization of CASC

Organization	Location	Products	Remarks
China academy of launch technology (CALT) aka 1st academy	Suburb of Nanyuan (Beijing)	Ballistic missiles satellite launch vehicles MRBM designs Subordinate institutes specialize in guidance, navigation, control subsystems, RVs, and launchers	Competes with CASIC 066 base and CASIC fourth academy 10th research institute focuses on hypersonic cruise vehicles
Academy of aerospace solid propulsion technology (AASPT)	Xian	Solid rocket motors of 2 m or more 1 Design department, 5 Research institutes, and 3 Production facilities	Aka as 4th academy
China academy of space technology (CAST)	Hiadan District, Beijing	Satellites Attitude control, onboard communications, sensor subsystems, cryogenic technologies, antenna systems	Aka as 5th academy
Academy of space propellant technology (AALPT)	Shaanxi	Liquid fuel propellant systems, Heavy lift launch vehicle development (LM5) 4 Research institutes 1 Factory	Aka as 6th academy and as academy of liquid propulsion technology or base 067
Sichuan academy of aerospace technology (SAAT)	Chengdu, Sichuan Province	Space systems and subsystems Navigation, guidance and control components, terminal guidance radar systems	Aka as the 7th academy or as 062 Base
Shanghai academy of space technology (SAST) aka as the 8th academy	Shanghai	Launch vehicle systems, oversees SAR, ELINT, Weather satellites development in 509th Research Institute	Largest. Competes with CASC first academy and CASC fifth academy
China academy of aerospace electronics technology (CAAET)	Haidan District, Beijing	Electronics, Inertial guidance system, onboard computers	Aka as 9th academy
China academy of aerospace aerodynamics (CAA)	Beijing	Aerodynamic testing	10th or 11th academy?

(continued)

Table 15.1 (continued)

Organization	Location	Products	Remarks
Shenzen academy of aerospace technology	Shenzhen	Digital communication system, RFID, GPS vehicle location	Joint venture between CASC, Shenzhen City Government and Harbin Institute of Technology

technology industry. CASIC owns seven academies, two scientific research and development bases, six public listed companies, and employs over 135,000 people. It is a major producer of short- and medium-range ballistic and cruise missiles. Through its subsidiaries it provides parts, components, and subsystems for the Lunar Exploration and the Navigation Satellite programmes. CASIC entities also supply most of the ground systems for TT&C purposes including radars for several applications. Table 15.2 provides an overview of the various organizations under it.

The Chinese Academy of Sciences (CAS)

Chapters 1 and 2 of this book provide details of the pivotal role played by the CAS after Quian returned to China from the US in 1955. Because of the problems created by the Cultural Revolution the satellite effort was moved out of the CAS and placed directly under the 7th Ministry. After passing through the vicissitudes of the Cultural Revolution and a period of consolidation under Deng, the CAS role in the space programme once again became important under the 863 Programme. Both the Human Space Flight effort as well as the Lunar Exploration Programme came largely through the efforts of various key scientists affiliated to the CAS in different ways. In 2013, the CAS came out with a report entitled "Vision 2020: The Emerging Trends in Science & Technology and China's Strategic Options" (Kulacki 2013). In this they outline the major challenges facing China and China's priorities. The CAS vision identifies key areas of focus and wants China to match the major science powers in terms of original and innovative report (see Sect. 12.5 for details).

The CAS came into existence immediately after the establishment of the People's Republic of China in 1949. It is a huge organization with over 60,000 people. About 124 institutions come directly under it. These are distributed all over China. Research institutes, universities, and companies with CAS investment all fall under its ambit. Several organizations and institutes dealing with different facets of the space programme also come under the CAS. These include the Institute of Remote Sensing and Digital Earth (RADI), the National Space Science Centre (NSSC), and the National Astronomical Observatories of China (NOAC). The Shanghai Institute of Microsystems and Information Technology is involved in a major way with the building of microsatellites including the Chuangxin series of satellites[9] (Aliberti 2015).

---

[9] pp. 20–22.

Table 15.2 The organization if CASIC

Organization	Location	Products	Remarks
Academy of information technology		Microsatellites, satellite applications, GPS inertial guidance units, MEMS-based guidance and navigation systems and ground stations	Aka the 1st academy
Changfeng mechanics & electronics technical academy	Western suburbs of Beijing	Air and Space Defense Systems. ASAT and associated radars 1 Design department, 10 Specialized research institutes, 3 Factories, 9 Independent commercial enterprises	Aka the 2nd academy or academy of defense technology
Haiying electromechanical technology	Yungang, suburb of Beijing	Cruise missiles 1 Design department, 10 Research institutes, 2 Factories	aka the 3rd academy or academy of cruise missile technology
Hexi chemical & machinery corporation	Hohhot inner mongolia	DF21 MRBMs System integration Collaborates with non-CASIC enterprises	Aka as the 4th Academy or Academy of Launch technology
Hexi Chemistry & Machinery Company or Academy of Propulsion Technology	Hohhot, inner mongolia	Solid rocket motors, casings, nozzles, grains, and igniters 4 Research institutes, 2 Factories	Aka as the 6th Academy Carved out of the CASC Fourth Academy to create competition in solid rocket motors
Sanjiang aerospace group	Xiaogan, North of Wuhan	Supply cruise missile components to the Third Academy DF-11, SRBM, TEL vehicles	Aka as the 9th Academy or 066 Base
Jiangnan aerospace group	Guizhou Province	Specialized missile components and software. Composite materials, gyroscopes, autopilot systems, and batteries 2 Research institutes, 20 Institutes and factories and 3 Technology centers	Aka as 061 Base
Human Space Bureau	Hunan's Shaoyang area	Specialty materials, magnets, antennas	Aka as 068 Base. or the Hunan Space Agency

Though the activities of many of these come directly under the CAS there is a complex and involved coordination system in place to ensure synergy and coherence. CAS seems to be having a greater voice in the conduct of space activities as indicated in Chap. 12 of this book.

15.8 Domestic Competition

China has in place a space industry ecosystem that provides for competition in different parts of the space value chain. CALT and SAST are both entities within CASC. They compete both in the launch vehicle and spacecraft domains. Like CAST they can also build any kind of satellite. The CAS can also build satellites and satellite payloads via various institutions under it. The Kuaizhou mobile launcher provided by the 066 Base under CASIC is also entering the launch services industry in China.

Industries directly and indirectly under these major state-sponsored agencies are also competing for shares in the domestic market. Section 9.15 that deals with remote sensing provides one illustration of industry involvement. In the Case Study under Sect. 16.7, another approach for realizing superior services via the international cooperation route for small satellites is detailed.

More recently, several private start-up companies dealing with satellites have been set up in China. Most of these are backed by universities and hedge funds. Their focus is services via the launch of microsatellites. Three companies provide an idea of likely future trends in this area (Lin and Singer 2016).

Onespace was founded in June 2015 with support from the National Defense Science and Industry Bureau. Their 59-tonne space launch vehicle is expected to launch a 500 kg payload in low earth orbit in 2018. Onespace also has ambitions eventually to build a manned space capsule.

Landspace technology is closely associated with Tsinghua University. Their aim is to build a medium space launch vehicle for manned and unmanned use. Their target markets are microsatellite launches for European and Southeast Asian countries. The Zhuque space launcher that failed in 2018 is their first attempt to become a launch services provider.

Three other small launchers made their debut in 2019 with some successes.

Shenzhen Yu Long Aerospace Science and Technology is another Chinese space start-up. It has performed work with an array of sounding rockets that reach suborbital heights. A rocket with a 165 kg payload reached an altitude of 35 km in January 2016. Shenzhen Yu Long has announced its hopes to fly a liquid-fueled rocket in 2020, and a manned launch in 2025.

More recently, Exspace was founded in 2016. The Chairman of this company Zhang Di is also a Deputy Director of the Fourth Academy of the China Aerospace Science and Industry Corporation (CASIC). Expace's target market is to launch small satellites for domestic and foreign customers.

HEAD Aerospace Technology Company is a service company founded in 2007 with its headquarters in Beijing. It has established collaborations with some aerospace companies in Europe. It is a leading space company in China, developing, producing, and selling advanced space products and technology for both domestic and international space markets. The Company opened its Dutch office in 2016, making it the first private company from the Chinese space industry to set up business in The Netherlands. According to the company profile available in their website, HEAD plans to acquire a constellation of 24 small Skywalker satellites. The first commercial Chinese Automatic Identification System (AIS) satellite "HEAD-1" was launched on an LM-4C launcher in November 2017. Several other HEAD AIS satellites have been launched since then.

15.9   Overall Trends

China has in place the technology, organization, and infrastructure for competing with the other major space powers in the global arena. It is possibly the only country in the world that can try and match the US. Given the increasingly important role of space in the conduct of world affairs the US-China space dynamic will drive the other space powers.

References

Aliberti M (2015) When China goes to the moon. European Space Policy Studies Institute (ESPI), Studies in Space Policy, vol 11. Springer
Dutton P, Erickson AS, Martinson R (eds) (2014) Chinese air and space-based ISR, integrating aerospace combat capabilities over the near seas. Naval War College Press, China's Near Seas Combat Capabilities Chapter 7
Harvey B (2013) (2013), China in space: the great leap forward. Springer Praxis Books, New York
Kulacki G (2013) Strategic options for Chinese space science and technology, a translation and analysis of the 2013 Report. Chinese Academy of Sciences http://www.ucsusa.org/sites/def ault/files/legacy/assets/documents/nwgs/strategic-options-for-chinese-space-science-and-tec hnology-11-13.pdf
Lin J, Singer PW (2016) Watch out spaceX: China's space start up industry takes flight private Chinese companies could launch microsatellites and one day. Taikonauts. http://www.popsci. com/watch-out-spacex-chinas-space-start-up-industry-takes-flight
Spaceflight (2016) China debuts Long March 7 Rocket from new Wenchang Satellite Launch Center. http://spaceflight101.com/long-march-7-maiden-launch/china-debuts-long-march-7-roc ket-from-new-wenchang-satellite-launch-center/
Weimin H (2013) Briefing on China's satellite communications progress and future. http://ieeexp lore.ieee.org/stamp/stamp.jsp?arnumber=6650313
Xinhua, (2016), China launches space tracking ship Yuanwang-7, July 12. http://en.people.cn/n3/ 2016/0712/c90000-9084868.html

Chapter 16
International Footprints of China's Space Programme

16.1 Background

As a relatively latecomer to the elite space club, China recognized that it had to catch up in many key areas related to space. It has very cleverly used a variety of approaches towards buying and selling various space products and services to achieve its purposes. It has also used its growing prowess in space as a part of its geopolitical strategy to achieve larger national aims and aspirations.

Historically Chinese capabilities in launchers have been ahead of their capabilities to build state-of-the-art satellites. Under the Deng regime there was a need to justify the space programme in terms of hard economic payoffs. This led directly to the creation of China's international arm for the marketing of its launch services, the China Great Wall Industries Corporation (CGWIC), in 1980 even before the first Geostationary Satellite Launcher the CZ 3 had been launched.

Given its relative backwardness in satellite-related technologies during the early days of the space programme, China decided to import communication satellites for meeting its domestic requirements. It has cleverly leveraged this domestic demand to strike agreements with major international satellite companies that facilitated China's acquisition of key technologies. It has combined this acquired knowledge with strong internal capabilities that is able to assimilate new knowledge and quickly provide indigenous substitutes and solutions.

Once relations with the US soured, China has also needed to get its state-of-the-art bought satellites launched by foreign launch service providers because of US export restrictions. It had to turn to other countries for help in its efforts to build state-of-art satellites.

In dealing with all these problems, China has combined commercial, political, and technology needs into a coherent strategy for furthering its space and national interests. As seen in the book, China has been able to catch up with the more advanced space powers through a judicious blend of indigenous development and selective imports. It is now poised to compete on a more equal footing with the more advanced

© National Institute of Advanced Studies 2022
S. Chandrashekar, *China's Space Programme*,
https://doi.org/10.1007/978-981-19-1504-8_16

space powers and become a much bigger player in the world space scene. The international component of its space strategy is particularly important for understanding how China has reached its current position in the world space order.

16.2   Satellites and Launch Services for International Customers

Chinese launchers have together launched 69 **satellites** and payloads as a part of China's international cooperative efforts. These satellites can be grouped into four categories. These include the following:

- Launch services for the satellites of other countries.
- In-orbit delivery of China-built satellites launched with Chinese launchers.
- In-orbit delivery of Chinese satellites for cooperative science and application programmes with other countries.
- Launch of piggyback satellites and payloads on various Chinese launchers on an availability basis.

China has provided 26 launchers for 42 satellites from 1990 to end 2020. These include 15 satellites for the US using nine launchers, 5 launches for Brazil under the CBERS programme, and 4 dedicated launchings for Australian satellites.

China has also provided services for single satellite launches by Europe, the Philippines, Indonesia, Turkey, dual launch of Saudi Arabian remote sensing satellites as well as one test satellite for Pakistan. The Pakistan test satellite was launched along with another remote sensing satellite built for Pakistan by China. In 2020, it provided a dedicated CZ 6 launcher for launching 10 remote sensing satellites for Argentina. It had earlier launched several small satellites for Argentina as piggyback payloads. For details, see Annexure 16.1.

Twelve satellites built by China and launched using Chinese launchers have been delivered in orbit to several countries. Nine of them are communication satellites delivered in GSO. These include two communication satellites for Nigeria and one satellite each for Laos, Pakistan, Belarus, Venezuela, Bolivia, and Algeria. A Palapa satellite to be delivered in orbit failed to reach orbit in 2020. Three remote sensing satellites have been delivered in orbit two for Venezuela and one for Pakistan. The details are in Annexure 16.2.

China has also built and launched two Doublestar satellites as a part of its contribution to a joint programme with the European Space Agency (ESA) to study the influence of solar activity on the radiation environment of the earth. In 2018, it launched the CFOSAT a joint French Chinese Ocean and weather research satellite.

It has also launched five Huanjing (HJ) remote sensing satellites carrying both optical and SAR payloads as its contribution to the joint programme that it has with the APSCO group of countries. Three launchers have been used for this purpose.

In 2019, it collaborated with France and launched a satellite for testing an ion thruster. Two other small satellites were placed in orbit to test Ka band communication technology in a CAS collaboration with Germany (Annexure 16.2).

China has also launched 26 satellites in a piggyback mode for several countries. A list of the payloads and the countries can be found in Annexure 16.3. Figure 16.1 provides a breakup of these cooperative efforts in satellites.

Commercial interests as dedicated or piggyback launches constitute 80 of the 91 satellites launched. They dominate China's space launch service connections with the countries of the world. When the piggyback satellites are taken out China has used a total of 43 launchers to place 61 satellites in various orbits for several countries as well as for providing Chinese satellite services for use by cooperating partner countries. Out of a total of 396 launches that have taken place till end 2020 the international component represents about 11% of all launches.

Figure 16.2 provides an overview of the launch services provided by China to various international customers.

This includes satellites launched as well as Chinese-built satellites delivered in orbit. It also includes satellites and launch services delivered as a part of international cooperation agreements. Since most of the launches are commercial, China has benefited substantially from its exports of launch services and in-orbit delivery of satellites.

During the heydays of the Cold War, relations between the US and China were good. Though the US did ration the number of commercial launchers for China, there were no major restrictions that prevented US satellite makers to use lower priced Chinese launch services to make their products more competitive. Recognizing these opportunities China also developed a new launcher the CZ 2E since its CZ 3 launcher could not place more than about 1200 kg satellites in GTO.

The US–China bonhomie did not last long. The Cox Committee Report put out in 1998 accused China of using its leverage over US satellite companies to acquire critical knowledge and technologies that had military implications. Restrictions of

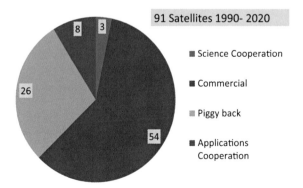

Fig. 16.1 Space cooperation—China 1970–2020

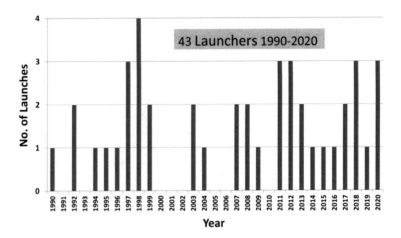

Fig. 16.2 Chinese launchers for international customers

various kinds led progressively to the creation of the International Trade Arms Regulation (ITAR) regime which required US clearance of any satellite carrying American parts before they could be launched on Chinese launchers. This measure effectively prevented Western satellite makers from using Chinese launch services. These US-led measures had a direct impact on China's share of the global launch services market.

16.3   Comsat Launch Services—Impact of US Embargo

Out of the 43 Chinese launches for international customers, 18 of them have placed the satellites of several countries into a Geostationary Transfer Orbit on their way to the GSO. Figure 16.3 provides a timeline of Chinese launches into GTO for various international customers.

From Fig. 16.3, it is evident that between the period 1998 and 2006 not a single foreign-built satellite was put into orbit by a Chinese launcher. This was a direct consequence of US actions on China arising out of the Cox Committee report. Just before this between 1990 and 1997 China had carried out four launches for Australia, two for the US, and one for the Philippines.[1]

Once the US had dried up as a source of help, China sought assistance from Europe. Through a combination of imported satellites as well as collaborations with key European satellite makers, it sought to bridge its gaps in satellite capabilities. It

---

[1] The CZ 2E launcher specifically developed for GSO satellites as well as the establishment of the more favorably located Xi Chang launch site was a part of this commercialization effort. China also conducted a dummy satellite test to prove its vehicle to the Australians before launching their satellites.

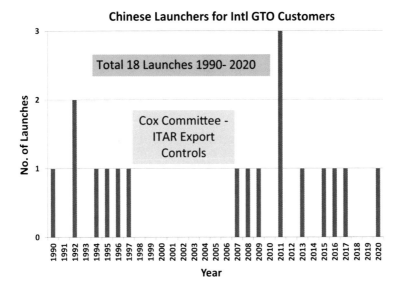

Fig. 16.3 Chinese launch services for GSO communication satellites

persuaded European satellite makers to make satellites that were ITAR free (carried no US components). Once these results fructified it has sought to bypass US restrictions by building satellites with no US parts and delivering them in orbit to its customers most of whom are from the developing world.

Since then, China has also built, launched, and delivered satellites in orbit to several countries. These include two satellites for Nigeria, one satellite for Pakistan, one satellite for Laos, one satellite for Bolivia, one satellite for Venezuela, one satellite for Belarus, one satellite for Algeria, and one for Indonesia. The trend data suggest that China will deliver more satellites in orbit to more countries. It has also provided launch services for the Indonesian Palapa satellite[2] as well for a European Eutelsat satellite.

One can see the results of these efforts in Fig. 16.3. From 2007 onwards, China has been successful in reclaiming some of its lost share of the global launch services market. By delivering satellites in-orbit it is also making inroads into the global GSO Communication Satellite Market.

As one can see from Fig. 16.3, US export restrictions have had a major impact on China's attempts to enter the launch services market. Based on the data, on an average, China has lost at least one launch service opportunity between the years 1998 and 2006. Assuming an average price of $30 to $ 55 million per launch, China has lost $ 270–$ 495 million dollars because of US sanctions[3] (Chen 1993).

---

[2] Some problems with the launcher placed it in a sub-optimal transfer orbit. However, the mission was retrieved in part by using the satellite propulsion system to raise it to the geostationary orbit.

[3] Most of the satellites would be in the heavy to very heavy launch category. These prices are on the lower side.

Reported Chinese prices for in-orbit delivery of satellites are in the range of $220–$250 million. China has therefore realized about $2 billion from its in-orbit delivery business model[4] (Reddy 2018). China also provides financing through its banks for satellite purchases. Through such efforts it is trying to become a bigger player in the global SATCOM and launch services market despite US restrictions.

16.4   Foreign Launch Services for Procured Satellites

Annexure 16.4 provides details of satellites bought by China that have been launched by both Chinese launchers and foreign launch service providers.

To get around export restrictions, China which had always bought state-of-the-art satellites but launched them with its own CZ 3 series of launchers had by necessity to buy launch services from other countries. Russia, Ukraine, and the US have provided these launch services for the various satellites procured by China.

Over the period 1990–2020, China has bought 21 communication satellites from other countries. Starting from 1997 onwards seven of them have been launched by foreign launchers. These seven bought launches also represent a net loss to China which could have otherwise used its own launchers to launch them.

In addition to the above communication satellites, China has also collaborated with the UK-based small satellite company SSTL for building state-of-the-art small satellites. These have been launched using Russian Kosmos 3 M launchers.

China has also launched an indigenously built Yinghuo satellite meant for Mars exploration on a Ukraine Zenit 3 F launcher. The mission was a failure. This is also listed as a foreign launch though the satellite was built by China. Another small satellite NUDsat was launched by the Indian PSLV in 2017.

Case Study 1

The Economic Impact of US sanctions on China

The trend data of Fig. 16.3 that shows Chinese launch of satellites to GSO for international customers helps to put some numbers on the economic impact of sanctions.

From Fig. 16.3, it is evident that China moved from an average launch rate of about one launch per year for international customers during the period 1990–1997 to zero launches during the period 1998–2006. Even at a reduced price of $35–$55 million for a heavy lift launcher like the CZ 3B this translates into a direct economic loss of between $ 270 and $ 495 million (Chen 1993).This estimated loss is on the lower side. If the market was a competitive market,

---

[4] The financing of the satellite may also have been facilitated by China.

China's cheaper launchers could have captured more than the earlier trend average of about one launch per year.

Because China could not use its own launchers even to launch its own bought satellites China has been required to pay for foreign launch services. For the period 1997 onwards, China has used foreign launch services for its procured satellites. The seven launches would have again cost China close to $350 million (Chen 1993).[5] If cheaper Chinese launchers were used instead China could have saved at least 50% of this amount. The additional cost of launch services because of sanctions can be estimated at about $175 million.

Taken together the direct economic cost of sanctions on China can be estimated to be $450–$670 million. These estimates of course do not account for the costs of lost launch opportunities and the indirect benefits that would have accrued to domestic organizations in China because of increased production. These are also likely to be significant.

16.5 Remote Sensing and Other Areas

China has launched a total of 42 satellites using 25 launchers into LEO and SSO orbits for many countries. These include remote sensing satellites as well orbiting Iridium communication satellites for the US. Figure 16.4 provides a timeline of various satellites launched into sun synchronous and low earth orbits by China for various countries.

In remote sensing, most developments in China have taken place indigenously. The early Chinese efforts were directed towards recoverable satellites that used straightforward photography. The first digital camera-based remote sensing satellite flown by China was possibly the first CBERS satellite developed in collaboration with Brazil in 1999.

This initial cooperative effort has been sustained in a major way with successive generations of CBERS satellites being launched on Chinese launchers. The CBERS series has seen six satellites and offers a package of services to those interested in receiving data from them.

China has also delivered two remote sensing satellites in orbit to Venezuela and one to Pakistan. It has provided launch services for remote sensing satellites built by Turkey and Pakistan. It has been the launch services contractor for a pair of Saudi Arabian satellites launched together on a CZ 2D launcher in 2018. These account for 12 of the 25 launches.

In the early part of its commercialization efforts when the US was not yet an acknowledged adversary, China also launched a total of 13 satellites (including one

---

[5] An Atlas rocket used to launch the Asiasat 4 would have cost $ 55 million p 51.

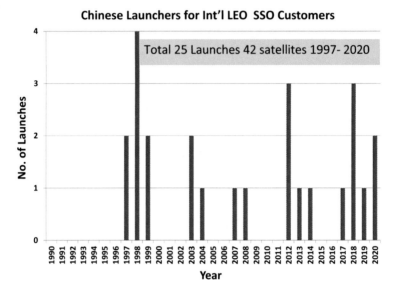

Fig. 16.4 Chinese launch services for SSO and LEO customers

dummy demonstration payload) on 7 CZ 2C-modified launchers for the Iridium orbiting constellation of communication satellites.

The CZ 2C launchers have also been used to place two Doublestar satellites in suitable orbits. These satellites, one in a polar orbit around the sun and the other in an equatorial orbit around the sun, were China's contribution to a joint programme with ESA to study the impact of the sun on the earth's environment. Together with data returned by the four ESA launched Cluster spacecraft they provide comprehensive coverage of the earth sun space.[6] In 2018, it also launched the CFOSAT a joint ocean weather-related scientific mission in cooperation with France.

China has initiated a major effort at trying to use its space capabilities for fostering economic and strategic links with its neighbors. The Asia–Pacific Space Cooperation Organization (APSCO) has been specifically set up by China for this purpose. Five Huanjing (HJ) remote sensing satellites through three launchers have been provided for this international initiative.

Case Study 2
APSCO and Other Regional Initiatives

China has also developed and launched five Huanjing satellites as a part of its contribution to the Asia–Pacific Space Cooperation Organization (APSCO).

---

[6] http://www.esa.int/Our_Activities/Space_Science/Cluster/Laurels_for_Cluster-Double_Star_t eams.

This international organization has been created by China to promote the use of space technology including remote sensing data by neighboring and other interested countries.

The Headquarters of the Organization is in Beijing. Member Countries in APSCO are Bangladesh, China, Iran, Mongolia, Pakistan, Peru, Thailand, and Turkey. Prior to the establishment of the Secretariat in 2006, workshops were held to define and organize cooperative efforts in the use of space technology. Indonesia was a part of the original nine APSCO countries but has not signed the convention. Turkey became a member a year later.[7]

The Huanjing constellation of satellites consists of four optical and one SAR satellite that have been launched as a part of China's contribution to APSCO.

APSCO has also routinely held annual meetings to define areas of common interest for cooperative work. Areas of study include small satellite buses for multiple missions, navigation, GSO communication satellites, and propagation losses for Ka band communications. Participant countries in APSCO meetings include Malaysia, Argentina, Ukraine, Indonesia, and Brazil.

China also has agreements with eight countries for receiving weather data from its meteorological satellites. These include Peru, Thailand, Bangladesh, Pakistan, Mongolia, Iran, and Indonesia. It also has in place a weather information dissemination service called Fengyuncast just like Eumetcast from Europe and Geonetcast from the US.

The APSCO arrangement maybe the precursor for a regional space-based service organization that includes remote sensing, weather, and navigation services. Apart from commercial motives, this initiative can be directly linked to China's efforts at countering US influence over the countries of the Asia–Pacific region.

16.6 Piggyback Satellites

Many countries have used Chinese launchers to place small satellites in orbit in a piggyback mode (Annexure 16.3). Most of them would have involved payments. European countries as well as Japan used China's recoverable capsules for experiments related to microgravity during the early years of the programme. Sweden also used a Chinese launcher for its piggyback Frejya satellite.

Countries that have launched piggyback satellites on Chinese launchers include Brazil, Pakistan, Turkey, Sweden, France, Luxemburg, Spain, Poland, Ecuador, Argentina, Germany, Denmark, Belarus, and Japan. In 2019, Sudan and Ethiopia were added to the list.

---

[7] http://www.apsco.int/AboutApsco.asp?LinkCodeN=1.

16.7   China's International Space Footprints and Strategy

Table 16.1 provides an overview of the various countries who have been involved in cooperative space activities with China.[8]

Table 16.1 shows that despite US sanctions, China and the US do have a strong relationship. Though the relationship has nose-dived after 1998, China continues to depend on US satellites to meet its own needs. Due to ITAR regulations, it has also needed to get these bought satellites launched from abroad. However, as the detailed assessments carried out in this book show, China has made significant progress in catching up on key satellite products that include communications, remote sensing, navigation, and weather satellites. As its ties with the US have deteriorated China has increasingly turned towards Europe for help in building state-of-the-art satellites.

Apart from help in key technologies for communications satellites, European companies such as Thales Alenia have also helped China bypass ITAR regulations by building satellites with no US parts. Such satellites built by European companies have been launched by China for several third world countries. Some aspects of the cooperation have therefore been of mutual benefit though there is awareness in Europe that China would use these opportunities to develop their own competitive products.

China had also contributed financially to the European Galileo navigation satellite project and might have benefited from key technology acquisitions from European companies.

Europe is also increasingly becoming aware that China may become a major competitor in the commercial space domain. Though this may lead to some problems, European economic needs and China's financial strength and internal markets would continue to provide the basis for good ties with some erosion in Europe's current dominant position.

The Doublestar Cluster Cooperation programme between Europe and China for space science studies represents a very successful partnership between two space powers[9] (Liu et al. 2005). China has also launched a payload for ESA on its recoverable satellite Shijian 10 in April 2016.[10] France and Germany are major players in the European Space Programme that also enjoy a strong relationship with China. France has played a key role in Chinese efforts to build state-of-the-art communication satellites and in its efforts to bypass the ITAR. The joint CFOSAT initiative between France and China for ocean and weather research suggests that both countries find it advantageous to work together.

China's relationship with the UK and its efforts to achieve parity in small satellites is illustrative of how China has cleverly leveraged its position to catch up. A separate Case Study 3 captures the dynamics of this evolving relationship.

---

[8] Data compiled from several Open and Public Sources.

[9] pp 2707–2712.

[10] https://directory.eoportal.org/web/eoportal/satellite-missions/s/shi-jian-10.

Table 16.1   China's international space footprints

Country	China sells			China buys		Comments
	Launchers	In-orbit delivery	Piggyback	Satellites	Launchers	
US	9	0	0	15	4	US dominant even after sanctions
Canada	0	0	1	0	0	Small satellite—nothing major
Europe	1	2	0	6	0	Strong ties—Europe upper hand
France	0	1	2	0	0	Major part of the European link
Germany	0	0	3	0	0	Major part of the European link
UK	0	0	0	2	0	DMC precursors SSTL collaboration
Spain	0	0	1	0	0	Small satellite—nothing major
Denmark	0	0	2	0	0	Small satellites—nothing major
Japan	0	0	1	0	0	No major relationship
Australia	4	0	0	0	0	Strong early ties—weak currently
Luxemburg	0	0	1	0	0	Flag of convenience
Sweden	0	0	1	0	0	Launch of Swedish satellite—history
Russia	0	0	0	0	4	Strong ties anti west—former enemies
Poland	0	0	1	0	0	No major relationship
Ukraine	0	0	0	0	2	Early launch services—pro-West
Laos	0	1	0	0	0	Dominant relationship—satellite supplier
Pakistan	1	2	1	0	0	Strong relationship Anti-India
Philippines	1	0	0	0	0	Launch services—weaker relationship
Indonesia	1		0	0	0	Launch services—weaker relationship

(continued)

Table 16.1 (continued)

Country	China sells			China buys		Comments
	Launchers	In-orbit delivery	Piggyback	Satellites	Launchers	
Belarus	0	1	1	0	0	In-orbit delivery—subsidized satellite
Turkey	1	0	1	0	0	Strong relationship remote sensing
Nigeria	0	2	0	0	0	In-orbit delivery—oil, materials
Brazil	6	0	2	0	0	6 CBERS satellites—strong ties
Argentina	1	0	8	0	0	CZ 6 launch of 10 satellites—strong ties
Bolivia	0	1	0	0	0	In-orbit delivery materials rich Lithium
Venezuela	0	3	0	0	0	In orbit delivery—strong ties—Oil
Ecuador	0	0	1	0	0	No major relationship
APSCO	0	5	0	0	0	Regional focus—US counter
Algeria	0	1	0	0	0	5000 kg COMSAT—commercial
Saudi Arabia	1	0	0	0	0	Two Saudi RS satellites—commercial
India	0	0	0	0	1	Minor deal no strong link
Sudan	0	0	1	0	0	Small satellite
Ethiopia	0	0	1	0	0	Small satellite
Total	**26**	**20**	**29**	**23**	**11**	

Case Study 3

International Relations and Technology Assimilation—The Case of Small Satellites

China's well-orchestrated strategy of collaboration with one of the world's leading small satellite companies illustrates China's approach towards catching up with the advanced space powers.

The 50 kg Tsinghua 1 satellite represented collaboration between Tsinghua University and Surrey Satellite Technology Limited (SSTL), a pioneering small satellite company functioning under the University of Surrey. This satellite was launched from Plestesk as a piggyback payload on a Kosmos 3 M rocket in June 2000. This was a small experimental remote sensing satellite.

Building upon this experience the Tsinghua 2 satellite was again launched from Plestesk in 2005. This was a 166 kg satellite supported by China's Ministry of Science & Technology (MOST) as well by the Beijing Landview Mapping Information Technology Company (BLMIT). The Tsinghua 2 was also built by SSTL and launched from Plestesk on a Russian rocket.

Internationally, this satellite was projected to be a part of a Disaster Management Constellation (DMC) of satellites that China was setting up. Built and operated on a budget of 101 million RMB, the satellite had both a 4-m resolution narrow swath capability as well as a broad swath capability with a coarser resolution. This satellite built by UK's SSTL and launched by Russia avoided US export controls since China was not directly involved. The satellite was operated and used by BLMIT during the period 2005 to 2010.

In parallel with these efforts, China and the UK were also exploring other possibilities for cooperation and trade. The Chinese Premier Wen Jin Bao and the UK Prime Minister David Cameron signed an agreement on space cooperation in 2011. This programme, budgeted at RMB 13.9 billion, involved a contract between SSTL and BLIMT under which SSTL will build and launch three electro-optical satellites with a resolution of 1 m for the Twenty First Century Aerospace Technology Limited, a subsidiary company under BLIMT.

The deal also envisaged that SSTL will own and operate the satellites but lease the entire capacity to Twenty First Century Aerospace Ltd. This is a new business model for remote sensing applications that is an extension of business arrangements used for satellite communications.[11]

To complete the picture SSTL contracted ISRO to place the satellites in Sun Synchronous Orbit (SSO) as a single constellation spaced out in the same orbit providing extensive and near-continuous coverage over the world.

On July 10, 2015, the PSLV Launcher placed a constellation of three identical DMC3 remote sensing satellites built by Surrey Satellite Technology Limited (SSTL) into a 647 km 98.06-degree inclination sun synchronous orbit.

Each of these 447 kg satellites carried an advanced electro-optical imaging instrument that can provide 4 m multispectral imagery in four bands as well as a 1 m PAN image.

The satellites are spaced 120 degrees apart in the orbit plane so that they can image any target area at least once in a day. The satellites do not carry

---

[11] https://directory.eoportal.org/web/eoportal/satellite-missions/d/dmc-3.

chemical propellants for attitude and orbit control. They use a state-of-the-art xenon thruster for this purpose.[12]

This is a classic case of how China has used international linkages and connections to further its own strategic agenda involving both civilian and military applications. Very early on China had identified SSTL as a key player in its small satellite programme. From the joint launch of the Tsinghua 1 in 2000 to the subsequent launch of the 166 kg Tsinghua 2 with various operational Chinese entities in 2005 to the final launch of the operational constellation by the PSLV in 2015, one can see a clear long-term strategy that combines technology trends, applications and use with technology absorption and indigenous product development.

The technology part of the programme is well integrated with its international relation postures at the level of the prime ministers. The strategy of buy and launch from abroad to avoid US sanctions as well as jointly build and manage strengthens internal technical capabilities that are then moved to the relevant entities within the Chinese Military industrial complex.

One can soon expect a Chinese equivalent of an SSTL that may offer a much cheaper product for international customers. Small satellite technology will also be leveraged to cater to military uses.

As US relations with China and Russia have plummeted, the two erstwhile adversaries are increasingly getting closer. To avoid US ITAR regulations, China has used Russian launchers for their bought satellites. For its human flight programme, China has turned to Russia for help in key areas of technology[13] (Harvey 2013).

Though the relationship is commercial, and Russia still has reservations on China, they are likely to be drawn closer because of a common anti-US sentiment. Internationally both at the Conference on Disarmament as well as in the Committee on the Peaceful Uses of Outer Space they are likely to take common positions.

China has also used launchers from Ukraine for some of its bought satellites. Given the current problems between the pro-West Ukrainian government and Russia this relationship may weaken. However, despite these irritants China may be able to source technology from Ukraine in case of need. Increasingly keeping in mind Chinese indigenous capabilities, China may not need Ukraine help for its space effort.

Moving away from the China's demand side towards the supply side Brazil stands out as a major partner in the realm of remote sensing satellites and their applications. This convergence of common interests is likely to continue both in bilateral and multilateral fora like BRICS.

---

[12] http://www.isro.gov.in/sites/default/files/pdf/pslv-brochures/PSLV-C28.pdf.
[13] pp 265–266.

The other countries for whom China has delivered satellites in orbit are Bolivia, Venezuela, Nigeria, Pakistan, Belarus, and Laos. Indonesia could also be added to this list though the Palapa satellite failed to reach orbit.

Most countries of Latin America are oil or resource-rich and these initiatives make both economic and strategic sense. Ties with Argentina have become stronger with the establishment of a deep space tracking station in the Nequen Province of Argentina that became operational in 2017. Since then, China has launched several remote sensing satellites for Argentina. The strong ties with Nigeria are also based on similar logic. Barring a major downturn in US-China relations these convergences of common interests are likely to continue.

Belarus is a close ally of Russia. It is a country whose geographic location is critical for Russia in countering western attempts to destabilize it. Belarus has also used Chinese space capabilities and bought a Chinese GSO communication satellite. Turkey is also a country that links Europe with Asia and Russia with Western Europe and US spheres of influence. It is also therefore a very important country strategically.

While Chinese influence over these countries through supply of space services may not be significant as of now, they provide a possible foothold for the future depending on how events in this region of the world pan out.

Pakistan is a China neighbor and an all-weather friend. The strong military and strategic relationship of which the space part has so far been relatively small will continue. China will provide needed help for Pakistan as and when it decides to enhance its capabilities in space technologies including satellites and launchers.

Laos is a close friend and ally of China. Chinese help is likely to continue and space cooperation with Laos will continue to feature as a major component of China's strategy.[14]

In the past, both the Philippines (1997) and Indonesia (2009) have used lower cost Chinese launchers for their foreign-built communication satellites. Australia too has used Chinese launch services to launch its first-generation communication satellites using a launcher (CZ 2E) specifically developed for them. US ITAR regulations may have forced them to shift to other launch service providers. US sanctions therefore do mean lost business opportunities for China.

In addition to the presence of ties, the absence of ties also provides us with some insights on China's approach on how it deals with the countries of the Asia–Pacific region where it is one of the major dominant powers. Among the ASEAN states, China has no major space links with Singapore and Malaysia.[15] Vietnam too has no connections. Though Indonesia and Philippines have used Chinese launch services and Indonesia even ordered an in-orbit delivery of a communication satellite, they may re-consider such involvements not only because of US sanctions but also because of China's increasingly intransigent behavior in the South China Sea.

---

[14] Cambodia a close ally of China may not yet be ready internally for receiving Chinese space largesse.

[15] Singapore has used the services of India's PSLV for launching small satellites. Indonesia, Japan, and South Korea are other countries in the Asia–Pacific Region that have used the PSLV for launching smaller satellites. On December 16, 2015, a dedicated PSLV launcher put six Singapore satellites into a 550 km sun synchronous orbit http://www.isro.gov.in/pslv-c29-teleos-1-mission.

Both Thailand and Indonesia are a part of a network of countries that receive Chinese weather data. This however may not be a significant factor that could influence the relationship.

Given this context China's efforts to use space services to promote and foster Chinese influence and to moderate at least to some extent US power over the region may run into problems.

It is also surprising that China and North Korea have no cooperative space activities. North Korea has emerged as a space power with the launch of two remote sensing satellites into SSO using its own indigenously developed launchers (Chandrashekar et al. 2016). Though China supports North Korea, it appears to be wary about sharing its space capabilities and technology with its neighbor.

Given current developments in the South China Sea as well as US imposed restrictions on China, it is unlikely that many of the countries in the Asia–Pacific region would use Chinese launch services in the future even though they may be lower priced. Japan too has been a traditional foe and the current impasse in space-related issues is likely to continue.

16.8   China and the United Nations

China became a member of the United Nations Committee on the Peaceful Uses of Outer Space (UNCOPUOS) in 1980 following Deng's visit to the US in 1979. It ratified the Outer Space Treaty in 1980. This was followed by the ratification of Astronaut Agreement, the Liability Convention, and the Registration Convention in 1988. China is therefore committed to being perceived as a responsible space power on the world stage. The recent uncontrolled reentries of the Tiangong 1 Space Laboratory and the spent stage of the CZ 5B rocket have raised international concerns of likely damage from Chinese space debris. As a signatory to the four major treaties on space, China will be bound to demonstrate to the world that it is indeed a responsible space power. One would expect that it will take all the necessary measures to mitigate, reduce, and eliminate the possibilities of space debris causing damage on the earth.

China also became a member of the International Astronautical Federation (IAF) and the International Telecommunications Union (ITU) in 1980. It has also been a member of the Conference on Disarmament (CD) whose mandate is to discuss the militarization and weaponization of space.

It has in collaboration with the Russian Federation tabled a Treaty at the CD on "The Prevention of an Arms Race in Outer Space (PAROS)". The meetings at the CD on the arms race in space have made no visible progress over the last 40 years due to the irreconcilable differences between the US and the Russia China combine.

Given China's rapid increase in Space Power and the common interests it has with Russia to keep the US in check the deadlock at the CD will continue. As discussed in Chap. 13 on China's military uses of space, the US-China rivalry will spill over into the increasingly important military domain of space.

China will also continue to use its space prowess to further its geopolitical ambitions as a responsible power on the world stage.

Case Study 4

United Nations and China Agreement for Increased Space Cooperation[16]

In 2016, China initiated a programme to facilitate international cooperation in the peaceful uses of outer space. After signing a Framework and a Funding Agreement with the UN Office of Outer Space Affairs (UNOOSA) China made a formal offer of cooperation in the peaceful uses of outer space to the Committee on the Peaceful Uses of Outer Space (COPUOS) at its 59th meeting held in Vienna during June 2016.

Under this framework, UNOOSA and the China Manned Space Agency (CMSA) will work together to enable UN Member States to conduct experiments on China's forthcoming Space Station. There will also be opportunities for countries to send astronauts and payload engineers to the space station. The programme also seeks to build capacities within Member States for undertaking space technology and applications projects. China will provide funding support through CMSA for these activities.

"Space exploration is the common dream and wish of humankind. We believe that the implementation of the agreements will definitely promote international cooperation on space exploration, and create opportunities for United Nations Member States, particularly developing countries, to take part in, and benefit from, the utilization of China's space station" (CMSA Deputy Director General Wu Ping).

China will also try to counter US and its allies in the Asia–Pacific region through a strengthening of its regional cooperation mechanisms such as APSCO. It will use both bilateral and multilateral initiatives for this purpose. Lower priced space products and space services will be a significant component of its regional and global strategies. On balance, however, China despite lower prices will find it difficult to compete with US space service offerings in areas like weather and navigation. Figure 16.5 which is based on the relationships outlined in Table 16.1 provides an overview of China's current international space footprints.

---

[16] http://www.unoosa.org/oosa/en/informationfor/media/2016-unis-os-468.html.

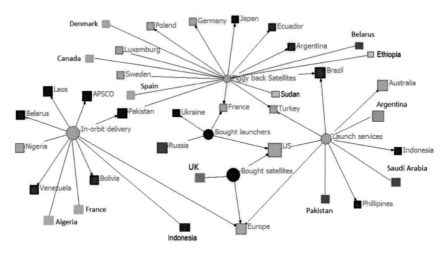

Fig. 16.5 China's international space commerce footprints—end 2020

Annexure 16.1: Chinese Launch Services to Other Countries—End 2020

Country	Satellite	Date of launch	Launcher	Orbit	Launch site	Nos	Comment
US	Echostar 1	12/28/1995	CZ 2E	GTO	Xichang	1	Commercial dual launch
	Intelsat 708	2/14/1996	CZ 3B	GTO	Xichang	1	*Commercial failure*
	Iridium dummy	9/1/1997	CZ 2C	620 km 86.5°	Taiyuan	1	Commercial
	Iridium 42	12/8/1997	CZ 2C	620 km 86.5°	Taiyuan	1	Commercial dual launch
	Iridium 44	12/8/1997	CZ 2C	620 km 86.5°	Taiyuan	1	Commercial dual launch
	Iridium 51	3/25/1998	CZ 2C	620 km 86.5°	Taiyuan	1	Commercial dual launch
	Iridium 61	3/25/1998	CZ 2C	620 km 86.5°	Taiyuan	1	Commercial dual launch
	Iridium 69	5/2/1998	CZ 2C	620 km 86.5°	Taiyuan	1	Commercial dual launch
	Iridium 71	5/2/1998	CZ 2C	620 km 86.5°	Taiyuan	1	Commercial dual launch

(continued)

(continued)

Country	Satellite	Date of launch	Launcher	Orbit	Launch site	Nos	Comment
	Iridium 76	8/19/1998	CZ 2C	620 km 86.5°	Taiyuan	1	Commercial dual launch
	Iridium 78	8/19/1998	CZ 2C	620 km 86.5°	Taiyuan	1	Commercial dual launch
	Iridium 88	12/19/1998	CZ 2C	620 km 86.5°	Taiyuan	1	Commercial dual launch
	Iridium 89	12/19/1998	CZ 2C	620 km 86.5°	Taiyuan	1	Commercial dual launch
	Iridium 93	6/11/1999	CZ 2C	620 km 86.5°	Taiyuan	1	Commercial dual launch
	Iridium 92	6/11/1999	CZ 2C	620 km 86.5°	Taiyuan	1	Commercial dual launch
Europe	Eutelsat W3C	10/7/2011	CZ 3B/E	GTO		1	Commercial
Australia	Aussat demo	*7/16/1990*	CZ 2E	GTO	Xichang	1	*Failure-piggyback Badr*
	Optus B1	8/13/1992	CZ 2E	GTO	Xichang	1	Commercial
	Optus B2	12/21/1992	CZ 2E	GTO	Xichang	1	Commercial
	Optus B3	8/27/1994	CZ 2E	GTO	Xichang	1	Commercial
Philippines	Agila 2	8/19/1997	CZ 3B	GTO	Xichang	1	Commercial
Indonesia	Palapa D1	8/31/2009	CZ 3B	GTO	Xichang	1	Commercial
Turkey	Gokturk-2	12/18/2012	CZ 2D	SSO-680 km	Jiuquan	1	Commercial
Brazil	CBERS 1	10/14/1999	CZ 4B	SSO-740 km	Taiyuan	1	Commercial-cooperation
Brazil	CBERS 2	10/21/2003	CZ 4B	SSO-740 km	Taiyuan	1	Commercial-cooperation
Brazil	CBERS 2B	9/19/2007	CZ 4B	SSO-740 km	Taiyuan	1	Commercial-cooperation
Brazil	CBERS-3	12/9/2013	CZ 4B	SSO-740 km	Taiyuan	1	Commercial-failure
Brazil	CBERS-4	12/7/2014	CZ 4B	SSO-740 km	Taiyuan	1	Commercial-cooperation
Brazil	CBERS 4A	12/20/2019	CZ 4B	SSO-630 km	Taiyuan	1	Commercial-cooperation
Pakistan	Pak Tes 1A	7/9/2018	CZ 2C	SSO-610 km	Jiuquan	1	Pakistan-built RS test satellite
Saudi Arabia	Saudisat 5A	12/7/2018	CZ 2D	SSO-550 km	Jiuquan	1	China launch for Saudi Arabia
Saudi Arabia	Saudisat 5B	12/7/2018	CZ 2D	SSO-550 km	Jiuquan	1	China launch for Saudi Arabia
Argentina	Nusat 9–18	6/11/2020	CZ 6	SSO-475 km	Taiyuan	**10**	Dedicated commercial launch
Total			**26**			**42**	

Annexure 16.2: In-Orbit Delivery and Satellites for Intl Cooperative Programmes—End 2020

Country	Satellite	Date of launch	Launcher	Orbit	Launch site	Nos	Comment
Europe	Double Star 1	12/29/2003	CZ 2C	79,000 × 570 km 28.2°	Xichang	1	Contribution to ESA study
	Double Star 2	7/25/2004	CZ 2C	38 568 × 560 km 90°	Taiyuan	1	Contribution to ESA study
Laos	LaoSat 1	11/20/2015	CZ 3B	GSO	Xichang	1	Satellite Built and Delivered
Pakistan	Paksat 1R	8/11/2011	CZ 3BE	GSO	Xichang	1	Satellite Built and Delivered
Pakistan	PRSS 1	7/9/2018	CZ 2C	SSO-610 km	Jiuquan	1	Satellite Built and Delivered
Belarus	Belintersat 1	1/15/2016	CZ 3B	GSO	Xichang	1	Satellite Built and Delivered
Nigeria	NIGCOMSAT-1	5/13/2007	CZ 3BE	GSO	Xichang	1	Satellite Built and Delivered
	NIGCOMSAT-1R	12/19/2011	CZ 3BE	GSO	Xichang	1	Satellite Built and Delivered
Venezuela	Simon Bolibar 1	10/29/2008	CZ 3BE	GSO	Xichang	1	Satellite Built and Delivered
	VRSS-1	9/29/2012	CZ 2D	SSO-640 km	Jiuquan	1	Satellite Built and Delivered
	VRSS 2	10/9/2017	CZ 2D	SSO-640 km	Jiuquan	1	Satellite Built and Delivered
Bolivia	Tupac Katari 1	12/20/2013	CZ 3B	GSO	Xichang	1	Satellite Built and Delivered
APSCO	Huanjing 1A	9/6/2008	CZ 2C	SSO-640 km	Taiyuan	1	Contribution to APSCO
	Huanjing 1B	9/6/2008	CZ 2C	SSO-640 km	Taiyuan	1	Contribution to APSCO
	Huanjing 1C	11/18/2012	CZ 2C	SSO-500 km	Taiyuan	1	Contribution to APSCO

(continued)

(continued)

Country	Satellite	Date of launch	Launcher	Orbit	Launch site	Nos	Comment
	Huanjing 2A, 2B	9/27/2020	CZ 4B	SSO-647 km	Taiyuan	2	Contribution to APSCO
Algeria	AlComsat	12/10/2017	CZ 3B	GSO	Xichang	1	Satellite Built and Delivered
France	CFOSAT	10/29/2018	CZ 2C	SS0-520 km	Jiuquan	1	Contribution to joint study
Indonesia	Palapa N1	4/9/2020	CZ 3BE	GSO	Xichang	1	Commercial in-orbit delivery
Total		**17**				**20**	

Annexure 16.3: International Piggyback Payloads and Satellites on Chinese Launchers—End 2020

Country	Satellite	Date	Launcher	Orbit	Site	Nos	Comment
France	FSW 0–9	8/5/1987	CZ 2C	420 × 172 km 62.9°	Jiuquan	1	Microgravity experiment FSW
France	Test Satellite	11/3/2019	CZ 4B	SSO-500 km	Taiyuan	1	Iodine ion thruster test
Germany	FSW 1–2	8/5/1988	CZ 2C	310 × 206 km 63°	Jiuquan	1	Microgravity experiment FSW
Germany	Orbcom Test	11/7/2019	KZ 1	SSO-1050 km	Jiuquan	2	CAS collaboration orbcom
Japan	FSW-2 3	10/20/1996	CZ 2D	342 × 171 km 63°	Jiuquan	1	Microgravity experiment FSW
Luxembourg	Vesselsat-2	1/9/2012	CZ 4B	SSO-480 km	Taiyuan	1	Orbcom AIS package ZY 3A
Spain	3 CAT-2	8/15/2016	CZ 2D	SSO 500 km	Jiuquan	1	Launched with QUESS satellite

(continued)

(continued)

Country	Satellite	Date	Launcher	Orbit	Site	Nos	Comment
Sweden	Freja	10/6/1992	CZ 2C	1763 × 596 km 63°	Jiuquan	1	Launched with FSW 1–4
Poland	BLITE PL	8/19/2014	CZ 4B	SSO-620 km	Taiyuan	1	Launched with Gefen 2
Pakistan	Badr-1	7/16/1990	CZ 2E	986 × 205 km 28.50°	Xichang	1	Launched with dummy Aussat
Turkey	TurkSat-3USat	4/26/2013	CZ 2D	SSO 640 km	Jiuquan	1	Launched with Gefen 1
Brazil	SAC 1	10/14/1999	CZ 4B	SSO-740 km	Taiyuan	1	Launched with CBERS 1
Brazil	Floripasat	12/20/2019	CZ 4B	SSO-625 km	Taiyuan	1	Launched with CBERS 4A
Argentina	CubeBug-1	4/26/2013	CZ 2D	SSO 640 km	Jiuquan	1	Launched with Gefen 1
Argentina	Nusat-1	5/30/2016	CZ 4B	SSO 500 km	Taiyuan	1	Launched with Ziyuan 3 02
Argentina	Nusat-2	5/30/2016	CZ 4B	SSO 500 km	Taiyuan	1	Launched with Ziyuan 3 02
Argentina	Nusat-3	6/15/2017	CZ 4B	552 X 542 km 43°	Jiuquan	1	Launched with HXMT
Argentina	Nusat-4	2/2/2018	CZ 2D	SSO 500 km	Jiuquan	1	Launched with Zhanzheng 1
Argentina	Nusat-5	2/2/2018	CZ 2D	SSO 500 km	Jiuquan	1	Launched with Zhanzheng 1
Argentina	Nusat-6 and 7	15/1/2020	CZ 2D	SSO 470 km	Taiyuan	2	Launched with Jilin
Ecuador	NEE 01pogos	4/26/2013	CZ 2D	SSO 640 km	Jiuquan	1	Launched with Gefen 1
Canada	KIPP	1/19/2018	CZ 11	1065 X 525 km SSO	Jiuquan	1	Launched with Jilin 07 and 08
Denmark	GOMX 4A	2/2/2018	CZ 2D	SSO 500 km	Jiuquan	1	Launched with Zhanzheng 1
Denmark	GOMX 4B	2/2/2018	CZ 2D	SSO 500 km	Jiuquan	1	Launched with Zhanzheng 1

(continued)

(continued)

Country	Satellite	Date	Launcher	Orbit	Site	Nos	Comment
Belarus	CubeBel 1	10/29/2018	CZ 2C	SSO-520 km	Jiuquan	1	Launched with CFOSAT
Sudan	SRSSS 1	11/03/2019	CZ 4B	SSO-510 km	Taiyuan	1	Launched with Gefen 7
Ethiopia	ETRSS 1	12/202/2019	CZ 4B	SSO-625 km	Taiyuan	1	Launched with CBERS 4A
Total						**29**	

Annexure 16.4

Procured satellites launched by China—end 2020

Country	Satellite	Launch date	Launcher	Orbit	Launch site	Comment
US	Asiasat 1	4/7/90	CZ-3	GTO	Xichang	Hughes 376 bus
	Apstar 1	7/21/94	CZ-3	GTO	Xichang	Hughes 376 bus
	Apstar 2	1/25/95	CZ-2E	GTO	Xichang	Hughes 601 bus failure
	Asiasat 2	11/28/95	CZ-2E	GTO	Xichang	LM-7000 bus
	Apstar 1A	7/4/96	CZ-3	GTO	Xichang	Hughes 376 bus
	Zhongxing 7	8/18/96	CZ-3	GTO	Xichang	Hughes 376 bus
	Apstar 2R	10/16/97	CZ-3B	GTO	Xichang	Loral 1300 bus
	Zhongwei 1	5/30/98	CZ-3B	GTO	Xichang	A2100A Lockheed bus
Europe	Sinosat 1	7/18/98	CZ-3B	GTO	Xichang	SB-3000 Thales/Aerospatiale bus
	Apstar 6	4/12/05	CZ 3B	GTO	Xichang	Alcatel/Thales Alenia bus
	Zhongxing 6B	7/5/07	CZ-3B	GTO	Xichang	Spacebus 4000 Thales
	Zhongxing 9	6/9/08	CZ-3B	GTO	Xichang	Spacebus 4000 Thales
	Apstar 7	3/31/2012	CZ 3B/E	GTO	Xichang	Spacebus 4000 Thales
	Zhonxing-12	11/27/2012	CZ 3B/E	GTO	Xichang	Spacebus 4000 Thales

(continued)

(continued)

Procured satellites launched by China—end 2020						
Country	Satellite	Launch date	Launcher	Orbit	Launch site	Comment
Procured Satellites Launched by Others—End 2020						
US	Asiasat 3	12/24/1997	Proton	GTO	Baikonur	Hughes 601 Russia launch
	Asiasat 3S	3/21/1999	Proton	GTO	Baikonur	Hughes 601 Russia launch
	Asiasat 4	4/12/2003	Atlas 3B	GTO	CCAS	Hughes 601 US launch
	Asiasat 5	8/11/2009	Zenit 3 SLB	GTO	Baikonur	SSL 1300 bus Ukraine launch
	Asiasat 7	11/25/2011	Proton ILS	GTO	Baikonur	SSL 1300 bus Russia
	Asiasat 8	8/5/2014	Falcon	GTO	CCAS	SSL bus US launch
	Asiasat 6	9/7/2014	Falcon	GTO	CCAS	SSL bus US launch
UK	Tsinghua 1	6/28/2000	Kosmos 3 M	690 km SSO	Plasetsk	SSTL Russia launch
	China-DMC + 4	10/27/2005	Kosmos 3 M	690 km SSO	Plasetsk	SSTL Russia launch
China-Built Satellite Launched by Others—End 2020						
China	Yinghuo-1	11/8/2011	Zenit 3F	Mars Mission	Baikonur	Failure Ukraine launch
	NUDsat	6/23/2017	PSLV	500 km SSO	SHAR	Piggyback launch
Total	**25**					

References

Chandrashekar S, Ramani N, Vishwanathan A (2016) Analysis of North Korea's February 2016 successful space launch. http://isssp.in/wp-content/uploads/2016/04/North-Korean-Feb-2016-Successful-Space-Launch.pdf

Chen Y (1993) China's space commercialization effort—Organization, policy and strategies. Space Policy

Harvey B (2013) China in space the great leap forward. Springer Praxis Books, New York

Liu ZX, Escoubet CP, et al (2005) The double star mission. In: Annales Gephysicae, vol 23. pp 2707–2712. http://www.ann-geophys.net/23/2707/2005/angeo-23-2707-2005.pdf

Reddy VS (2018) China's design to capture regional satcom markets, ORF Special Report. https://www.orfonline.org/wp-content/uploads/2018/07/ORF_SpecialReport_70_China_SatCom.pdf